The First Computers

History of Computing

Memories That Shaped an Industry, Emerson W. Pugh, 1984

The Computer Comes of Age: The People, the Hardware, and the Software, R. Moreau, 1984

Memoirs of a Computer Pioneer, Maurice V. Wilkes, 1985

Ada: A Life and Legacy, Dorothy Stein, 1985

IBM's Early Computers, Charles J. Bashe, Lyle R. Johnson, John H. Palmer, and Emerson W. Pugh, 1986

A Few Good Men from Univac, David E. Lundstrom, 1987

Innovating for Failure: Government Policy and the Early British Computer Industry, John Hendry, 1990

Glory and Failure: The Difference Engines of Johann Müller, Charles Babbage and Georg and Edvard Scheutz, Michael Lindgren, 1990

John von Neumann and the Origins of Modern Computing, William Aspray, 1990

IBM's 360 and Early 370 Systems, Emerson W. Pugh, Lyle R. Johnson, and John H. Palmer, 1991

Building IBM: Shaping an Industry and Its Technology, Emerson W. Pugh, 1995

A History of Modern Computing, Paul Ceruzzi, 1998

Makin' Numbers: Howard Aiken and the Computer, edited by I. Bernard Cohen and Gregory W. Welch with the cooperation of Robert V. D. Campbell, 1999

Howard Aiken: Portrait of a Computer Pioneer, I. Bernard Cohen, 1999

The First Computers—History and Architectures, edited by Raúl Rojas and Ulf Hashagen, 2000

The First Computers—History and Architectures

edited by Raúl Rojas and Ulf Hashagen

The MIT Press
Cambridge, Massachusetts
London, England

First MIT Press paperback edition, 2002
© 2000 Massachusetts Institute of Technology

This book was set in Times Roman and Helvetica by the editors.

Library of Congress Cataloging-in-Publication Data

The first computers: history and architectures / edited by Raúl Rojas and Ulf Hashagen.
    p. cm.—(History of computing)
  Includes bibliographical references and index.
  ISBN 978-0-262-18197-6 (hc.: alk. paper)—978-0-262-68137-7 (pbk.: alk. paper)
  1. Computers—History. 2. Computer architecture—History. I. Series. II. Rojas, Raúl,
  1955–   . III. Hashagen, Ulf.

QA76.17.F57   2000
004'.0921—dc21

                                                                        99-044811

Contents

Part III: The German Scene

Part IV: The British Scene

Part V: Early Japanese Computers

Preface

We are proud to present this volume to all programmers, computer scientists, historians of science and technology, and the general public interested in the details and circumstances surrounding the most important technological invention of the twentieth century — the computer. This book consists of the papers presented at the International Conference on the History of Computing, held at the Heinz Nixdorf MuseumsForum in Paderborn, Germany, in August 1998. This event was a satellite conference of the International Congress of Mathematicians, held in Berlin a week later. Using electronic communication, the contributions for this volume were discussed before, during, and after the conference. Therefore, this is a collective effort to put together an informative and readable text about the architecture of *the first computers* ever built.

While other books about the history of computing do not discuss extensively the structure of the early computers, we made a conscious effort to deal thoroughly with the architecture of these machines. It is interesting to see how modern concepts of computer architecture were being invented simultaneously in different countries. It is also fascinating to realize that, in those early times, many more architectural alternatives were competing neck and neck than in the years that followed. A thousand flowers were indeed blooming — data-flow, bit-serial, and bit-parallel architectures were all being used, as well as tubes, relays, CRTs, and even mechanical components. It was an era of *Sturm und Drang*, the years preceding the uniformity introduced by the canonical von Neumann architecture.

The title of this book is self-explanatory. As the reader is about to discover, attaching the name "world's first computer" to any single machine would be an over-simplification. Michael R. Williams makes clear, in the first chapter in this volume, that any of these early machines could stake a claim to being a first in some sense. Speaking in the plural of the *first computers* is therefore not only a diplomatic way around any discussion about claims to priority, it is also historically correct. However, this does not mean that our authors do not strongly push their case forward. Every one of them is rightly proud of the intellectual achievement materialized in the machines they have studied as historians, rebuilt as engineers, or even designed as pioneers. And this volume has its share of all three kinds of writers. This might well be one of the strengths of this compilation.

Why study old architectures?

Some colleagues may have the impression that nothing new can be said about the first computers, that everything worth knowing has already been published somewhere else. In our opinion, this is not the case; there is still much to be learned from architectural comparisons of the early computers. A good example is the reconstruction of Colossus, a machine that remained classified for many years, and whose actual design was known to only a small circle of insiders. Thanks to Tony Sale, a working replica of Colossus now exists, and full diagrams of the machine have been drawn. However, even when a replica has been built, the internal structure of the machine has sometimes remained undocumented. This was the case with Konrad Zuse's Z1 and Z3, reconstructed for German museums by Zuse himself. Since he did not document the machines in a form accessible to others, we had the paradox in Germany of having the machines but not knowing exactly how they worked. This deficit has been corrected only in recent years by several papers that have dissected Zuse's machines.

Another example worth analyzing is the case of the Harvard Mark I computer. Every instruction supplies a source and a destination: numbers are moved from one accumulator to another, and when they arrive they are added to the contents of the accumulator (normal case). The operation can be modified using some extra bits in the opcode. This architecture can be streamlined by defining different kinds of accumulators, which perform a different operation on the numbers arriving. Thus, one accumulator could add, the other subtract, and yet another just shift a number. This is exactly the kind of architecture proposed by Alan Turing for the ACE, a computer based on the single instruction MOVE. We notice only the similarity between both machines when we study their internal organization in greater depth.

It is safe to say that there are *few* comparative architectural studies of the first computers. This volume is a first step in this direction. Moreover, we think that this book can help motivate students of computer science to look at the history of their chosen field of study. Courses on the history of computing can be made more interesting for these students, not always interested in the humanities or history in itself, by showing them that there is actually much to be learned from the successes and failures of the pioneers. Some kinds of computer architectures even reappear when the architectural constraints make a comeback. The Connection Machine, a supercomputer of the 1980s, was based on bit-serial processors, because they were cheap and could be networked in massive amounts. Reconfigurable hardware is a new buzzword among the computer science community, and the approach promises to speed up computations by an order of magnitude. Could it be that the microchips of the future will look like the ENIAC, like problem-dependent rewireable machines?

Those who do not know the past are condemned to live it anew, but the history of computing shows us that those who know the past can even put this knowledge to good use!

Structure of the book

Part I deals with questions of method and historiography. Mike Mahoney shows that computer science arose in many places simultaneously. He explains how different theoretical schools met at the crossroads leading to the fundamental concepts of the discipline. Robert Seidel then discusses the relevance of reconstructions and simulations of historical machines for the history of science. New insights can be gained from those reconstruction efforts. In the next chapter, Andreas Brennecke attempts to bring some order to the discussion about the invention of the first computers, by proposing a hierarchical scheme of increasingly flexible machines, culminating in the stored program computer. Finally, Harry Huskey, one of the pioneers at the conference, looks at the constraints imposed on computer architectures by the kind of materials and logical elements available during the first decades following World War II.

Part II of the book deals with the first American computers. John Gustafson, who led the reconstruction of Atanasoff's machine, describes the detective work that was necessary in order to recreate this invention, destroyed during the war and considered by some, including a federal judge, to be the first computer built in the U.S. He addresses the limitations of the machine but also explains how it could have been used as a calculator. I. Bernard Cohen, whose Aiken biography is the best study of a computer pioneer published up to now, contributed a chapter which sheds light on the architectural solutions adopted by Aiken and clarifies why he did not build an electronic machine. Professor Jan Van der Spiegel and his team of students performed the feat of putting the ENIAC on a single chip. Their paper provides many details about the operation of the machine and discusses its circuits in depth. Their description is the best and most comprehensive summary of ENIAC's architecture ever written. William Aspray and Paul Ceruzzi review later developments in the computer arena in their contributions and show us how the historian of computing can bring some order in this apparent chaos.

Part III looks at the other side of the Atlantic. For the first time, a single book written for the international public discusses the most important early German computers: the Z1, Z3, and Z4, as well as the electronic machines built in Göttingen. Raúl Rojas, Ambros Speiser, and Wilhelm Hopmann review all these different machines, discussing their internal operation. In his contribution Hartmut Petzold looks at the emergence of a computer industry in Germany and the role played by Konrad Zuse. Friedrich L. Bauer, a well-known German pioneer, looks again at the high-level programming language invented by Zuse, the *Plankalkül* (calculus of programs), which he considers his greatest achievement. Friedrich Kistermann and Thomas Lange analyze the structure of two almost forgotten, yet very important machines, the DEHOMAG tabulator and the first general-purpose analog computer, built by Helmut Hoelzer in Germany. Hoelzer's analog machines were used as on-board computers during the war.

The first British computers are explained in Part IV. Tony Sale describes the reconstruction of Colossus, which we mentioned above. Brian Napper and Chris Burton analyze the architecture and reconstruction of the Manchester Mark I, the world's first stored-program computer. Frank Sumner reviews the Atlas, a real commercial spin-off of the technological developments that took place in Manchester during those years. In the final chapter of this section, Martin Campbell-Kelly, editor of Babbage's Collected Works, takes a look at the EDSAC, the computer built in Cambridge, and tells us how much can be learned from a software simulation of a historical machine.

Finally, Part V makes information available about the first Japanese computers. Seiichi Okoma reviews the general characteristics of the early Japanese machines and Eiiti Wada describes the PC-1 in more depth, a computer that is very interesting from a historical viewpoint, since it worked using majority logic. The same kind of circuits had been studied in the U.S. by McCulloch and Pitts, and also had been used by Alan Turing in his written proposal for the ACE machine. Apparently, the only hardware realization was manufactured in Japan and used for the PC-1.

Acknowledgments

The *International Conference on the History of Computing* could not have been held without the financial support of the Deutsche Forschungsgemeinschaft (DFG), the Heinz-Nixdorf MuseumsForum in Paderborn and the Freie Universität Berlin. The HNF took care of all the logistics of a very well organized meeting, and Goetz Widiger from FU Berlin managed the Web site for the conference. Zachary Kramer, Philomena Maher, and Anne Carney took care of correcting our non-native speakers' English. We thank them all. Our gratitude also goes to all contributors to this volume, who happily went through the many revisions and changes needed to produce a high-quality book. The Volkswagen Foundation provided Raúl Rojas funding for a sabbatical stay at UC Berkeley, where many of the revisions for the book were made.

Raúl Rojas and Ulf Hashagen

The First Computers

A Preview of Things to Come: Some Remarks on the First Generation of Computers

Michael R. Williams

Abstract. The editors of this volume have asked me to prepare this introduction in order to "set the scene" for the other papers. It is often difficult to know just how much knowledge people have about the early days of computing – however you define that term. If one reads a sophisticated description which details some small aspect of a topic, it is impossible to follow if your intention was simply to learn some basic information. On the other hand, if you are an historian that has spent your entire working life immersed in the details of a subject, it is rather a waste of time to carefully examine something which presents the well known facts to you, yet again. This means that, no matter what I include here, I will almost certainly discuss things of no interest to many of you! What I do intend to do is to review the basics of early computer architecture for the uninitiated, but to try and do it in a way that might shed some light on aspects that are often not fully appreciated – this means that I run the risk of boring everyone.

1. Classifications of computing machines

As a start, let us consider the word "computer." It is an old word that has changed its meaning several times in the last few hundred years. Coming, originally, from the Latin, by the mid-1600s it meant "someone who computes." It remained associated with human activity until about the middle of this century when it became applied to "a programmable electronic device that can store, retrieve, and process data" as *Webster's Dictionary* defines it. That, however, is misleading because, in the context of this volume, it includes all types of computing devices, whether or not they were electronic, programmable, or capable of "storing and retrieving" data. Thus I think that I will start by looking at a basic classification of "computing" machines.

One can classify computing machines by the technology from which they were constructed, the uses to which they were put, the era in which they were used, their basic operating principle, analog or digital, and whether they were designed to process numbers or more general kinds of data.

Perhaps the simplest is to consider the technology of the machine. To use a classification which was first suggested to me by Jon Eklund of the Smithsonian, you can consider devices made from five different categories:

- Flesh: fingers, people who compute – and there have been many famous examples of "idiot savants" who did remarkable calculations in their head, including one that worked for the Mathematics Center in Amsterdam for many years;
- Wood: devices such as the abacus, some early attempts at calculating machines such as those designed by Schickard in 1621 and Poleni in 1709;
- Metal: the early machines of Pascal, Thomas, and the production versions from firms such as Brunsviga, Monroe, etc.;
- Electromechanical devices: differential analyzers, the early machines of Zuse, Aiken, Stibitz, and many others;
- Electronic elements: Colossus, ABC, ENIAC, and the stored program computers.

This classification, while being useful as an overall scheme for computing devices, does not serve us well when we are talking about developments in the last 60 or 70 years.

Similarly, any compact scheme used for trying to "pigeon-hole" these technological devices will fail to differentiate various activities that we would like to emphasize. Thus, I think, we have to consider any elementary classification scheme as suspect. Later in this volume there is a presentation of a classification scheme for "program controlled calculators" which puts forward a different view.[1]

2. Who, or what, was "first"

Many people, particularly those new to historical studies, like to ask the question of "who was really first?" This is a question that historians will usually go to great lengths to avoid. The title of this volume (*The First Computers – History and Architectures*) is certainly correct in its use of the word *first* – in this case it implies that the contents will discuss a large number of the early machines. However, even the subtitle of this introduction – "Some Remarks on the First Generation of Computers" – is a set of words full of problems. First, the use of the word "computer" is a problem as explained above. Second, the words "first generation" have many different interpretations – do I include the electromechanical machines of Zuse, Stibitz, and

---

[1] See in this volume: A. Brennecke, "A Classification Scheme for Program Controlled Calculators."

Aiken (which were certainly "programmed") or am I limiting myself to the modern "stored program" computer – and even then, do I consider the first generation to begin with the mass production of machines by Ferranti, UNIVAC, and others, or do I also consider the claims of "we were first" put forward by the Atanasoff-Berry Computer (ABC), Colossus, ENIAC, the Manchester Baby Machine, the EDSAC, and many more?

Let me emphasize that there is no such thing as "first" in any activity associated with human invention. If you add enough adjectives to a description you can always claim your own favorite. For example the ENIAC is often claimed to be the "first electronic, general purpose, large scale, digital computer" and you certainly have to add all those adjectives before you have a correct statement. If you leave any of them off, then machines such as the ABC, the Colossus, Zuse's Z3, and many others (some not even constructed such as Babbage's Analytical Engine) become candidates for being "first."

Thus, let us agree, at least among ourselves, that we will not use the word "first" – there is more than enough glory in the creation of the modern computer to satisfy all of the early pioneers, most of whom are no longer in a position to care anyway. I certainly recognize the push from various institutions to have their people declared "first" – and "who was first?" is one of the usual questions that I get asked by the media, particularly when they are researching a story for a newspaper or magazine.

In order to establish the ground rules, let us say that there are two basic classes of machines: the modern stored program, digital, electronic computer, and the other machines (either analog or digital) that preceded, or were developed and used after the invention of the stored program concept.

During the recent celebrations of the 50th anniversary of the creation of the Manchester Baby Machine, one of the speakers remarked that "You don't go into a pet store and ask to buy a cat and then specify 'I would like one with blood please' – similarly, you don't buy a computer and ask for it to have a memory, you just assume that it will be part of the machine." The possession of a large memory for both instructions and data is a defining characteristic of the modern computer. It is certainly the case that the developers of the modern computer had a great deal of trouble finding devices that would make a suitable memory for a stored program computer, so it is with this topic that I would like to begin my more detailed remarks.

3. Memory systems

It is quite clear where the concept of the stored program computer originated. It was at the Moore School of Electrical Engineering, part of the University of Pennsylvania, in the United States. What is not so clear is who invented the concept. It was formulated by the group of people who were, then, in the middle of the construction of the ENIAC and was a response to the problems they were beginning to see in the design of that machine – principally the very awkward control system which required the user to essentially "rewire" the computer to change its operation. It is clear that the concept had been discussed before John von Neumann (who is often thought of as its inventor) was even aware of the ENIAC's existence, but which of the ENIAC team members first suggested it as a potential solution is unknown. This embryonic concept required several years of research and development before it could be tested in practice – and it was even later before the implications of its power were fully appreciated. Von Neumann, and others, certainly took part in this aspect of the concept's development.

While many people appreciated the elegance of a "stored program" design, few had the technological expertise to create a memory device which would be:

- inexpensive
- capable of being mass produced in large quantities
- had low power consumption
- was capable of storing and retrieving information rapidly

Indeed, these criteria were not all to be satisfied until the commercial development of the VLSI memory chip. It was certainly impractical to attempt to construct a large memory from the types of technology (relays and vacuum tubes) that had been the memory elements in the earlier computing machines.

Many different memory schemes were suggested – one pioneer even describing his approach to the problem as "I examined a textbook on the physical properties of matter in an attempt to find something that would work." Obvious candidates were various schemes based on magnetism, electrical or heat conductance, and the properties of sound waves in different media. The ones used for the first computers were modifications of work that had been done to aid in the interpretation of radar signals during World War II. The most successful memory schemes fall into two different categories: delay line mechanisms, like those used for Turing's Pilot ACE (Fig. 1),[2] and electrostatic devices, like those used for the Manchester "Baby" (Fig. 2).[3] For a complete description of the mechanisms of each of these, the interested reader should refer to texts on the history of computing.[4]

---

[2] See in this volume: Harry D. Huskey, "Hardware Components and Computer Design."

[3] See in this volume: R.B.E. Napper, "The Manchester Mark 1 Computers."

[4] See, for example, Michael R. Williams, *A History of Computing Technology*, second edition, (IEEE Computer Science Press, 1997); or, for a more detailed treatment of early memory systems, see J. P. Eckert, "A Survey of Digital Computer Memory Systems," *Proceedings of the IRE*, October, 1953, to be reprinted in the 20-4 issue of *Annals of the History of Computing*.

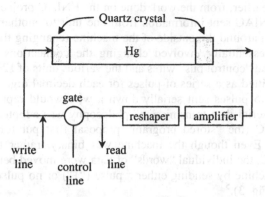

Figure 1: Diagram of the operation of a typical mercury delay line

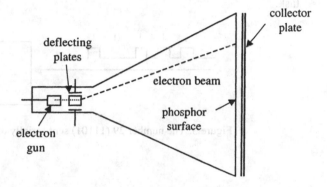

Figure 2: Diagram of the operation of a typical electrostatic memory tube, in this case a "Williams tube"

These two different memory schemes were intimately connected with the basic computer architecture of the first machines and it is now time to briefly examine a few aspects of that topic before we progress further.

4. Elementary architecture of the first machines

The first of the modern computers can be considered to be divided into two different classes depending on how they transferred information around inside the machine. The idea for the stored program computer originated, as stated earlier, from the work done on the ENIAC project in the United States. The ENIAC sent information from one unit to another via a series of wires that ran around the outside of the machine (changing the job the ENIAC was doing essentially involved changing the connections between these "data bus" and "control bus" wires and the various units of ENIAC). Numbers were transmitted as a series of pulses for each decimal digit being moved, for example, 5 pulses sent serially down a wire would represent the digit 5, etc. This "serial data transmission" philosophy was adopted in the design of the EDVAC (the "stored program" proposal first put forward by the ENIAC team). Even though the machine was binary, rather than decimal like the ENIAC, the individual "words" of data were moved between various parts of the machine by sending either a pulse ("1") or no pulse ("0") down a single wire (Fig. 3).[5]

Many of the early computers used this form of data transmission because of two factors: a) it required fewer electronic components to control the signals, and b) it was already known how to design circuits to accomplish this task.

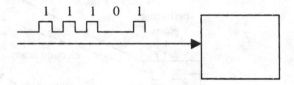

Figure 3: The number 29 (11101) sent serially down a wire

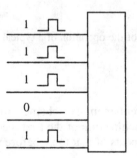

Figure 4: The number 29 (11101) sent down a number of parallel wires

5 See in this volume: Jan Van der Spiegel et al., "The ENIAC: History, Operation, and Reconstruction in VLSI."

The problem with serial transmission is that it is slower than attempting to transmit data via a number of parallel wires – to transmit n bits in a word usually took n clock pulses. When some groups were attempting to create a very high performance machine, they wanted to take advantage of the increase in speed given by transmitting all the data pulses in parallel – a mechanism which would allow n bits to be transmitted in only one clock pulse (Fig. 4).

The first stored program computer project to adopt a parallel transmission scheme was the IAS computer being developed at the Institute of Advanced Study by the team led by von Neumann. This project took much longer to become operational than most of the early machines, simply because the parallel nature of the architecture required the electronic circuits to be much more precise as to the timing of pulses. The additional problem with parallel data paths is that the memory must be able to provide all n data bits of a word at one time.

Delay lines, by their very nature, are serial memory devices – the bits emerge from the delay line one at a time. If you were to incorporate a delay line memory into an, otherwise, parallel machine, you would have to store all 40 bits of a word (in the case of the IAS machine) in 40 different delay lines. Even then it would be awkward because delay lines do not have accurate enough timing characteristics to allow this to be easily engineered. What was needed was the more exact (and higher speed) electronic system of an electrostatic memory. It was still necessary to store one bit of each word in a different electrostatic tube, but at least it was a solution to the problem.

Figure 5: John von Neumann and the IAS computer

The illustration above, of von Neumann standing beside the IAS machine, clearly shows 20 cylindrical devices in the lower portion of the machine – these were one half of the 40 tubes that made up the memory (the other half were on the other side of the machine). Each tube stored 1,024 bits – the first tube stored the first bit of each of the 1,024 words, the second tube contained the second bit, etc.

Of course, it was still possible to use the electrostatic storage tubes in a serial machine as was done with the first machine at Manchester University and the subsequent commercial versions produced by Ferranti. In this case a single word would be stored on one "line" of dots on one tube and the individual bits would be simply sent serially to a computer when required.

When one looks at the history of the early computers, it is often the case that the famous "family tree" diagram (first produced in a document from the U.S. Army) is mentioned (Fig. 6). If you examine that classification scheme you will note that a number of factors are missing.

This categorization of computers obviously takes a very American view of the situation and also leaves out any of the pre-electronic developments that led up the creation of the ENIAC. A better, but still flawed, version was created by Gordon Bell and Allen Newell[6] (Fig. 7). Here, at least, some of the precursors to the modern computer are acknowledged and the major difference between serial and parallel machines are noted. They also include the early British developments at Cambridge, Manchester, the National Physical Laboratory, and have an acknowledgement of the work of Konrad Zuse.

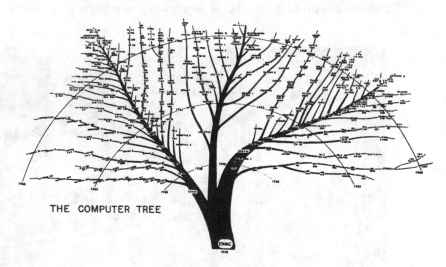

THE COMPUTER TREE

Figure 6: The original U.S. Army "family tree"

---

[6] Gordon Bell and Allen Newell, *Computer Structures, Reading and Examples* (McGraw-Hill, 1971).

Figure 7: The Bell and Newell "family tree"

A more practical approach to listing the early machines might be to group them in some form that will illustrate the times during which they were developed and used. For this task the usual "timeline" is perhaps the best choice of visual device (Fig. 8). There were, however, about a thousand machines created between the years 1930 and 1970 which deserve some consideration in a chart like this and that number prohibits a reasonable representation on anything that will fit into a page. Thus I will suggest that only a few of the most important early machines can be noted in this way – even so, the diagram soon becomes so crowded that it is difficult to see.

There are still a number of thing that can be easily gained from that diagram. It is possible, for example, to understand at a glance that a great deal of very inventive work was done just about the time of the Second World War – most of it, of course, inspired and paid for by the military. The timeline is approximately (but not completely) arranged so that increasing technical sophistication goes from the lower projects to the upper ones. While not a surprise, it certainly does indicate that the faster, more complex, devices were based on the experience gained in earlier experiments.

Another interesting chart, but unfortunately one too complex to show here, would be this timeline with arrows between the projects showing the sources of inspiration, technical advice, and even the exchange of technical personal – the chart would be too complex because almost all the events shown (with the exception of the work of Zuse) relied heavily on one another in these matters.

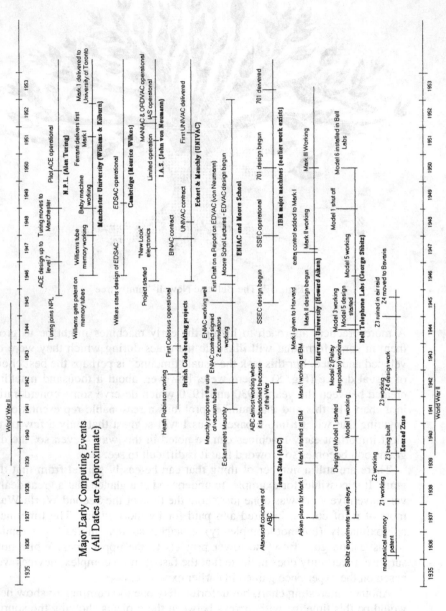

Figure 8: A timeline of major early computer projects

The machines in this timeline are the subject of many of the papers in this volume – some discuss the technical details, some the uses to which they were put, and others refer to the "down stream" effects these developments had on other machines and people. I hope this timeline will provide a handy reference to help you keep the temporal order straight.

Another guide to the novice might well be some of the technical details of the machines themselves. Rather than go into a lengthy description of the different architectures, I propose to offer the chart in Fig. 9 which, I hope, will help to do the job. It would certainly be wrong for anyone to rely on the information contained in this table because it was mostly constructed from my memory – other papers in this volume will offer more detailed information on individual projects.

A glance down any column will show the very wide range of projects and the tremendous increase in complexity as the teams gained experience. For example, the 3 years between the creation of the Bell Labs Model 2 and the Model 5 (1943-1946) saw an increase of complexity from 500 relays to over 9,000; the control systems expand from a simple paper tape reader to one containing 4 problem input/output stations, each with 12 paper tape readers; the control language developing from an elementary "machine language" to one in which instructions were given in a form recognizable today ("BC + GC = A"); and the physical size of each machine increases to the point where the Model 5 required two rooms to house its 10 tons of equipment.

M-mechanical; R - relays; V - vacuum tubes; PT - punched paper tape; B - binary; D - decimal; C - Complex Numbers; FP - floating point; DL - delay line; W - Williams tube; SP - stored program

Machine		Technology	Control system	Memory size	Data Input/Output	Operations per second
Zuse Z1	1938	M	PT (movie film)	16 M, B, FP	keyboard/lamps	+ 1/3, × 1/5
Z2	1939	R	same	same	same	
Z3	1941	R - 2,600	same	64 R, B, FP		× 1
Z4		R	PT	1,000 M, B, FP	3 ttys	+ 3, × 1
The Z3 was the first machine to properly execute a program (on paper tape) - Dec 7, 1941						
Bell	1	1940 R - 450	3 ttys	1? C	PT	
Labs	2	1943 R - 500	PT	5 D	4 PT	
	3	1944 R - 1,300	PT	10 D	4 PT	
	4	1945 R - 1,300	PT	10 D	48 PT	
	5	1946 R - 9,000	PT	44 D		+ 3, × 1/6
Aiken	I	1943 M	PT	72 D	PT/TTY	+ 3, × 1/6
	II	1947 R	PT	50 D, FP	PT/TTY	+ 8, × 1
ABC		V - 600	fixed job	60 B	cards	Designed to solve systems of linear equations
SSEC	1948	R - 23,000, V - 13,000	66 PT	158 D	PT	+ 250
IBM 701	1952	V	SP	4,096 B, W	cards	+ 17,000, × 2,200
ENIAC	1945	V - 18,000	plug wires	20 D	cards	+ 5,000
Colossus	1943	V	fixed job		PT	
Baby	1948	V	SP	32 B	switches/crt	
The Manchester 'baby' machine was the first ever to execute a stored program (1948)						
EDSAC	1949	V - 3,000	SP	512 B	PT/TTY	+ 700, × 225
EDSAC was the first really usable stored program computer (1949)						

Figure 9: Some technical details of early computer projects

5. Conclusions

The distinction between "programmable calculators" and "stored program computers" is seen to be one which can not be readily made on any technological basis. For example, the memory size of the Zuse Z4 machine (a "calculator") is many times larger than either the first (the Manchester "Baby") or second (Cambridge EDSAC) stored program computers. Similarly the massive amounts of technology used on either the IBM SSEC or the ENIAC were far in excess of that used on any of the early stored program computers. The distinction also can not be made on the basis of a date by which any particular project was started or finished – many different machines controlled by punched paper tape were begun after the first stored program computers were created. Any one attempting to casually indicate that project X was "obviously" the first computer on the basis of only a few considerations can be easily proved wrong. As I indicated in my opening remarks: there is more than enough glory in the creation of this technology to be spread around all the very innovative pioneers.

About the only simple conclusion that can be noted is that the problem of creating a memory for the different types of machines was the main stumbling block to the development of computing technology. Until this problem had been solved the computer remained a device which was only available to a few. Now that we have the size and the cost of all the components reduced to almost unimaginable levels, the computer has become a universal instrument that is making bigger and faster changes to our civilization than any other such development – it is well worthwhile knowing where, and by whom, these advances were first made and this volume will certainly help in telling this story.

MICHAEL R. WILLIAMS obtained a Ph.D. in computer science from the University of Glasgow in 1968 and then joined the University of Calgary, first in the Department of Mathematics then as a Professor of Computer Science. It was while working at Glasgow that he acquired an interest in the history of computing. As well as having published numerous books, articles, and technical reviews, he has been an invited lecturer at many different meetings, and has been involved in the creation of 8 different radio, television, and museum productions. During his career he has had the opportunity to work for extended periods at several different universities and at the National Museum of American History (Smithsonian Institution). Besides his work as Editor-in-Chief for the journal *Annals of the History of Computing*, he is a member of several editorial boards concerned with publishing material in the area of the history of computing.

Part I: History, Reconstructions, Architectures

Computers and Mathematics, 1950–1970

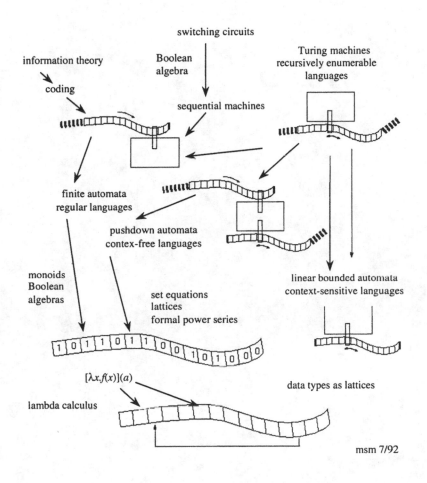

switching circuits

information theory

Boolean
algebra

Turing machines
recursively enumerable
languages

coding

sequential machines

finite automata
regular languages

pushdown automata
contex-free languages

monoids
Boolean
algebras

set equations
lattices
formal power series

linear bounded automata
context-sensitive languages

$[\lambda x, f(x)](a)$

data types as lattices

lambda calculus

msm 7/92

The Structures of Computation

Michael S. Mahoney

Abstract. In 1948 John von Neumann decried the lack of "a properly mathematical-logical" theory of automata. Between the mid-1950s and the early 1970s such a theory took shape through the interaction of a variety of disciplines, as their agendas converged on the new electronic digital computer and gave rise to theoretical computer science as a mathematical discipline. Automata and formal languages, computational complexity, and mathematical semantics emerged from shifting collaborations among mathematical logicians, electrical engineers, linguists, mathematicians, and computer programmers, who created a new field while pursuing their own. As the application of abstract modern algebra to our dominant technology, theoretical computer science has given new form to the continuing question of the relation between mathematics and the world it purports to model.

1. History and computation

The focus of this conference lies squarely on the first generation of machines that made electronic, digital, stored-program computing a practical reality. It is a conference about hardware: about "big iron," about architecture, circuitry, storage media, and strategies of computation in a period when circuits were slow, memory expensive, vacuum tubes of limited life-span, and the trade-off between computation and I/O a pressing concern. That is where the focus of the nascent field and industry lay at the time. But, since this conference is a satellite conference of the International Congress of Mathematicians, it seems fitting to consider too how the computer became not only a means of doing mathematics but also itself a subject of mathematics in the form of theoretical computer science. By 1955, most of the machines under consideration here were up and running; indeed one at least was nearing the end of its productive career. Yet, as of 1955 there was no theory of computation that took account of the structure of those machines as finite automata with finite, random-access storage. Indeed, it was not clear what a mathematical theory of computation should be about. Although the theory that emerged ultimately responded to the internal needs of the computing community, it drew inspiration and impetus from well beyond that community. The theory of computation not only gave mathematical structure to the computer but also gave computational structure to a variety of disciplines and in so doing implicated the computer in their pursuit.

As many of the papers show, this volume is also concerned with how to do the history of computing, and I want to address that theme, too. The multidisciplinary origins and applications of theoretical computer science provide a case study of how something essentially new acquires a history by entering the histories of the activities with which it interacts. None of the fields from which theoretical computer science emerged was directed toward a theory of computation per se, yet all became part of its history as it became part of theirs. Something similar holds for computing in general. Like the Turing Machine that became the fundamental abstract model of computation, the computer is not a single device but a schema. It is indefinite. It can do anything for which we can give it instructions, but in itself it does nothing. It requires at least the basic components laid out by von Neumann, but each of those components can have many different forms and configurations, leading to computers of very different capacities. The kinds of computers we have designed since 1945 and the kinds of programs we have written for them reflect not the nature of the computer but the purposes and aspirations of the groups of people who made those designs and wrote those programs, and the product of their work reflects not the history of the computer but the histories of those groups, even as the computer in many cases fundamentally redirected the course of those histories.

In telling the story of the computer, it is common to mix those histories together, choosing from each of them the strands that seem to anticipate or to lead to the computer. Quite apart from suggesting connections and interactions where in most cases none existed, that retrospective construction of a history of the computer makes its subsequent adoption and application relatively unproblematic. If, for example, electrical accounting machinery is viewed as a forerunner of the computer, then the application of the computer to accounting needs little explanation. But the hesitation of IBM and other manufacturers of electrical accounting machines to move over to the electronic computer suggests that, on the contrary, its application to business needs a lot of explanation. Introducing the computer into the history of business data processing, rather than having the computer emerge from it, brings the questions out more clearly.

The same is true of theoretical computer science as a mathematical discipline. As the computer left the laboratory in the mid-1950s and entered both the defense industry and the business world as a tool for data processing, for real-time command and control systems, and for operations research, practitioners encountered new problems of non-numerical computation posed by the need to search and sort large bodies of data, to make efficient use of limited (and expensive) computing resources by distributing tasks over several processors, and to automate the work of programmers who, despite rapid growth in numbers, were falling behind the even more quickly growing demand for systems and application software. The emergence during the 1960s of high-level languages, of time-sharing operating systems, of computer graphics, of communications between computers, and of artificial intelligence increasingly refocused attention from the physical machine to abstract models of computation as a dynamic process.

Most practitioners viewed those models as mathematical in nature and hence computer science as a mathematical discipline. But it was mathematics with a difference. While insisting that computer science deals with the structures and transformations of information analyzed mathematically, the first Curriculum Committee on Computer Science of the Association for Computing Machinery (ACM) in 1965 emphasized the computer scientists' concern with effective procedures:

> The computer scientist is interested in discovering the pragmatic means by which information can be transformed to model and analyze the information transformations in the real world. The pragmatic aspect of this interest leads to inquiry into effective ways to accomplish these at reasonable cost.[1]

A report on the state of the field in 1980 reiterated both the comparison with mathematics and the distinction from it:

> Mathematics deals with theorems, infinite processes, and static relationships, while computer science emphasizes algorithms, finitary constructions, and dynamic relationships. If accepted, the frequently quoted mathematical aphorism, 'the system is finite, therefore trivial,' dismisses much of computer science.[2]

Computer people knew from experience that "finite" does not mean "feasible" and hence that the study of algorithms required its own body of principles and techniques, leading in the mid-1960s to the new field of computational complexity. Talk of costs, traditionally associated with engineering rather than science, involved more than money. The currency was time and space, as practitioners strove to identify and contain the exponential demand on both as even seemingly simple algorithms were applied to ever larger bodies of data. Yet, as central as algorithms were to computer science, the report continued, they did not exhaust the field, "since there are important organizational, policy, and nondeterministic aspects of computing that do not fit the algorithmic mold."

---

[1] "An Undergraduate Program in Computer Science – Preliminary Recommendations," *Communications of the ACM*, 8, 9 (1965), 543–552; at 544.

[2] Bruce W. Arden (ed.), *What Can Be Automated?: The Computer Science and Engineering Research Study* (COSERS) (Cambridge, MA: MIT Press, 1980), 9.

Thus, in striving toward theoretical autonomy, computer science has always maintained contact with practical applications, blurring commonly made distinctions among science, engineering, and craft practice, or between mathematics and its applications. Theoretical computer science offers an unusual opportunity to explore these questions because it came into being at a specific time and over a short period. It did not exist in 1955, nor with one exception did any of the fields it eventually comprised. In 1970, all those fields were underway, and theoretical computer science had its own main heading in *Mathematical Reviews*.

2. Agendas

In tracing its emergence and development as a mathematical discipline, I have found it useful to think in terms of *agendas*. The *agenda*[3] of a field consists of what its practitioners agree ought to be done, a consensus concerning the problems of the field, their order of importance or priority, the means of solving them, and perhaps most importantly, what constitutes a solution. Becoming a recognized practitioner means learning the agenda and then helping to carry it out. Knowing what questions to ask is the mark of a full-fledged practitioner, as is the capacity to distinguish between trivial and profound problems; "profound" means moving the agenda forward. One acquires standing in the field by solving the problems with high priority, and especially by doing so in a way that extends or reshapes the agenda, or by posing profound problems. The standing of the field may be measured by its capacity to set its own agenda. New disciplines emerge by acquiring that autonomy. Conflicts within a discipline often come down to disagreements over the agenda: what are the really important problems?

As the shared Latin root indicates, agendas are about action: what is to be *done*?[4] Since what practitioners do is all but indistinguishable from the way they go about doing it, it follows that the tools and techniques of a field

---

[3] To get the issue out of the way at the beginning, a word about the grammatical number of *agenda*. It is a Latin plural gerund, meaning "things to be done." In English, however, it is used as a singular in the sense of "list of things to do." Since I am talking here about multiple and often conflicting sets of things to be done, I shall follow the English usage, thus creating room for a non-classical plural, *agendas*.

[4] Emphasizing action directs attention from a body of knowledge to a complex of practices. It is the key, for example, to understanding the nature of Greek geometrical analysis as presented in particular in Pappus of Alexandria's *Mathematical Collection*, which is best viewed as a mathematician's toolbox. See my "Another Look at Greek Geometrical Analysis," *Archive for History of Exact Sciences* 5 (1968), 318–348.

embody its agenda. When those tools are employed outside the field, either by a practitioner or by an outsider borrowing them, they bring the agenda of the field with them. Using those tools to address another agenda means reshaping the latter to fit the tools, even if it may also lead to a redesign of the tools, with resulting feedback when the tool is brought home. What gets reshaped and to what extent depends on the relative strengths of the agendas of borrower and borrowed.

There are various examples of this from the history of mathematics, especially in its interaction with the natural sciences. Historians speak of Plato's agenda for astronomy, namely to save the phenomena by compounding uniformly rotating circles. One can derive that agenda from Plato's metaphysics and thus see it as a challenge to the mathematicians. However, one can also – and, I think, more plausibly – view it as an agenda embodied in the geometry of the circle and the Eudoxean theory of ratio. Similarly, scientific folklore would have it that Newton created the calculus to address questions of motion. Yet, it is clear from the historical record, first, that Newton's own geometrical tools shaped the structure and form of his *Principia* and, second, that once the system of the *Principia* had been reformulated in terms of the calculus (Leibniz', not Newton's), the mathematical resources of central-force mechanics shaped, if indeed it did not dictate, the agenda of physics down to the early nineteenth century.

Computer science had no agenda of its own to start with. As a physical device it was not the product of a scientific theory and hence inherited no agenda. Rather it posed a constellation of problems that intersected with the agendas of various fields. As practitioners of those fields took up the problems, applying to them the tools and techniques familiar to them, they defined an agenda for computer science. Or, rather, they defined a variety of agendas, some mutually supportive, some orthogonal to one another. Theories are about questions, and where the nascent subject of computing could not supply the next question, the agenda of the outside field provided its own. Thus the semigroup theory of automata headed on the one hand toward the decomposition of machines into the equivalent of ideals and on the other toward a ring theory of formal power series aimed at classifying formal languages. Although both directions led to well defined agendas, it became increasingly unclear what those agendas had to do with computing.

3. Theory of automata

Since time is limited, and I have set out the details elsewhere, a diagram will help to illustrate what I mean by a convergence of agendas, in this case leading to the formation of the theory of automata and formal languages.[5] The core of the field, its paradigm if you will, came to lie in the correlation between four classes of finite automata ranging from the sequential circuit to the Turing machine and the four classes of phrase structure grammars set forth by Noam Chomsky in his classic paper of 1959.[6] With each class goes a particular body of mathematical structures and techniques, ranging from monoids to recursive function theory.

As the diagram shows by means of the arrows, that core resulted from the confluence of a wide range of quite separate agendas. Initially, it was a shared interest of electrical engineers concerned with the analysis and design of sequential switching circuits and of mathematical logicians interested in the logical possibilities and limits of nerve nets as set forth in 1943 by Warren McCulloch and Walter Pitts, themselves in pursuit of a neurophysiological agenda.[7] In some cases, it is a matter of passing interest and short-term collaborations, as in the case of Chomsky, who was seeking a mathematical theory of grammatical competence, by which native speakers of a language extract its grammar from a finite number of experienced utterances and use it to construct new sentences, all of them grammatical, while readily rejecting ungrammatical sequences.[8] His collaborations, first with mathematical psychologist George Miller and then with Bourbaki-trained mathematician Marcel P. Schützenberger, lasted for the few years it took to determine that phrase-structure grammars and their automata would not suffice for the grammatical structures of natural language.

---

[5] For more detail see my "Computers and Mathematics: The Search for a Discipline of Computer Science," in J. Echeverría, A. Ibarra and T. Mormann (eds.), *The Space of Mathematics* (Berlin/New York: De Gruyter, 1992), 347–61, and "Computer Science: The Search for a Mathematical Theory," in John Krige and Dominique Pestre (eds.), *Science in the 20th Century* (Amsterdam: Harwood Academic Publishers, 1997), Chap. 31.

[6] Noam Chomsky, "On Certain Formal Properties of Grammars," *Information an Control* 2, 2 (1959), 137–167.

[7] Warren S. McCulloch and Walter Pitts, "A Logical Calculus of the Ideas Immanent in Nervous Activity," *Bulletin of Mathematical Biophysics* 5 (1943), 115–33; repr. in Warren S. McCulloch, *Embodiments of Mind* (MIT, 1965), 19–39.

[8] "The grammar of a language can be viewed as a theory of the structure of this language. Any scientific theory is based on a certain finite set of observations and, by establishing general laws stated in terms of certain hypothetical constructs, it attempts to account for these observations, to show how they are interrelated, and to predict an indefinite number of new phenomena. A mathematical theory has the additional property that predictions follow rigorously from the body of theory." Noam Chomsky, "Three Models of Language," *IRE Transactions in Information Theory* 2, 3 (1956), 113–24; at 113.

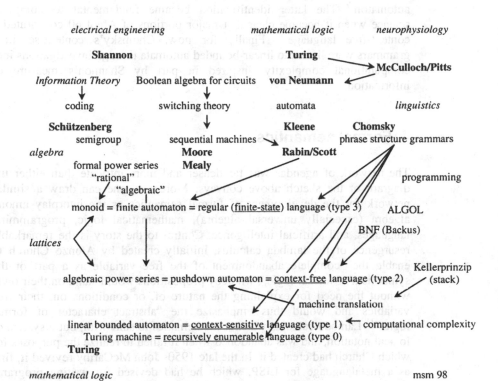

Figure 1: The Agendas of Computer Science

Schützenberger, for his part, came to the subject from algebra and number theory (the seminar of Bourbakist Pierre Dubreil) by way of coding theory, an agenda in which Benoit Mandelbrot was also engaged at the time. It was the tools that directed his attention. Semigroups, the fundamental structures of Bourbaki's mathematics, had proved unexpectedly fruitful for the mathematical analysis of problems of coding, and those problems in turn turned out to be related to finite automata, once attention turned from sequential circuits to the tapes they recognized. Pursuing his mathematical agenda led Schützenberger to generalize his original problem and thereby to establish an intersection point, not only with Chomsky's linguistic agenda, but also with the agenda of machine translation and with that of algebraic programming languages. The result was the equivalence of "algebraic" formal power series, context-free languages, and the pushdown (or stack) automaton.[9] The latter identification became fundamental to computer science when it became clear that major portions of Algol 60 constituted a context-free language.[10] Finally for now, Chomsky's context-sensitive grammars were linked to linear-bounded automata through investigations into computational complexity, inspired in part by Shannon's measure of information.

4. Formal semantics

The network of agendas was far denser and more intricate than either the diagram or the sketch above conveys. Moreover, one can draw a similar network for the development of formal semantics as the interplay among algebra (especially universal algebra), mathematical logic, programming languages, and artificial intelligence. Central to the story is the remarkable resurgence of the lambda calculus, initially created by Alonzo Church to enable the "complete abandonment of the free variable as a part of the symbolism of formal logic," whereby propositions would stand on their own, without the need for explaining the nature of, or conditions on, their free variables and would thus emphasize the "abstract character of formal logic."[11] Lambda calculus was not mathematics to start with, but a system of logical notation, and was abandoned when it failed to realize the purposes for which Church had created it. In the late 1950s John McCarthy revived it, first as a metalanguage for LISP, which he had devised for writing programs emulating common-sense reasoning and for mechanical theorem-proving, and then in the early 1960s as the basis of a mathematical theory of computation focused on semantics rather than syntax.

"Computer science," McCarthy insisted, "must study the various ways elements of data spaces are represented in the memory of the computer and how procedures are represented by computer programs. From this point of view, most of the work on automata theory is beside the point."[12] In

---

9   At about the same time, but apparently quite independently, Robert Rosen brought the semigroup model of coding into the agenda of mathematical biophysics at the University of Chicago in "The DNA-Protein Coding Problem," *Bulletin of Mathematical Biophysics* 21(1959), 71–95.

10   Seymour Ginsburg and H. Gordon Rice, "Two Families of Languages Related to ALGOL," *Journal of the ACM* 9 (1962), 350–371.

11   Alonzo Church, "A Set of Postulates for the Foundation of Logic," *Annals of Mathematics*, 2nd ser., 33 (1932), 346–66.

12   "Towards a Mathematical Science of Computation," *Proc. IFIP Congress 62* (Amsterdam: North-Holland, 1963), 21–28; at 21. Automata theory stayed too close to the machine, he explained: "... the fact of finiteness is used to show that the automaton will eventually repeat a state. However, anyone who waits for an

McCarthy's view, programs consisted of chains of functions that transform data spaces. Automata theory viewed functions as sets of ordered pairs mapping the elements of two sets and was concerned with whether the mapping preserved the structures of the sets. McCarthy was interested in the functions themselves as abstract structures, not only with their equivalence but also their efficiency. A suitable mathematical theory of computation, he proposed, would provide, first, a universal programming language along the lines of Algol but with richer data descriptions;[13] second, a theory of the equivalence of computational processes, by which equivalence-preserving transformations would allow a choice among various forms of an algorithm, adapted to particular circumstances; third, a form of symbolic representation of algorithms that could accommodate significant changes in behavior by simple changes in the symbolic expressions; fourth, a formal way of representing computers along with computation; and finally a quantitative theory of computation along the lines of Shannon's measure of information.

Except for the last item, which in the mid-1960s became the focus of the rapidly developing field of computational complexity, McCarthy's agenda for a formal semantics initially attracted little support in the United States. It did catch on in England, however, where under Christopher Strachey's leadership P.J. Landin pursued the lambda calculus approach to programming language semantics and, where R.M. Burstall, seconded by Donald Michie's Machine Intelligence Unit at Edinburgh, attempted to link it to universal algebra as a means of proving the correctness of programs. Strachey himself pursued the peculiar problem posed by the storage of program and data in a common memory, which in principle allowed unrestricted procedures which could have unrestricted procedures as values; in particular a procedure could be applied to itself.

To see the problem, consider the structure of computer memory, represented mathematically as a mapping of contents to locations. That is, state σ is a function mapping each element l of the set L of locations to its value $\sigma(l)$ in V, the set of allowable values. A command effects a change of state; it is a function γ from the set of states S into S. Storing a command means that γ can take the form $\sigma(l)$, and hence $\sigma(l)(\sigma)$ should be well defined. Yet, as Dana Scott insisted in his "Outline of a mathematical theory of computation" in 1970, "[t]his is just an insignificant step away from the self-application problem $p(p)$ for 'unrestricted' procedures p, and it is just as hard to justify mathematically."

IBM 7090 to repeat a state, solely because it is a finite automaton, is in for a very long wait." (*Ibid.*, 22).

[13] Cobol, he noted, suffered from its attachment to English, and Uncol was "an exercise in group wishful thinking" (*Ibid*, 34).

Computers and Mathematics, 1950–1970

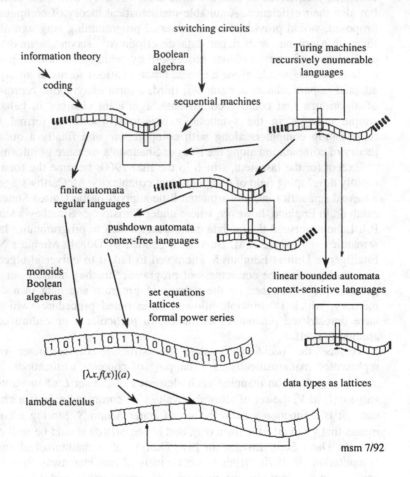

Figure 2: Machines and Languages

The fixpoint operator of the lambda calculus seemed to point a way around the problem, but that made it clear that the lack of a mathematical model for the lambda calculus threatened to undermine the enterprise.

To date, no mathematical theory of functions has ever been able to supply conveniently such a freewheeling notion of function except at the cost of being inconsistent. The main *mathematical* novelty of the present study is the creation of a proper mathematical theory of functions which accomplishes these aims (consistently!) and which can be used as the basis for the *metamathematical* project of providing the "correct" approach to semantics.[14]

By creating a model in the form of continuous lattices with fixpoints, Scott not only made the lambda calculus the foundation for denotational, or mathematical, semantics, but also added a new item to the agenda of abstract lattice theory. How did lambda calculus become mathematics? It is a question of interest, of getting onto the mathematical agenda. Computers gave the lambda calculus mathematical meaning, because it served to give a mathematical account of computation.

But giving mathematical structure to the lambda calculus in turn pushed mathematical semantics toward a focus on abstract functions and hence toward a recent branch of mathematics that had seemed, even to one of its creators, Samuel Eilenberg, of limited applicability to theoretical computer science, namely category theory. The interaction in the 1970s and 1980s of semantics with universal algebra, in particular Omega-algebras, and then categories parallels that of algebra and automata in the 1960s. By 1988, Saunders Maclane's *Categories for the Working Mathematician* had a counterpart in Andrea Asperti's and Giuseppe Longo's *Categories, Types, and Structures: An Introduction to Category Theory for the Working Computer Scientist.*

5. Computers and mathematics

What does this all add up to? It is in part a story of how a subject becomes mathematical, and one can tell it as an example of the traditional view of the relation of mathematics to its applications. The concepts are created for "internal" reasons and then applied. But there is intriguing evidence to suggest a more complex interaction. Let me turn to three mathematicians to help me make the point. In 1948, John von Neumann said concerning the theory of automata:

There exists today a very elaborate system of formal logic, and, specifically, of logic as applied to mathematics. This is a discipline with many good sides, but also with certain serious weaknesses. This is not the occasion to enlarge upon the good sides, which I certainly have no intention to belittle. About the inadequacies, however, this may be said:

---

[14] *Ibid.,* 4–5.

Everybody who has worked in formal logic will confirm that it is one of the technically most refractory parts of mathematics. The reason for this is that it deals with rigid, all-or-none concepts, and has very little contact with the continuous concept of the real or of the complex number, that is, with mathematical analysis. Yet analysis is the technically most successful and best-elaborated part of mathematics. Thus formal logic is, by the nature of its approach, cut off from the best cultivated portions of mathematics, and forced onto the most difficult part of the mathematical terrain, into combinatorics.

The theory of automata, of the digital, all-or-none type, as discussed up to now, is certainly a chapter in formal logic. It will have to be, from the mathematical point of view, combinatory rather than analytical.[15]

Von Neumann subsequently made it clear he wanted to pull the theory back toward the realm of analysis, and he did not expand upon the nature of the combinatory mathematics that might be applicable to it.

In reviewing the role of algebra in the development of computer science in 1969, Garrett Birkhoff, whose lattice theory, once thought useless, was proving a fundamental tool of the new field, remarked that finite Boolean algebras had held no interest for him as a mathematician because they were all equivalent up to isomorphism. But as Boolean algebra was applied to the analysis and design of circuits, it led to problems of minimization and optimization that proved both difficult and interesting. The same held true of the optimization of error-correcting binary codes. Together,

> [these] two unsolved problem in binary algebra ... illustrate the fact that *genuine applications can suggest simple and natural but extremely difficult problems*, which are overlooked by pure theorists. Thus, while working for 30 years (1935–1965) on generalizing Boolean algebra to lattice theory, I regarded finite Boolean algebras as trivial because they could all be described up to isomorphism, and completely ignored the basic "shortest form" and "optimal packing" problems described above.[16]

---

[15] John von Neumann, "On a Logical and General Theory of Automata" in *Cerebral Mechanisms in Behavior – The Hixon Symposium*, ed. L.A. Jeffries (New York: Wiley, 1951), 1–31; repr. in *Papers of John von Neumann on Computing and Computer Theory*, ed. William Aspray and Arthur Burks (Cambridge, MA/London: MIT Press; Los Angeles/San Francisco: Tomash Publishers, 1987), 391–431; at 406.

[16] Garrett Birkhoff, "The Role of Modern Algebra in Computing," *Computers in Algebra in Number Theory* (American Mathematical Society, 1971), 1–47, repr. in his *Selected Papers on Algebra and Topology* (Boston: Birkhäuser, 1987), 513–559; at 517; emphasis in the original.

Earlier in the article, Birkhoff had pointed to other ways in which "the problems of computing are influencing algebra." To make the point, he compared the current situation with the Greek agenda of "rationalizing geometry through constructions with ruler and compass (as analog computers)."

> By considering such constructions and their optimization in depth, they were led to the existence of irrational numbers, and to the problems of constructing regular polygons, trisecting angles, duplicating cubes, and squaring circles. These problems, though of minor technological significance, profoundly influenced the development of number theory.
>
> I think that our understanding of the potentialities and limitations of algebraic symbol manipulation will be similarly deepened by attempts to solve problems of optimization and computational complexity arising from digital computing.

Birkhoff's judgment, rendered at just about the time that theoretical computer science was assigned its own main heading in *Mathematical Reviews*, points to just one way in which computer science was opening up a new realm of mathematical interest in the *finite* but *very large*. Computational complexity was another way.

Several years later, Samuel Eilenberg, who had collaborated with Saunders Maclane in the creation of category theory, decided that automata and formal languages had progressed individually and in tandem to the point where they could be placed on a common mathematical foundation. The current literature, though algebraic in content and approach, reflected the specific interests that had motivated them. "It appeared to me," wrote Eilenberg in the preface of his intended four-volume *Automata, Languages, and Machines*,

> that the time is ripe to try and give the subject a coherent mathematical presentation that will bring out its intrinsic aesthetic qualities and bring to the surface many deep results which merit becoming part of mathematics, regardless of any external motivation.[17]

Yet, in becoming part of mathematics the results would retain the mark characteristic of their origins. All of Eilenberg's proofs were constructive in the sense of constituting algorithms.

> A statement asserting that something exists is of no interest unless it is accompanied by an algorithm (i.e., an explicit or effective procedure) for producing this "something."

---

[17] *Automata, Languages, and Machines* (2 vols., NY: Columbia University Press, 1974), Vol. A, xiii.

In addition, Eilenberg held back from the full generality to which abstract mathematicians usually aspired. Aiming at a "restructuring of the material along lines normally practiced in algebra," he sought to reinforce the original motivations rather than to eradicate them.

Both mathematics and computer science would benefit from his approach, he argued:

> To the pure mathematician, I tried to reveal a body of new algebra, which, despite its external motivation (or perhaps because of it) contains methods and results that are deep and elegant. I believe that eventually some of them will be regarded as a standard part of algebra. To the computer scientist I tried to show a correct conceptual setting for many of the facts known to him (and some new ones). This should help him to obtain a better and sharper mathematical perspective on the theoretical aspects of his researches.

Coming from a member of Bourbaki, who insisted on the purity of mathematics, Eilenberg's statement is all the more striking in its recognition of the applied origins of "deep and elegant" mathematical results.

What is particularly important about the formation of theoretical computer science as a mathematical discipline is that in the application of mathematics to computation the traffic traveled both ways. While providing mathematical grounding for the powerful techniques embedded in current programming tools, theoretical computer science gave "physical" meaning to semigroups, lattices, Omega-algebras, categories, thus placing some of the most abstract, "useless" concepts of modern mathematics at the heart of modern technology. In doing so, it motivated their further analysis as mathematical entities, bringing out unexpected properties and relationships among them.

What has it done for computing? That is a trickier question. Despite the elegant theory and the powerful tools based on it, computer science is still a long way from possessing the sort of mathematical theory McCarthy envisioned, and certainly from the practical goal he set for it. In a discussion on the last day of the second NATO Conference on Software Engineering held in Rome in October 1969, Christopher Strachey, Director of the Programming Research Group at Oxford University, lamented that "one of the difficulties about computing science at the moment is that it can't demonstrate any of the things that it has in mind; it can't demonstrate to the software engineering people on a sufficiently large scale that what it is doing is of interest or importance to them."[18] Almost two decades later, the situation had not changed much. C.A.R. Hoare, in his inaugural lecture as Professor of Computation at Oxford in 1985 told his audience that he supposed as a matter

---

[18] Peter Naur, Brian Randell, and J.N. Buxton (eds.), *Software Engineering: Concepts and Techniques. Proceedings of the NATO Conferences* (NY: Petrocelli, 1976), 147.

of principle that computers are mathematical machines, computer programs are mathematical expressions, a programming language is a mathematical theory, and programming is a mathematical activity. "These are general philosophical and moral principles, and I hold them to be self-evident – which is just as well, because all the actual evidence is against them. Nothing is really as I have described it, neither computers nor programs nor programming languages nor even programmers."[19] Given the mathematical sophistication of theoretical computer science by 1985, that seems a remarkable statement. Given the traditional role of mathematics in modern science, it is a statement worthy of the attention of historians and one rich in historiographical possibilities.

To see why brings us around to a historiographical theme of this conference. The history of science has until recently tended to ignore the role of technology in scientific thought, though perhaps less so in Germany than elsewhere, as indeed this conference testifies. The situation has begun to change with recent work on the role and nature of the instruments that have mediated between scientists and the objects of their study, ranging from telescopes and microscopes in the 17th century to bubble chambers in the 20th. But, outside of the narrow circle of people who think of themselves as historians of computing, historians of science (and indeed of technology) have ignored the instrument that by now so pervades science and technology as to be indispensable to their practice. Increasingly, computers not only mediate between practitioners and their subjects but also replace the subjects with computed models. One might argue that no instrument since the 17th century has shaped the practice of science to the extent that the computer has done. Some time soon, historians are going to have to take the computer seriously as an object of study, and it will be important, when they do, that they understand the ambiguous status of the computer itself.

MICHAEL S. MAHONEY is Professor of History in the program in history of science at Princeton University, where he earned his Ph.D. in history in 1967. A specialist on the development of the mathematical sciences from antiquity to 1700, he turned in the early 1980s to the history of computing, where his research has focused on the formation of theoretical computer science as a mathematical discipline and on the origins and development of software engineering. He has worked as consultant to Bell Labs both on software de-

---

[19] C.A.R. Hoare, "The Mathematics of Programming," in his *Essays in Computing Science* (Hemel Hempstead: Prentice Hall International, 1989), 352.

velopment and on an oral history of UNIX. He also served in 1990–1991 as Chair of an OTA Advisory Panel for "Computer Software and Intellectual Property: Meeting the Challenges of Technological Change and Global Competition." Mahoney has been Editor of the ACM Press's History Series, a member of the Conference Committee, and a member of the Advisory Committee for SIGGRAPH's "Milestones: The History of Computer Graphics." He served as historian for the second ACM Conference on the History of Programming Languages (Cambridge, 1993).

Reconstructions, Historical and Otherwise: The Challenge of High-Tech Artifacts

Robert W. Seidel

Abstract. I examine the reconstruction of artifacts by museums, the use of artifacts in reconstructions of history by historians, and the potential of virtual reconstruction for historical purposes based upon past efforts to display and to explain the development of high-technology objects like the particle accelerator, laser, and computer at the University of California's Lawrence Hall of Science, Los Alamos National Laboratory, the Smithsonian Institution. The historical reconstruction of the development of the cyclotron and of the laser shows the importance of teamwork between historians, scientists, and engineers in formulating accurate historical reconstructions.

1. Introduction

The current interest in the simulation, reconstruction and reactivating of early computers reflects an enthusiasm on the part of the practitioners. The historian's interest in artifacts is different from the practitioner's. The difference between their perspectives creates a tension between the historian's use of artifacts and the practitioner's reconstruction of them that should be reconciled if both are to profit from such reconstructions. I would like to reflect on the nature of such a reconciliation, not least because it goes to the heart of the nature of the museum and of the history of technology.

A brief review of the history of science museums and the disciplinary construction of the history of technology indicates some of the major difficulties historians, scientists, and museum professionals have had in interpreting artifacts. Although artifacts exist, and can be validated using historical techniques, the task of the historian/curator in interpreting their meaning is fraught with pitfalls.[1] The historian's interpretation is no longer (if it ever was) privileged, but it is "authoritative" in a constrained sense.

---

[1] Cf., *inter alia*, Steven Lubar and W. David Kingery, *History from Things: Essays on Material Culture* (Washington & London: Smithsonian Institution Press, 1993).

As Paul Forman[2] has argued, the task of the historian requires critical independence, but an exchange of views between historians and their subjects can be mutually instructive.[3] Historians of science and technology have deferred to their subjects as the valuators of "historic" accomplishments, and have used their representations of historical reality, although their interpretations involve critical assessment of historical testimony, text and artifact. It is in the careful use of his critical tools that the historian discovers history.

This is evident in museums, where historians have interpreted technology for centuries. Such colossi as the Deutsches Museum, the Chicago Museum of Science and Industry, the National Air and Space Museum, and the Science Museum of London are testimonies to the accomplishments of modern technology, richly funded by government and industry, overflowing with artifacts, only a small percentage of which can be displayed, and staffed by museum curators whose expertise ranges from undergraduate study to advanced degrees in history. Their disciplinary interests, when blended in the crucible of exhibit development with the institutional interests of museum administrators in patronage and popularity, are often diluted. However, when the artifact is subjected to an interpretation that takes into account not only its construction, function, and the details of its invention and development, but also its political, social, and economic contexts, both the historian and the participant can take pride in the accomplishment.[4]

In historical publication, as opposed to display, the three-dimensional aspect of the artifact must give way to a two-dimensional graphic representation or a verbal description. In such research, particularly in the mature areas of the history of technology, the artifact seems to recede into the background as the context within which it developed swells to fill the mental picture that the historian paints. The use of the artifact as a primary source, however, may answer crucial questions about that development, and new media, like virtual reconstructions in cyberspace, may enhance the historian's use of that information.

Although some historians portray the business, military, and scientific contexts of computers, most history of computing still focuses on artifacts and their makers. As in the case of other artifacts, large sums are still available for their celebration and display, often from the makers themselves. Hence, the Computer Museum in Boston is the work of the same individuals who built Digital Equipment Corporation. The Microsoft, Intel, Magnavox, and DEC museums represent and celebrate the accomplishments of these

---

[2] Paul Forman, "Independence, Not Transcendence, for the Historian of Science," *Isis 82* (1991), 71–86.

[3] Cf., e.g. Roger Stuewer, ed., *Nuclear Physics in Retrospect: Proceedings of a Symposium on the 1930s* (Minneapolis: University of Minnesota Press, 1979), 318–322.

[4] Edward Tenner, "Information Age at the National Museum of American History," *Technology and Culture* 33 (Oct. 1992), 780–87.

corporations. The Smithsonian Institution relies heavily on the private sector for funding and the raw materials of their exhibits. As in other forms of patronage, questions of autonomy and emphasis arise.

Recent controversies between scientists, veterans, congress, and the curators of Smithsonian exhibits have cast a pall over historical interpretation of science and technology. The disputes over the *Enola Gay* and "Science in American Life" exhibits have made it clear that historians employed by museums do not enjoy the academic freedom guaranteed to their academic colleagues, and that the director who relies too heavily on revisionist history will be toppled by the powers that be. More traditional historians may face the criticism of postmodernists, feminists, animal rights advocates, political activists, and other "politically correct" special interest groups in their attempts to reconstruct the past. The historian may well feel safer writing a monograph than facing the political consequences of building a museum exhibit that expresses the same interpretations.

Within the museum, moreover, artifacts dominate the representation of history. Text, when used, serves primarily to describe the artifact. While the broader contexts of development may be suggested by the grouping of artifacts in exhibits or displays with thematic unities which, like those of the dramatic arts, suggest the time, place, and circumstance within which their construction took place, but, as one might imagine, the suggestion of technological determinism by the dominance of the artifact is seldom balanced by an account of the determination of the technology by its environment. As historians of technology have moved toward an understanding of the sociological, economic, and other environmental determinants, practitioners and possessors of artifacts have often refused to follow.

The historical reconstruction of computers by practitioners also privileges the artifact, although often in virtual form. In the past twenty years, the volume of "hardware" history has grown as the computer itself shrank from gigantic proportions to the desktop and the microchip processor. As the artifacts of modern computing become invisible, older, larger computers supply a symbol of computing to practitioners, the public, and patrons which is not only visible, but comprehensible.

The inherent lack of interesting visual clues has plagued the interpretation of the computer from ENIAC to the present. The visual presentation of computers has required "special effects" enhancements ranging from the Ping-Pong ball hemispheres used to magnify the blinking lights of the ENIAC, to the gigantic and elaborate movie computers of *Colossus: The Forbin Project*, and other films. While simulation of the operation of actual computers can represent the functionality in more significant ways, substituting software representations for glorified hardware, it is unclear how this serves the purposes of display, the act of historical interpretation, or the antiquarian passions that have fueled interest in artifacts in the past.

The reconstruction of the artifact can help the reconstruction of the past. However, the use of artifacts for historical reconstruction requires the same critical apparatus that has informed the study of texts. Critical questions should be posed in the design of projects to reconstruct artifacts, but seldom are. In what follows, I compare museum, historical, and virtual reconstructions to illuminate this process.

2. Museum reconstructions

The museum was once, for the professional scientist, the laboratory within which he conducted his experiments and, quite literally, sought his "Muse." Museums deconstructed reality. The identification, classification, and organization of nature were their scientific *raison d'être*.[5] Whatever assistance the carefully arranged and labeled collections of museums gave to memory, it was not through the reconstruction of the reality they purported to represent. The act of classification itself does violence to reality in order to make it accessible to reason. Indeed, most reconstructions of memory were literary, from journals and diaries of observant travelers who brought their own impressions of the noble savage and the heart of darkness to their readers in the form of structured narrative.[6]

Historians built their discipline on textual criticism. In formulating their narratives of the past, they seek to interpret the evidence of the past to construct an intelligible story for the present. The physical analogue of this effort is the restoration of historic sites. The ruins of ancient Anasazi pueblos in the southwest, industrial cities like Lowell in the northeast, and historical Williamsburg in the mid-Atlantic United States provide an experience of the past that is more "authentic" than outdoor museums that assemble buildings from other locales. These reconstructions limited by the imagination and knowledge of the curators and exhibit staff, the materials available, and the interpretation by guides.

Indoor museums present a different sort of problem for those interested in reconstructing the past. A focus on the design, function, performance, or operating characteristics of an artifact, without regard for intellectual, economic, social, political, technical, and other influences or effects, may help visitors to understand a machine and appreciate its technical evolution but not *why* it came into being when it did, looked like it did, or was used as it was. Similarly, celebratory exhibits that present the "myth of progress" and the "heroic inventor" as sufficient explanation for the origins, development, and

---

5   Pickstone, John V., "Museological Science? The Place of the Analytical/Comparative in 19th-Century Science, Technology and Medicine," *History of Science* 32 (1994), 111–138.

6   Justin Stagl, *A History of Curiosity: The Theory of Travel, 1550–1800*, (Chur, Switzerland: Harwood Academic Publishers, 1995).

impact of technology give short shrift to the underlying historical forces that determine them.[7]

In an effort to go beyond this stage, historians of technology, whose sympathy for internal history of artificial devices is greater than many of their colleagues, have undertaken to review museum exhibits to answer questions such as "whether the exhibit has a unifying theme of purpose. If it does, is it clearly stated, ... valid in the context of other historical work [and] innovative? ... Is the theme argued effectively? ... Is there a structure to the exhibit that leads the visitor through the development of the theme?"[8]

Within this framework, artifacts may be used as evidence of scale, of use, of origins, of inventive style, and of cultural and social context. "Academic historians," one curator warns, "are not familiar with the study of three-dimensional objects and have seldom if ever used museum exhibits as sources for research." Reviews in *Technology and Culture* and *American Heritage of Invention and Technology* suggest that the best sources for such research are the records of research conducted by curators in the construction of the exhibit. These should be saved and made accessible to scholars, and, "at the very least, there should be an annotated copy of the exhibit script – including a list of artifacts."[9]

Yet, the reconstruction of history should include involvement with the material culture of the past, just as the reconstruction of the artifact should include an understanding of history. In order to understand why this is not yet a common practice, I want to turn now to an exploration of the kind of reconstructions historians have done.

3. Historical reconstruction

I found little interest in material culture among most historians of nuclear science and technology at the Bradbury Science Museum. To my knowledge, in the years since it has systematically collected and documented artifacts of the atomic age in its warehouse, no scholar has asked to examine that collection. I attempted to stimulate such interest in the conventional manner by convening a symposium[10] on postwar technology transfer. It included the first

---

7   Joseph J. Corn, "Interpreting the History of American Technics," in *History Museums in the United States: A Critical Assessment,* ed. Warren Leon and Roy Rosenzweig (Urbana and Chicago, 1989), 237–261.

8   Bernard S. Finn, "Exhibit Reviews: Twenty Years After," *Technology and Culture.* 20:4 (Oct. 1989), 996–998.

9   Ibid., p. 1002.

10 Robert W. Seidel and Paul Henriksen, *Proceedings of the Symposium On The Transfer of Technology from Wartime Los Alamos to Peacetime Research* (Los Alamos, Bradbury Science Museum, 1989). Among the attendees were David Allison, Bill Aspray and Peter Galison.

public display of the neutrino detector with which Fred Reines conducted his Nobel Prize-winning detection of the free neutrino. Peter Galison, already engaged in the study of "the material culture of microphysics" made use of the occasion.[11]

This reluctance to use material culture seemed to change in the early 1990s when Steven Lubar and W. David Kingery published *History from Things*. A number of scholars who had used artifacts in their studies of history contributed to the collection.[12] Further evidence of this interest could be found in monographs like Robert Smith's *The Space Telescope*.

From Homer to Hayden White, history has used and created texts. There is nothing in this procedure that rules out reading artifacts as texts:

> In the terminology of history, artifacts are primary sources: Several scholars have observed that any artifact ... is a historical event ... An artifact is something that happened in the past, but, unlike other historical events, it continues to exist in our own time. Artifacts constitute the only class of historical events that occurred in the past but survive into the present. They can be re-experienced: they are authentic, primary, historical material available for first-hand study. Artifacts are historical evidence.[13]

What have these kinds of reconstructions to tell historians? Bern Dibner wrote of the monumental efforts required to move Egyptian obelisks to Rome, Paris, London, and New York. They have, he maintained, "been chiseled, raised, lowered and moved again by methods revealing to our engineers ... we are fortunate to have clear records of the mechanics used in the moving and erection of the Vatican obelisk in 1586 ... by means that must have, in some measure, resembled those used by the Roman engineers, if not by the Egyptians themselves."[14] Although the intended audience is engineers, the intent of the study is to reveal what engineers have failed to do:

> Not only did the Egyptian engineers not have such modern aids but the cutting and finishing of the hard granite, its transportation over hundreds of miles, and its erecting, were accomplished by these ancients with a modesty that has kept such deeds from being adequately recorded. Whereas there exist thousands of sculptures, bas-reliefs, gems, paintings, papyri, and models of the religious, regal, and domestic life of the Egyptians, their advanced technology is illustrated by extremely few known

---

[11] Peter Galison, *Image and Logic: The Material Culture of Microphysics* (Chicago: University of Chicago Press, 1997), 461.

[12] Lubar, *History from Things* (note 1.)

[13] Jules David Prown, "The Truth of Material Culture: History or Fiction," *Ibid*, 2–3.

[14] Bern Dibner, *Moving the Obelisk* (Cambridge: MIT Press, 1970), 7–8.

examples. We must therefore *reconstruct their tools* and methods from the results they achieved.[15]

Reverse engineering of past techniques provides a way to "fill in the gaps" in the text. It can also substitute for the text when "technological processes cannot be adequately described with words. Nonliterate peoples have carried out complex technological processes with such skill and sophistication that duplicating them has proved to be a challenging task for modern practitioners." Even literate scientists and engineers have not necessarily recorded their methods and techniques in forms accessible to the historian.[16]

Historians of science tend to focus on scientific instruments, rather than the means of production in craft, manufacture, industry and government. The recent volume on instruments by Robert Bud and Deborah Warner shows the value of this focus, as have the earlier studies of Cohen, Daumas, E. G. R. Taylor, Gerald L. E. Turner, and others.[17] Their use of material culture in these studies has varied. Often, the instruments have served as inspiration for historical research, which in turn enriches the understanding of the instrument. The reconstruction of these instruments is rarer, in part because of the historian's preference for texts over techniques as secondary sources, and in part because he or she does not have the required skills. Clearly, they are fascinated with machines.[18] In order to indicate how this use of material culture has been successful in the historiography of science and technology, I will examine two familiar cases.

---

[15] *Ibid.* Emphasis added.

[16] Robert B. Gordon, "The Interpretation of Artifacts in the History of Technology," in Lubar (note 1), p. 74.

[17] Robert Bud and Deborah Jean Warner, eds., *Instruments of Science: An Historical Encyclopedia* (New York: Garland, 1998); Bud and Susan E. Cozzens, eds., *Invisible Connections: Instruments, Institutions, and Science* Bellingham, Wash.: SPIE Optical Engineering Press, 1992); Gerard L.E. Turner, *Nineteenth-Century Scientific Instruments* (London: Sotheby Publications; Berkeley: University of California Press, 1983); E. G. R. Taylor, *The Mathematical Practitioners of Hanoverian England, 1714–1840,* (London, Cambridge University Press, 1966), *The Mathematical Practitioners of Tudor & Stuart England* (Cambridge: University Press, 1954); Maurice Daumas, *Les Instruments Scientifiques aux XVIIe et XVIIIe Siecles.* (Paris: Presses universitaires de France, 1953) Trans. and ed. Mary Holbrook, *Scientific Instruments of the Seventeenth and Eighteenth Centuries* (New York, Praeger, 1972); I. B. Cohen, *Some Early Tools of American Science; An Account of the Early Scientific Instruments and Mineralogical and Biological Collections in Harvard University* (Cambridge: Harvard University Press, 1950).

[18] Otto Mayr, *Philosophers and Machines* (New York: Science History Publications, 1976), 1–4.

4. The Antikythera mechanism

Derek de Solla Price's investigation of the Antikythera mechanism required over 20 years of work. The device, which he convincingly dates from the first quarter of the first century BC, was discovered in a ship wreck near the island of Antikythera at the beginning of the present century. His studies are illustrative of the kind of investigation a well-trained historian of science can make of an artifact. He thoroughly analyzed the evidence using the most modern techniques, and constructed a solid historical argument that challenges accepted views of the past.

Price's first step was to determine the provenance of the artifact. This process determines the chain of evidence that places the object in space and time. Like its legal analogue, the reconstruction of the chain of evidence is essential to authenticate its place in history. Price did so by determining the circumstances of the discovery of the pieces of the artifact by sponge divers in 1901. He included the precise location, the process of recovery, and the handling of the artifacts by the Athens Museum. He did this, of necessity, from accounts made by others, including curators and archaeologists, like Gladys David Weinberg, who dated them to 80–50 B.C. Amongst these accounts he found a number of hypotheses that had to be reconsidered. None of them were satisfactory, in his view, after a painstaking review of the evidence presented in their support.[19]

Price's next step was to examine photographs of the artifact from its discovery to the 1950s. Curators had discovered pieces of the mechanism as the wood that had encased them dried and shrank away. The evidence that had appeared and disappeared due to cleaning and handling over the years enabled Price to identify the fragments of an early astronomic computer.[20]

Price traveled to Athens to examine the fragments with the assistance of the Greek epigrapher George Stamites, who, deciphered almost twice as many characters as had been previously read, strengthening Price's hypothesis that the device had been used to calculate the motion of the moon and planets. Price returned to Greece in June 1961 to check the inscriptions and joins for a final reconstruction, but found there was not enough visible of the gearing or dial work to make one. Having exhausted the bibliographic, iconographic, and epigraphic resources at his disposal, Price turned next to physics. In 1971, he became aware of *Isotopic Methods of Examination and*

---

[19] Gladys Davidson, et al., *The Antikythera Shipwreck Reconsidered* (Philadelphia, American Philosophical Society, 1965). Radiocarbon dating showed the ship dated from 220 ± 43 BC.

[20] Price, "Clockwork before the Clock," *Horological Journal* (Dec. 1955, Jan. 1956); cf. Price, *A History of Technology 3* (Oxford, 1957), 618.

Authentication in Art and Archaeology, and contacted the Greek AEC, which prepared gamma-radiographs and x-radiographs of the instrument.[21]

Price used this evidence to build, not a reconstruction of the computer, but an historical argument. He inferred a much higher level of technology in ancient times than had previously been recognized, and confuted Benjamin Farrington's assertion that the ancients had not engaged in technological pursuits because it was slave's work. Price suggested a new interpretation of the history of scientific instruments:

> With the Antikythera mechanism, whatever its function, we are evidently concerned with the rather different phenomenon of High Technology, ... specially sophisticated crafts and manufactures that are in some ways intimately associated with the sciences, drawing on them for theories, giving to the instruments and the techniques that enable men to observe and experiment and increase both knowledge and technical competence.

Price argued, that "the roots of those special skills and qualities which were balanced between the sciences and the crafts and were to become the crucial element in giving the world the Scientific and Industrial Revolutions and the recent age of High Technology" were to be found in ancient high technology:[22]

> Thus, though not sufficient, the tradition of clock-making can be seen to have been crucial to the emergence of our modern world. So much of present-day machinery derives from it that it has become commonplace to use the term 'clockwork' for anything with gear wheels – as in clockwork toy trains for example. The timekeeping, ticking mechanical clock itself can be traced back only to the thirteenth or fourteenth century, but the wider history of clockwork goes back beyond the extraordinary emergence of the clock to a long prior period which includes the lines that lead also to such diverse developments as the concept of perpetual motion, the design of calculating machines and computers, to automata and robots, and to magnetic compasses.[23] It is in this story that the Antikythera mechanism provides us with dramatic new evidence and the earliest relic of such a distinguished main line in technology.

---

[21] Price, "On the Origin of Clockwork, Perpetual Motion Devices and the Compass," *Contributions from the Museum of History and Technology* (Washington, D.C.: Smithsonian Institution, 1959); "An Ancient Greek Computer," *Scientific American* (June 1959), 60. Oak Ridge National Laboratory Report IIC–21 (Oak Ridge, October 1970).

[22] Price, *Gears,* 51–53.

[23] Price, "On the Origin of Clockwork, Perpetual Motion Devices, and the Compass," *Contributions from the Museum of History and Technology,* Smithsonian Institution Bulletin 218, No. 6 (1959): 81–112.

Was this the first computer? Price calls it "a Calendar Computer," and the "earliest relic" of the tradition leading to the computer, but this was before the issues arose that are the substance of technical and legal debates about "firsts," which have engendered a plethora of multi-adjectival "firsts." If a computer is a device that computes, one would have to look further back into the past, Price suggests, for Archimedes' sphere, which performed a similar computation based on an earlier understanding of astronomy. The Antikythera device performed calculations, Price believed, that took into account the eccentric orbits of planets.

We may be surprised that Price did not seek to assign priority for the invention of the computer. Priority is a product of rise of the scientific journal in the 17th century. It is more interesting for Price to see the Antikythera device as further evidence of the convergence of ancient mathematical techniques from Babylon and Hellenic Greece to produce modern Western science.[24]

The sudden efflorescence of the technology at the end of World War II also shows the confluence of two mathematical traditions, one related to business and the control of information in an industrial society, and the other relating to mathematical and scientific computation. Like the Babylonian and Greek mathematical traditions, these traditions were complementary, but proceeded according to different conceptions of calculating, of which we can, as Ceruzzi has shown, see traces in the early architecture of electronic computers.[25]

We do not, however, have the advantage of hindsight in evaluating the historical significance of this confluence. Price looked back on the history of scientific technology from the modern perspective, that of Galileo and Newton, through the Renaissance, Middle Ages, And Late Antiquity, to a device over 2 millennia old. We can have no idea, after only 50 years, where electronic computing technology is taking us, just as the anonymous maker of the Antikythera device could not foresee its impact on western civilization. Indeed, he had no idea that the end of the millennium in which he lived would see an eclipse of these techniques, to be revived only early in this millennium. Our millennium bug and the productivity paradox may greatly alter the future of electronic computing, even as the decline of the Roman republic and empire altered the development of their technology.

This leads us to yet another problem. Price was able to deduce the function and structure of the Antikythera mechanism with the help of modern techniques, a profound knowledge of the science of the time, and some inferences for ancient documents. He had no text to guide him or even to suggest a priori the existence of the technology. How much more pressed might our de-

---

[24] Derek de Solla Price, *Science since Babylon* (New Haven: Yale University Press, 1975), 40–50.

[25] Paul Ceruzzi, "Crossing the Divide: Architectural Issues and the Emergence of the Stored Program Computer," *IEEE AHC* 19:1 (Jan.–Mar. 1997), 5–9.

scendants be to discover the nature of the electronic digital computer 2000 years from now, without documents and only scattered artifacts to go by? No visual inspection would do. Indeed, unlike the mechanisms of the Antikythera device, the solid state componentry of contemporary computers would probably not survive with their micro-circuitry in any recognizable form, even with powerful microscopes. Earlier vacuum-tube computers of the type reconstructed recently, give more visual clues, but only to those acquainted with early 20th-century electronic technology. I think it unlikely, therefore, that the future archaeologist or historian could decipher the purpose of the computer without guidance from texts.

For the short-term future, however, we can expect that there will be plenty of books. Already, the popular market for books explaining the operation of applications software is crowded with paperbacks explaining the operations to everyone from dummies to systems analysts. The rapid evolution of software makes it unlikely, however, that most of these books will long survive. The most popular word-processing application of the 1980s, *WordStar*, is represented by some 500 entries in the INSPEC database, of which only a score were written after 1990. Lotus 1-2-3 is represented by about a thousand entries, 80% of which were published before 1990.[26] How many of these will survive is anyone's guess, but certainly our landfills will be filled with them.

Will computer programs and data also survive? As has been amply demonstrated, old programs can be migrated to new platforms and operating systems. These can be in turn widely distributed via the World Wide Web. Whether or not this activity will survive the textual record is still in doubt. The Digital Libraries are still under construction, and the Web is an impermanent medium. Paper, although fragile, has survived the ages, as have inscriptions in stone and metal. Microfilm has shown staying power, although audio and video-tapes have not. In considering the preservation of computer software, it might make sense to save all of these forms, with an emphasis on those which are more permanent. Our reliance on such documentation is exemplified by studies of the Difference Engines of the 19th century.

5. The Babbage Engine

The various constructions and reconstructions of the Babbage engines have relied on paper more than parts, and here, I think, the critical resources of the historian are essential to a proper understanding of the machine. The sources upon which many secondary accounts have relied are other secondary accounts. The primary resources, however, are to be found in Babbage's papers in the British Museum, the Science Museum of London, and the Royal Soci-

---

[26] The Charles Babbage Foundation has commissioned a task force to look at the history of software and suggest useful approaches to it.

ety of London. These have formed the basis for historical studies of the Difference Engines and the Analytical Engine by both computer scientists and historians of technology.[27]

Lindgren's extensive comparative study of early difference engines used literary sources to describe Babbage's engine, and the wooden and metal model of the Scheutz difference engine that he discovered in the bowels of a Stockholm museum.[28] The comparison of the two engines illuminated their differences in design, construction, and function. However, Lindgren went beyond these points to investigate the context provided by politics, science, society and economics.

The claims about the capacity of British industry for precision machine-building advanced by these authors are of particular interest to the historian of technology. The rise of the American system of manufactures in the first half of the 19th century surprised British politicians, especially since it had produced a technology for producing weapons with interchangeable parts. A number of Babbage's biographers have argued that he could not build his engines because British machine technology was not up to the task. Reconstruction of his early machines has led some to argue that the level of British technology employed was more advanced.[29] Lindgren and Bromley's analyses, based upon their historical and actual reconstructions suggest that while the technology was capable of producing Babbage's engines, his engineering skills were not.[30]

Lindgren further suggests that Babbage and his engineer, Clement, over-designed the Difference Engine, making it too difficult to build. The Scheutzs, working in an older, cut-and-try tradition, succeeded because they were willing to accept a cruder finish to their engine. In a sense, this reflects Andrew Carnegie's maxim that "pioneering don't pay," as well as the strictures of modern systems engineering as to the amount of new technology permissible in complex constructions.

---

[27] Bromley, Allan G., "Charles Babbage's Analytical Engine, 1838," *AHC* 4 (1982), 196–217; Bromley, Allan G., "The Evolution of Babbage's Calculating Engines," *AHC v. 9*, 1987, 113–36; Bruce Collier, "The Little Engines that Could've: The Calculating Machines of Charles Babbage," Ph.D. Thesis, Harvard University, August 1970; Michael Lindgren, *Glory and Failure: The Difference Engines of Johann Müller, Charles Babbage and Georg and Edvard Scheutz*, trans. Craig G. McKay (Cambridge: MIT Press, 1990).

[28] Lindgren (note 27), 76 ff.; 139 ff.

[29] Doron Swade, *Charles Babbage and His Calculating Engines*. (London: Science Museum, 1991), Bromley, "Analytical Engine, 1838," and " Evolution," (note 27).

[30] Lindgren *Glory* (note 27), 262–266; Bromley, "Evolution," (note 27), 113. The failure of Babbage's partnership with his engineer, Frederick Clement, was singled out as an important factor in the failure of the Difference Engine number 1 construction by Bruce Collier, "The Little Engine that Could've: The Calculating Machines of Charles Babbage," Ph. D. Dissertation, Harvard University, 1970), ch. 2.

Notwithstanding these technical issues, however, Lindgren finds the principal cause of failure of the Difference Engine to be the lack of a market for automated table-making.[31] This reminds us that technical virtuosity is, like other virtues, its own reward. The market may make a multi-billionaire of a Rockefeller or Gates, regardless of the technical excellence of their accomplishments. Xerox PARC exemplifies the lesson in our own time.[32]

This conclusion is the result of the two-stage process: the first step, which Price and Lindgren follow explicitly, is artifact analysis by archaeometry (the study of internal structure, formal analysis including reproductions and engineering interpretations, surficial marking, and physical properties). The second step is contextual analysis (the study of backward linkages to materials used, human and social components of design and manufacture as well as forward linkages to users and observers. "Although archaeometric analysis is essential for a complete investigation, the most significant history results from contextual interpretations," and "use technological artifacts to illustrate the development and application of material culture interpretation in a historical context."[33]

6. Challenges of high-tech artifacts

The historical reconstruction of high-tech artifacts might seem to require less attention to historical context than the examples I have cited. Yet, my investigations of accelerators and lasers and reconstructions of high-tech artifacts for museum displays suggest otherwise. Several examples may serve to show why.

The first "atom-smasher" is replicated in the exhibit "Lawrence and his Laboratory" at the Lawrence Hall of Science. Ernest Rutherford's tiny brass cylindrical chamber housed the first laboratory transformation of an element in 1917. The LHS replication was not a reproduction of the original chamber preserved in the Cavendish Laboratory at Cambridge, but of a copy at the

---

[31] Lindgren, *Glory* (note 27), 283–287.

[32] Popular journalistic accounts have provided a number of suggestions as to why it was Steven Jobs who derived the greatest commercial benefit from the work at Xerox PARC, but have not addressed the institutional and social questions that are posed by basic research in a market-driven economy.

[33] Lubar, *History from Things* (note 12), pp. xii–xiv. In contrast, Allan Bromley's studies of the Antikythera mechanism and the difference engines are internalist and presentist and present no interest whatsoever to the historian whose interests lie in actual rather than counterfactual history. The "logical" reconstruction according to modern engineering is particularly prone to error, since there is no evidence that ancient engineers held the same values or premises as modern ones. Cf. Dibner, *Moving* (note 14).

Smithsonian.[34] To simulate the experiment, an electronic analog allowed visitors to "fill the chamber" with air, oxygen, and nitrogen. This showed that the nuclear reaction, in which an alpha particle strikes a nitrogen nucleus and produces an isotope of oxygen as well as a proton, occurs in atmospheric air and nitrogen, but not in oxygen. This was also displayed in the formula:

$$_2^4\text{He}+_7^{14}\text{N}\rightarrow_8^{17}\text{O}+_1^1\text{H}$$

The object itself gave little clue as to its use or importance, apart from displaying its miniscule size compared to the gigantic accelerators that succeeded it. Indeed, the most intensive study of this experiment of which I am aware uses Rutherford's laboratory notebooks as the basis of the reconstruction, even though the organization of these documents is chaotic.[35] High-tech artifacts often challenge the imagination because their use is not evident from their form, and without textual documentation would be lost with the memories of their builders and users.[36]

At Los Alamos, in the Bradbury Science Museum a flow cytometer and a Cray 1A computer are displayed. The flow cytometer is a working reconstruction that displays cell-sorting by laser-induced florescence and electromagnetic separation. In its initial form, however, it was too authentic. Because of the very great sensitivity of the eye to the wavelength of the argonion laser used in the sorting, it was replaced with a helium-neon frequency shifted laser. Although this undermined the authenticity of the display, it increased its safety and preserved its functionality.[37]

The difficulties of interpreting artifacts in the history of computing, which Paul Ceruzzi has discussed elsewhere, were similar.[38] The Bradbury Science Museum's exhibit, "Theoretical Design Of Nuclear Weapons," displayed the

---

[34] Correspondence between P. W. Bishop and R. W. Seidel, 1971, CBI.

[35] David B. Wilson, *Rutherford: Simple Genius* (Cambridge: MIT, 1983), 395–405.

[36] Cf. E.g., Ernest Rutherford, *The Collected Papers of Lord Rutherford of Nelson, O.M., F.R.S.* Published under the scientific direction of James Chadwick. (Cambridge University Press, 1962); *Radiations from Radioactive Substances* (Cambridge University Press, 1951). Laboratory notebooks are tradition manuscript sources of greatly varying quality. Among the most useful I have seen are those of the Lawrence Berkeley National Laboratory, especially those of Eugene Gardner, who manufactured pions in the 184-inch cyclotron in 1948.

[37] Two nuclear rocket engines fell prey to the same concern for safety: as detectors of residual radioactivity become more sensitive, they reveal previously undetectable hazards of radiation exposure. Like the Harvard cyclotron that was one of the first to be employed at Los Alamos and radioactive glass from Trinity, they must be stored away from the public until the decay of radioactive isotopes outstrips the pace of detector development. The "nuclear weapons" on display are in fact training devices that never contained the critical assemblies.

[38] Paul Ceruzzi, " 'The Minds Eye' and the Computers of Seymour Cray," paper presented at the Second Science and Technology Museums Forum, National Museum of American History, Washington, D.C.

FERMIAC, parts of MANIAC, and images of several early computers that had been used to design successive generations of nuclear devices. In 1987, the C-Division leader donated the first Cray 1A computer that the laboratory had acquired in 1977. Its appearance suggested a circular park bench. In order to add interest, we had panels removed and a section made to reveal its inner workings, which, in turn, were explained by labels and voice-overs. Shielded by tinted glass, it was revealed at an appropriate time and in appropriate places by a son-et-lumiere arrangement that leant movement and drama to what was otherwise a static display. The use of a stop-motion video of the construction of a Cray computer, which compressed the action into less than a minute, provided a hyperdynamic experience of its reconstruction.

In all of the efforts described above, objects illustrated history and history illuminated objects. High-tech artifacts are black boxes. Los Alamos Laboratory photographs – some 750,000 in number and increasing proved complete photographic documentation of many objects, but the photographers' notebooks are inadequate to identify many of them. Retired laboratory staff have been hired to identify these pictures, which are not always worth a thousand words.

The high-artifact or its reconstruction gives the museum visitor a grasp of scale. Paul Forman's exhibit "Atom Smashers" at the National Museum of American history used different generations of accelerators to showed their evolution from the cyclotron to the Tevatron.[39] However, their very size worked against their use for research as part of the Smithsonian's collections. Indeed, they were disassembled and given away when "Atom Smashers" gave way to the "Information Age" Exhibit.[40]

When the Los Alamos Laser Fusion facility, Antares, shut down in 1987, the Bradbury Science Museum acquired one of the largest lasers in the world, 60 feet long and 14 feet in diameter. The Smithsonian had turned down as similar laser offered to them by AVCO-Everett on the grounds that they had inadequate space to store it. My earlier investigation of the history of military laser research and development made me aware of the historical importance of these lasers in the military and energy programs of the United States. Few museums, however, have the luxury of 43 square miles, the size of the Los Alamos National Laboratory. It is doubtful that large, high-tech artifacts will survive unless they are made Historic American Engineering Record or National or State Historical Sites and are preserved as have other recipients of this distinction have been. Accelerators, computers and lasers are more difficult to interpret than those artifacts, whose form reveals their function.[41]

---

[39] "Atom Smashers ... 50 Years," exhibit brochure. I am indebted to Michael Meo, Forman's curatorial assistant, for a copy of the brochure.

[40] Personal communication from Paul Forman.

[41] *America Preserved: A Checklist of Historic Buildings, Structures, and Sites Recorded by the Historic American Buildings Survey and the Historic American Engineering Record* (Washington: Library of Congress, 1995).

In the case of accelerators, M. Stanley Livingston, the co-inventor of the cyclotron and strong-focusing, elucidated the technical design and operation of particle accelerators, using scientific papers, manuscript materials, and a working knowledge of these machines. This work is fundamental to the historical interpretation of the development of accelerators in *Lawrence and his Laboratory*.[42] The reasons for failure and success in the environment of nuclear physics in the 1920s and 1930s were not only technical, however. Entrepreneurship, the availability of funding and cheap personnel, and accidents played a role. As an example, the source of inspiration for the cyclotron happened to be available to Lawrence only because of a coincidence. Lawrence's entrepreneurial activity secured financial support from diverse philanthropies interested in the medical applications of neutron and x-rays and artificially produced radioactive isotopes. The effort required to understand the technical development took us to two continents and through thousands of published and unpublished documents.[43]

A technical understanding of the operation and design of particle accelerators and a historical understanding of the context within which they were or were not used informed our answers to questions like: Why did the cyclotron "win" the race for high-voltages? Or did it? What environmental factors, both local and global, influenced its development and diffusion? Why did some resist using the cyclotron and turn to alternative technologies? Why were so many discoveries in nuclear physics that could have been made using the machine made elsewhere with simpler and smaller devices? How did the cyclotron contribute to the development of the technology of particle acceleration and, in particular, to the rise of the synchrotron concept after World War II? These questions, and others like them, enabled us to reconstruct the history of the particle accelerators with an understanding of the roles played by many individuals, organizations, and contingencies.

Although the LBL history project involved three historians and a half-dozen research assistants, the Laser History Project incorporated a Laser History Council, composed of three Nobel Laureates in the field and leaders of important industrial and academic laboratories, an Advisory Committee of historians, archivists, research administrators in the field and working scientists, four professional societies, two history centers, and the Laser Institute of America. The Advisory Committee Chairman delegated questions of historical interpretation and method to the historians, and charged his colleagues

---

[42] *High-energy accelerators* (New York, Interscience Publishers, 1954), *Particle Accelerators; A Brief History* (Cambridge, Mass., Harvard University Press, 1969); *The Development of High-Energy Accelerators* (New York: Dover, 1966); Livingston and John Blewett, *Particle Accelerators* (New York, McGraw-Hill, 1962)

[43] J. L. Heilbron and Robert W. Seidel, *Lawrence and His Laboratory: Volume 1 of a History of the Lawrence Berkeley Laboratory* (Berkeley: University of California Press, 1989), 82–83.

to identify and facilitate access to sources of information. Neither the Council nor the Committee exercised any editorial role over the products, although, of course, they were asked for advice and reviews in some, though not all cases. Advisory committee members also arranged funding from a number of private, professional, and government sources, including the Department of Defense, to support two historians and travel for research and interview purposes throughout the United States.

Artifacts like the laser, the computer, and the particle accelerator and detector are often the results of team efforts in academic, industrial, and government laboratories, and the documentation of these artifacts often requires also a team. The cooperation of scientists, engineers and historians is fruitful when each contributes his own knowledge and expertise. It is barren when the historian is regarded as a chronicler or amanuensis for the participant.

7. Virtual reconstructions

Simulation techniques can be useful to an historical understanding of high-tech artifacts. The virtual reconstruction of an artifact too large for display in an ordinary museum can effectively substitute for the more traditional model, picture, or video. Holography can supply more reality to the visual reconstruction of such an artifact. The virtual worlds that are accessible to anyone owning a Nintendo 64 or Sony Playstation suggest more ambitious used. Computer-Aided Visual Environments (CAVES), are used by several firms for training of operators, just as flight simulators have been. The Ars Electronica Center in Linz, Austria, has a CAVE that shows computer simulations of mathematical equations and virtual universes.[44]

Historians have begun to use such virtual reconstructions to facilitate research and education. The Perseus Project links graphics, maps, and ancient texts so that the student may read the text, view the context, and browse hypertext links relating to the entire Greek corpus of manuscripts, modern CIA maps of Greece, and many photographic and historical resources. Other attempts to put primary documents on line, such as the papers of Robert Oppenheimer and Thomas Edison Papers have been attempted but have not succeeded for reasons that are seldom announced: failure is an orphan, and success has many fathers.

The high costs, fluid technology, and limited market for such resources will probably make them far less common than print alternatives for some

---

[44] A description of the CAVE is available on the center's website at the address http://www.aec.at/center/proj/cave.html. Other CAVES have been built at NCSA for the Visualization of chaotic systems, at General Motors and Catapillar for design studies and prototyping. Lisa Picarille "Archival Rivals," *Computerworld* 31:3 (Jan. 20, 1997) 83ff. Cf. Karl Gerbel und Hannes Leopoldseder, *Die Ars Electronica Kunst im Zeitsprung* (Linz: Landesverlag, 1989.)

time to come, especially if copyright law is not amended. Our hopes of putting archival materials, rather than mere finding aids, on the Internet, have been deflated by the cost of scanning these materials by hand, which archivists find preferable to batch scanning for obvious reasons. These costs approach those of microfilming, which provides a more permanent archival record, but is done only in the case of major collections where resources are available.

The use of computer simulations in teaching history either in museums or classrooms, can assist in role-playing and discussion of historical issues, as the Oregon Trail program and the experiences of some schools have demonstrated.[45] The CD-ROM market is replete with recreations of historical battles and historical encyclopedias, which are seldom more than pouring old wine into new bottles.[46] The quality of the wine is not always up to the cost of the bottles, and for the most part, the term of art used to describe these items – "shovelware" – is appropriate: most of them are shoveled into the dumpster after sitting on shelves for many days.

Computers have proven to be very useful tools for historians, not in virtually reconstructing history, but in real historical reconstruction of texts through computer enhancement, statistical analysis through the use of spreadsheets, databases, and other more traditional applications like word processing. A cadre of historians devoted to the development of these uses of computing has been quite active over the past 18 years. Like other tools that have been made available by modern science and technology, such as neutron activation analysis and radiography, the utility of the computer is greater when it assists in answering, rather than posing, historical questions.

What sort of questions might be posed to virtual reconstructions of history such as we have seen at this conference? I can suggest a few, but, as a human, I suspect each historian would have his or her own set of questions for those fortunate enough to possess the technical and financial means to create reconstructions based upon historical data. The first of these is: Why did a given object come into existence? Why was it more successful in its environment than competing objects? Why does technical excellence not guarantee success? Great historical events, like the Industrial Revolution of the 19th century and World War II, provide a macrocosmic stimulus to historical change, but their effects upon technology are complex. The easy explanations in terms of "heroic inventors," seldom take their interactions with the wider world into

---

45 Jonah H. Peretti, Mark Cowett, Casey Charvet, "Historical Role Playing in Virtual Worlds: VRML in the History Curriculum and Beyond," *Computer Graphics* 31 (Aug. 1997), 64–65; http://www.newmand.k12.la.us/civilwar.

46 E.g. *The Day after Trinity*, which features the movie originally shown on the public television network in the United States, *Critical Mass*, which is based on the Los-Alamos sponsored book *Critical Assembly* by Lillian Hoddeson, Roger Meade, Catherine Westfall, and Paul Henriksen, and *Einstein*, which incorporates Ronald Clark's biography in full.

account. Modern technological innovations commonly arise from teams of scientists, engineers, and patrons. To emphasize one set of actors while neglecting others is a flaw of the heroic inventor tradition. The interaction of two or more individuals in the development of a new technology like the computer introduces more complexity.[47] Furthermore, the cultivation of financial, environmental and social resources is usually a part of the process, and Hughes has spoken of "systems builders" as those who organize these resources for technological development.[48]

A second question is: how? This is a dangerous question, because one must forget the obvious answers that spring to mind. It is very easy to read contemporary methods or technologies into the past, rather than finding it there. Historians are trained to make critical use of sources and to overcome projections about the uses of a technology into the past. For example, the question of how a technology *was used* is not answered historically by stating how it would be used today, as the examples of the cyclotron and laser show. Unless we adopt a very naïve technological determinism, we must admit that the values of a given time will influence the uses it will make of its tools.[49] We know, for example, that the first use of ENIAC was to make calculations related to thermonuclear weapon design. This was not its intended use, and the details of that use remain classified. We could not infer the design of the ENIAC from that use, nor could we predict the commercial success of the computer based upon the uses of other early computers. Yet, without that success, we would not be interested in the history of the computer. It would be of no interest how it worked, as the Babbage engines were of no great interest before the success of the Mark I moved Howard Aiken to identify them as a general source of inspiration. On the other hand, once the historical importance of an artifact is established, we may have reason to investigate how it actually worked, if only to dispel the clouds of hyperbole in which successful new technologies in general, and computers in particular, are enshrouded. Like the legend that Eli Whitney and Isaac Singer mass-produced muskets and sewing machines used interchangeable parts, which has been refuted by historians who have examined the artifacts and shown that they were not interchangeable,[50] the question of how a technology did or did not

---

[47] Cf. Lorraine Daston "Enlightenment Calculations," *Critical Inquiry 21* (Autumn, 1994) 182–202, and Simon Schaffer, "Babbage's Intelligence: Calculating Engines and the Factory System," ibid., 203–227.

[48] Thomas Parke Hughes, *American Genesis: A Century of Invention and Technological Enthusiasm* (Penguin, 1989).

[49] For a discussion of the historiography of technological determinism, see Merritt Roe Smith and Leo Marx, *Does Technology Drive History: The Dilemma of Technological Determinism* (Cambridge: MIT, 1994).

[50] See, for example, Woodbury, "Eli Whitney and the Legend of Interchangeable Parts," *Technology and Culture I:3* (1959), 318–350. David Hounshell, *From the American System to Mass Production* (Baltimore and London: Johns Hopkins, 1984), Appendix 1.

work may benefit from the actual physical reconstruction of the technology by knowledgeable craftsmen, if they do not assume that their prejudices about how it worked necessarily apply. On the other hand, attempts to prove those prejudices by reconstruction of artifacts can only lead to error.

8. Conclusion

The use of artifacts and reconstructions in museums, history, and virtual time and space can enhance and illuminate historical interpretations and perspectives. The use of artifacts for promotion, priority claims, entertainment, or antiquarian purposes should not be confused with history. Whether or not historical use of reconstructions, as opposed to artifacts, is possible depends on the quality of the reconstruction. If artifacts are primary sources, reconstructions are secondary, since they must embody the intent, interpretation, and interests of those who fund, conduct, and support this activity. Like any secondary source, these are subject to the same external and internal criticisms that historians have developed.[51] The examples of Price and Lindgren illustrate the value of that critical methodology in application to real historical objects. Reproductions can result from such an analysis, just as can monographs or articles. In dealing with past objects, like past texts, we ignore hundreds of years of historical practice and scholarship at our peril.

ROBERT W. SEIDEL, Director of the Charles Babbage Institute, is a historian of modern science and technology. His research focuses upon the history of modern physical science and related technologies. He is writing a history of the Lawrence Berkeley Laboratory (formerly the Radiation Laboratory of the University of California), of which the first volume, *Lawrence and his Laboratory* (co-authored by John Heilbron) has appeared. He has also written a popular history of the wartime atomic bomb program, *Los Alamos and the Making of the Atomic Bomb*, and published articles on the history of particle accelerators, computers in high-energy physics, military lasers, and the origins of radar.

---

[51] For a useful summary for non-historians, one may consult Jacques Barzun and Henry Graff, *The Modern Researcher*.

A Classification Scheme for Program Controlled Calculators

Andreas Brennecke

Abstract. What constitutes a computer? This question is of interest to historians trying to decide whether a specific device can be called a computer or not, and to computer scientists wanting to gain a better understanding of the essential concepts of computer architecture. In both cases, the answer must encompass more than just a definition. Many concepts and features have influenced the development of the modern computer. But which of them are necessary to realize the computer's full capabilities and which are the most important ones?

When considering historical machines, it is important to pay attention to their practical use and to identify the human activities involved. A new classification scheme will be proposed in this contribution that is derived from the historical analysis of computers and which cannot be obtained from theoretical concepts alone.

1. Introduction

The history of *program controlled calculators* – especially the question of which concepts and features are essential characteristics of a computer – is not only of interest to historians but also to computer scientists who want to obtain a better understanding of basic concepts:

> By attempting to define and classify, we come to understand more clearly the essential character of the computing machines with which history presents us when stripped of all technological and accidental differences and reduced to their fundamental logical design.[1]

---

[1] Allan G. Bromley, "What Defines a 'General-Purpose' Computer?" *Annals of the History of Computing* 5, no. 3 (1983): 303–305, on 305.

However, certain concepts seem to be so obvious today that their actual meaning remains opaque even to professionals of computing. Other concepts have become too complex to be easily understood today. As John A. N. Lee argued:

> One of the other benefits of retrospection is the occasion to identify the fundamentals of concepts which have now become perceptually complex.[2]

Hence, it came as no surprise to me that graduate students taking part in a course on the history of computers at our university were unable to give a proper definition of program control. They had some vague ideas but could not say whether Jacquard looms or analog computers are program controlled or not. Other terms such as *von Neumann architecture, universal computer, stored program computer* or *general purpose computer* were often used by them without proper understanding of the underlying ideas and mechanisms. Obviously, some clarification is needed.

This contribution proposes a classification scheme for program controlled calculators from the viewpoint of a computer scientist. It will be argued that the scheme may lead to a better understanding of fundamental concepts and that it may help historians to gain a better understanding of the course of events in the computer field.

2. Principles and practice

When classifying computers it is not sufficient to consider theoretical concepts only. Theoretical computer science has developed notions like *μ-recursive functions, primitive recursive functions* and the *Chomsky hierarchy* to describe the potential of computing devices. However, it was important for the constructors of program controlled calculators, especially in the early years, to design machines that could be implemented with the technology at the time. Theoretical concepts offer limited help here, because they do not sufficiently differentiate between what can be done in principle and what can be accomplished in practice.

Raúl Rojas has shown that Konrad Zuse's Z3 could be called a *universal computing device* if it is provided with indirect addressing.[3] This means that

---

2   John A. N. Lee, "Those Who Forget the Lessons of History Are Doomed To Repeat It – or, Why I Study the History of Computing," *Annals of the History of Computing* 18, no. 2 (1996): 54–62, on 58.

with indirect addressing and with no limitation on the size of memory, the Z3 could have calculated every computable function. In a later publication Rojas shows how to simulate indirect addressing for a finite memory using only the original instruction set of the Z3, namely by using *conditional executable code segments*. Thus, the Z3 is as universal as any modern computing device.[4] However, this theoretical result does not fully reflect the practical power of the Z3. Although the length of the simulation program is not important from a theoretical point of view, it is relevant for practical purposes.

Universal computation in the sense of theoretical computer science depends on having unlimited storage, which real machines obviously cannot have. Modern computers have a relatively large memory so that results from the theoretical models can still be extrapolated. But early computing machines had very few memory cells to store results. Konrad Zuse's Z3, for example, could store only 64 values, the ENIAC (Electronic Numerical Integrator and Computer) had 20 accumulators[5] and the ASCC (Automatic Sequence Controlled Calculator or Harvard Mark I) had 72.[6] Can one view these machines as universal machines when they are provided with an unlimited storage?

In the early days of program controlled machines, universality did not play an important role for developers or users. Nevertheless, considerations about the potentials of the computers are as old as the early program controlled devices themselves. Charles Babbage wrote about his Analytical Engine:

> These two memoirs [the memoir of General L. F. Menabrea and the notes made by Ada Augusta, Countess of Lovelace] taken together furnish, to those who are capable of understanding the reasoning, a complete demon-

---

[3] See Raúl Rojas, "Conditional Branching is not Necessary for Universal Computation in von Neumann Computers," *Journal of Universal Computer Science* 2, no. 11 (1996): 756–768.

[4] Rojas does not show that the Z3 is equivalent to the most powerful theoretical computing models such as the Turing machine, but that the computing model of the Z3 is equivalent to any existing computer which can only have a limited memory. Cf. Raúl Rojas, "How to Make Zuse's Z3 a Universal Computer," *Annals of the History of Computing* 20, no. 3 (1998): 51–54.

[5] See Herman H. Goldstine and Adele Goldstine, "The Electronic Numerical Integrator and Computer (ENIAC)," *Mathematical Tables and Other Aids to Computation* 2 (1946): 97–110. Reprint in: Brian Randell, ed., *The Origins of Digital Computers – Selected Papers*, 2nd ed. (Berlin: Springer-Verlag, 1982), 359–373. Reprint in: *Annals of the History of Computing* 18, no. 1 (1996): 10–16.

[6] See Howard H. Aiken and Grace M. Hopper, "The Automatic Sequence Controlled Calculator," *Electrical Engineering* 65 (1946): 384–391, 449–454, 522–528. Reprint in: Randell, 203–222.

stration – *that the whole of the developments and operations of analysis are now capable of being executed by machinery.*[7]

John V. Atanasoff, who built the ABC (Atanasoff-Berry Computer) remarked that his method of electronic computing could be extended to handle other tasks:

> Once our prototype had proved successful, we both [John V. Atanasoff and Clifford E. Berry] knew that we could build a machine that could do almost anything in the way of computation.[8]

The machine built by Atanasoff and Berry could only be used to solve systems of linear equations. Other problems had to be manually reduced to linear equations.

The ENIAC was used to solve problems in different fields.[9] Even assuming that the ENIAC contained all elements necessary for universal computation, and that it could be used for many tasks, it could not be used for very complex problems involving many calculations in practice. The ENIAC had to be rewired for each type of problem. During complex computations, the intermediate results had to be punched on cards and read again. The operator had to keep track of the cards and their contents and so, for very complex problems, the human intervention required was overwhelming – Stan Augarten remarks about the ENIAC that its programming was "a one-way ticket to the mad house."[10] However, theoretical computer science deals with the formal aspects of problems and theoretical machines only and never considers human activities – an insufficiency also mentioned by Allan G. Bromley, when he discusses the term *general purpose*:

---

[7]   Charles Babbage, *Passages from the Life of a Philosopher*, (London, 1864). Reprint: Martin Campbell-Kelly, ed., *Charles Babbage – Passages from the Life of a Philosopher.* (New Brunswick, N.J.: Rutgers University Press, 1994), 102. Charles Babbage viewed *analysis* as a special, but in his time enormously important domain: William J. Ashworth, "Memory, Efficiency, and Symbolic Analysis – Charles Babbage, John Herschel, and the Industrial Mind," *ISIS* 87 (1996): 629–653. Note that the term analysis had a different meaning in the 19th century.

[8]   John V. Atanasoff, "Advent of Electronic Digital Computing," *Annals of the History of Computing* 6, no. 3 (1984): 229–282, on 247.

[9]   See W. Barkley Fritz, "ENIAC – A Problem Solver," *Annals of the History of Computing* 16:1 (1994): 25–45.

[10] Stan Augarten, *Bit by Bit – An Illustrated History of Computers* (New York: Ticknor & Fields, 1984), 128.

Complexity theory gives some measure of how difficult a task is for machines of a given type, but gives no measure of how difficult it is to program the machine for that task – the measure that I feel is relevant here.[11]

In order to clarify the conceptual power of the machines and the work that must be done by human operators to solve certain problems, it is helpful to develop a classification scheme that considers both theoretical principles and practical implementations. This contribution presents such a classification, identifying several stages in the development of *sequence controlled machines*.[12]

3. Analog and digital

According to some dictionaries of computer science,[13] *analog* variables are continuous and digital variables take discrete values only. A clock with hands is called analog, whereas a clock with a numerical display is referred to as digital. But the second hand of some modern analog clocks moves in discrete steps. Therefore, it is not correct to use the term *analog* as a synonym of *continuous* and the term *digital* as a synonym of *discrete*.

In most publications, the terms analog and digital are used to distinguish between the representation of numbers by *physical quantities* or by *numerals*. In the example of the clock, this is an adequate description. The angle of the hands of the analog clock represents the current time, whereas the numerical display of a digital clock shows numbers.

There is another interpretation which comes closer to the intuitive meaning of the term analog.[14] There is a similarity between the rotation of the hour hand of the clock and the movement of the sun in the heavens as a consequence of the rotation of the earth. The clock builds a *physical analogy* to the rotation of the earth. This interpretation can be found in early papers on com-

---

[11] Bromley (n. 1 above), on 304.

[12] The history of sequence control is well described in Brian Randell, "The Origins of Computer Programming," *Annals of the History of Computing* 16, no. 4 (1994): 6–14.

[13] For example see *IEEE Standard Computer Dictionary – A Compilation of IEEE Standard Computer Glossaries* (New York, 1991) or *Duden Informatik – Ein Sachlexikon für Studium und Praxis* (Mannheim: Duden-Verlag, 1989).

[14] Analog: n. analogue ... analogue: 1. something having analogy to something else ... analogy: 1. a partial similarity between like features of two things, on which a comparison may be based: the analogy between the heart and a pump ... . *Webster's Encyclopedic Unabridged Dictionary of the English Language* (New York/Avenel, 1989).

puting machinery, in which analog and digital computers are compared.[15] In later publications, when *analog computing devices* had lost their significance, the definition based on representation by physical quantities was used more often. [16]

Applying the term analog to computers implies that physical quantities are used to represent numbers within a calculation process, e.g. the angle of a shaft or a disc. Analog computers embody a physical analogy of the problem. For example, a system of mechanical integrators, adders, etc., connected by shafts, constitutes an analogy to the ballistic motion of a projectile. Digital computers represent their numbers using numerals (digits) and they implement an abstract method (algorithm) to solve a problem. So far, this characterization tells us nothing about the control of analog and digital computers.

4. Program control is more than sequence control

The concepts *sequence control* and *program control* seem indistinguishable and the term sequence control is part of the name of early program controlled machines, e.g. the Automatic Sequence Controlled Calculator (ASCC) or the Selective Sequence Controlled Calculator (SSCC).[17] Frequently, the term program is used to refer to every sort of *sequence controlled machine* that has been built since the earliest known automata by Heron of Alexandria (Fig. 1). However, this makes it impossible to underline the conceptual differences between Heron of Alexandria's automata, early musical automata, Jacquard looms, sequence controlled calculating machines, and a modern computer.

Nevertheless, there seems to be some difference in the use of the terms sequence control and program control. Brian Randell[18] uses the term sequence control for all sorts of control mechanisms, but, when discussing the first

---

[15] For example see John von Neumann, "Entwicklung und Ausnutzung neuerer mathematischer Maschinen," Lecture in Düsseldorf to the Arbeitsgemeinschaft für Forschung des Landes Nordrhein-Westfalen on Sept. 9, 1954. Reprint in: Schuchmann, Hans-Rainer, ed., *Computertechnik im Profil* (Munich, 1984) or John V. Atanasoff, "Computing Machine for the Solution of Large Systems of Linear Algebraic Equations," Unpublished memorandum (Ames, Iowa, 1940). Reprint in: Randell (n. 5 above), 315–335. Atanasoff claimed "to be the first to use the word *analog* for computers". Atanasoff (n. 8 above), on 234. In this later publication Atanasoff used the term analog in the sense of physical quantity.

[16] For example see Karl Steinbuch and Wolfgang Weber, *Taschenbuch der Informatik – Band I: Grundlagen der Technischen Informatik* (Berlin: Springer-Verlag, 1974).

[17] Wallace J. Eckert, "Electrons and Computation" *The Scientific Monthly* 67:5, Nov. (1948): 315–323. Reprint in: Randell (n. 5 above), 223–232.

[18] See Randell (n. 12 above).

sequence controlled calculators, he prefers the term program control. From the author's point of view, program control denotes a specific quality compared to simple sequence control. This difference will be made clear now using the terms analog and digital as they are discussed above.

I use the term *analog* to characterize all sorts of *control mechanisms* that *represent physical analogies*, even though the representation of numbers is not analog, i.e. in the form of physical quantities. Such mechanisms are determined by the construction of the machine and always run the same way, although sometimes parts, such as a pegged cylinder, are removable. In contrast to the following definition of digital control, each *bit* of the control medium, e.g. a pegged cylinder, directly controls a mechanism which corresponds to this bit, e.g. to play a single tone. Analog control sequences are a part of the construction of the machine and can be illustrated by drawings of the controlled machine, as shown in Figs. 1, 2, 3 and 5.[19]

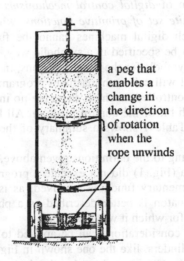

a peg that enables a change in the direction of rotation when the rope unwinds

Figure 1: The moving temple of Bacchus, an early automaton designed by Heron of Alexandria, in which a rope was wound around an axis with pegs.[20]

---

[19] Rolf Todesco called machines semi-automata if their functionality can be fully described by drawings, in contrast to automata which need a symbolic description. R. Todesco, *Technische Intelligenz oder Wie Ingenieure uber Computer sprechen*. (Stuttgart-Bad Cannstatt: Fromman-Holzboog, 1992).

[20] W. Walter, "Die gespeicherten Programme des Heron von Alexandria," *Elektronische Rechenanlagen* 15:3 (1973): 113–118.

Table 1: Use of the terms analog and digital to specify several aspects of computing devices

	analog	digital
application	physical analogy to a given problem	abstract method (algorithm)
representation of numbers	by physical quantities	by numerals
sequence control	determined by the construction of the machine which can only execute the same type of process	a finite set of basic functions that can be arranged in any desired sequence
description of the control process	with diagrams	symbolic

The notion of *digital control mechanisms* refers to a sequence of choices from a *finite set of primitive functions,* which can be arranged in arbitrary order. Such digital machines cannot be fully represented using diagrams. They must be specified in a symbolic way, for example using a sequence of arbitrary symbols for the basic functions, in the order in which they are to be executed. I will call such machines program controlled. In the sequence of a program controlled machine, there are no independent functions corresponding to each bit of the control medium. All bits are interpreted as a common function. Table 1 shows a summary of the distinction between analog and digital.

According to the definitions given above, the automaton built by Heron of Alexandria (Fig. 1) did not execute a program, because there were no individual elementary functions involved, as is necessary for program control. This automaton is better described as a physical analogy to the sacrificial ceremony for which it was designed.

Similar considerations can be applied to musical automata controlled by pegged cylinders, like the one shown in Fig. 2. The pegged cylinders enable the instrument concerned to play the desired melody as a sequence of single tones or combinations, called chords. Most of the combinations of peg positions correspond to disharmonies. In principle, it would be possible to describe all combinations of tones as a set of basic functions for a musical automaton, but in practice one would not do so.

The Jacquard loom (Fig. 3) is another example. The process of weaving does not depend on the cards that control the selection of the warp threads. The loom has only one state and always executes the same single function: raise a set of warp threads before bringing in the filling thread. Therefore the loom weaves a pattern corresponding to the holes in the cards. The cards can be considered an analogy to the weaving pattern and not a program.

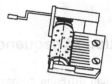

Figure 2: A simple musical automaton. Each peg of the turning cylinder touches one tooth of the comb directly producing the respective tone.[21]

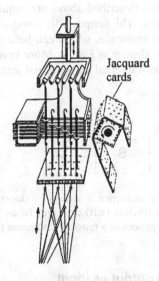

Jacquard cards

Figure 3: The mechanism of a Jacquard loom, the cards on the right control the raising and lowering of the warp threads[22]

The cylinders of most musical automata, like the sets of cards in the Jacquard loom, are interchangeable parts; another melody can be played or another pattern woven quite easily. So the control sequence can be either an inherent part of the machine or put on an external medium. However, and this is my point here, pegged cylinders and Jacquard cards can be used for program

---

21 Ibid.
22 Almut Bohnsack, *Der Jacquard-Webstuhl* (München, 1993).

control in other machines, but their use does not necessarily imply that the machines are program controlled.

5. A classification of sequence and program control

In this section, I present different computational models that can be summarized in a classification scheme for program control.

Executing a fixed sequence

The machines described above are sequence controlled but they do not process any input. The output will always remain the same, corresponding directly to the sequence, which can be a part of the machine or on an external medium, as shown in Fig. 4. More complex mechanisms may also be controlled by both internal and external sequences.

Figure 4: The sequence S always produces the same output. The sequence can be fixed by construction (left) or it may be an interchangeable part of the machine, e.g. the pegged cylinder of a musical automaton (right).

Using the output as input

The structure of Charles Babbage's Difference Engine[23] clarifies another point. The calculation of the machine is controlled by cams fixed on disks, whose functions are similar to those of pegged cylinders (Fig. 5).

The Difference Engine was a *special purpose machine* and the cams implemented the method of finite differences. However, in contrast to the mechanisms described previously, the Difference Engine's calculation depends on previous results, while the process of the musical automata or the Jacquard loom is fixed and the output is always exactly the same. The output

---

[23] See Allan G. Bromley, "The Evolution of Babbage's Calculating Engines," *Annals of the History of Computing* 9:2 (1987): 113–136.

of the Difference Engine is the input for the next step in the process of calculation.

There is no significant difference between program control without feed-back and analog sequence control, since the same output is always produced. For longer or more complex calculating processes, it is required that results already calculated be used again as input (Fig. 6). As long as the machine is unable to automatically feed the output back to the input, this must be done manually, as was done by human *computers* using desk calculators.

A *storage* facility is needed for automatic reuse of previously calculated results. The Difference Engine, for example, has registers to store the intermediate results. The number of intermediate results needed is limited by the highest order difference. Babbage designed his Difference Engine No. 2 to calculate seventh order differences with eight registers.

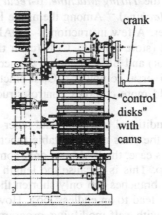

crank

"control disks" with cams

Figure 5: The control mechanism of the Difference Engine[24]

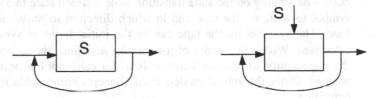

Figure 6: Using the output as input. The sequence S can either be fixed by the construction or be stored on an interchangeable external medium.

---

[24] Doron D. Swade, "Der mechanische Computer des Charles Babbage," *Spektrum der Wissenschaft*, April (1993), 78–84.

As previously stated, the size of the storage was a crucial factor in early program controlled machines. Thus, scarcely relevant for today's computers, an important question for historical calculators is how the output of the computer can be used as input in consecutive steps.

Using the output to influence control

Another very important feature, from a theoretical point of view, is *conditional branching*. Here the result of a calculating step is used to decide where the program sequence is to continue.

Conditional branching is a fundamental concept of universal computing. In theoretical computer science, several models for universal computing are used, such as the *Turing Machine*, μ-*recursive functions* or the *RAM (Random Access Machine)*.[25] Among them, the RAM comes closest to a conventional computer. A few instructions – LOAD (load a value into the accumulator), STORE (store a value), CLR (clear the accumulator), INC (increment the accumulator) and BRZ (branch if the accumulator is zero) – are sufficient to obtain a universal computer.[26] On a practical level, the conditional branch is used to build more complex control structures such as WHILE and FOR loops.

Without conditional branching and other extensions to simulate the conditional branch, the class of computable functions of the RAM is considerably reduced. In this case, the RAM can only compute outputs of finite length or it will never stop. This is because the length of the program has to be finite. Unconditional branches can only shorten the execution path in the program, otherwise they lead to an endless loop. However, *conditional branching* can be simulated with self-modifying programs or by indirect addressing, as shown by Rojas.[27] Hence, a more general description is that the output can influence control (Fig. 7).

The Turing machine reads a symbol from its tape during each step and *decides* – depending on the state transition table – which state to assume, which symbol to write on the tape and in which direction to move the read-write head. The symbols on the tape can be the initial input or symbols already calculated. Within the theory of μ-*recursive functions,* the μ- operator offers the opportunity to execute iterative function calls until a desired value is reached. Other theoretical models need concepts comparable to conditional branching.

---

[25] For example see John E. Hopcroft and Jeffrey D. Ullman, *Introduction to automata theory, languages, and computation* (Reading, Mass.: Addison-Wesley, 1979).

[26] See Rojas (n. 3 above).

[27] Ibid.

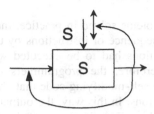

Figure 7: The output can be used to influence control and to decide where to continue the sequence S.

The importance of conditional branching had already been mentioned by Babbage in the context of his Analytical Engine,[28] but he never fully described how this mechanism could be built. In general, it is difficult to implement conditional branching in card or tape controlled machines. Nevertheless, this has been achieved. One example was the Bell Laboratories' Model V. The branches in this machine were described as "to 'hunt' for the beginning of any desired section."[29] The ASCC could not do branching, but it was able to execute iterations with a fixed number of loops. Additionally, it had a special register (the automatic check counter, register No. 72) which could be used to stop the machine if a tolerance in an iteration reached a given value.[30] In this way, the ASCC was able to make a simple decision depending on its results, but the human operator had to decide where to continue the process of calculation.

Finally, in early program controlled machines, conditional branching was rarely used and cannot be compared with conditional branching in today's computers. If output can influence the sequence of instructions, e.g. in the form of conditional branching in combination with other instructions mentioned above, a machine can execute all computable functions. Thus, from a theoretical point of view, no additional machine class is necessary.[31] Nevertheless, another class will be introduced which has more practical consequences.

---

[28] See Babbage (n. 7 above).

[29] See Franz L. Alt, "A Bell Telephone Laboratories' Computing Machine," *Mathematical Tables and Other Aids to Computation* 3 (1948): 1–13, 69–84. Reprint in: Randell (n. 5 above), 263–292, on 273.

[30] See Aiken, Hopper (n. 6 above).

[31] Every Turing Machine can be simulated by a GOTO-Program which uses conditional branches and only three registers containing natural numbers of unlimited size. Neither program modification nor indirect addressing is required for this theoretical simulation. Cf. Uwe Schöning, *Theoretische Informatik kurz gefaßt* (Mannheim: BI-Wiss.-Verlag, 1992), 102–104.

Using the output for control

For several problems solved in practice, the machine had to be able to execute the same sequence of instructions by using different memory locations, and the instructions had to be executed with addresses that had just been calculated. Therefore, the programmers of early machines, such as the EDVAC, wrote self-modifying code that changed the addresses coded inside its own instructions. In this way, the output of the program was also used to control the machine (Fig. 8). For example, the execution of loops with different addresses was used to solve linear systems of equations. Such problems could later be solved more elegantly by using index registers, which were used for the first time in the Manchester MARK I.

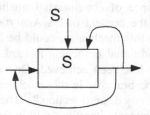

Figure 8: The output can be used as a control sequence S.

Furthermore, obtaining the output of calculations as sequences of instructions allows automatic translation of assembler code or high-level languages into executable programs. Therefore, it was possible to write programs in an easier and more natural manner. Without this capability, the computer would never be used for very complex problems, because programming the software becomes the crucial factor as the problem increases in size.

6. Conclusion

The classification scheme proposed here represents the view of a computer scientist. It is not the only possible classification, but it indicates the different stages which can be found in the history of sequence controlled calculators. The four classes are summarized in Table 2. The scheme can now be used to discuss stages in the development of the computer. Often a *stored program* machine is taken as synonym of computer. From the theoretical point of view, a stored program is not necessary for universal computation.[32] Tech-

---

[32] Cf. Rojas (n. 3 above).

nologically, a stored program allows fast access to the instructions and the instructions are stored in the same memory as the data (Table 2).

Compared with the classification scheme, the stored program concept offers an easy way for output to influence the calculating process, as well as for output to be taken as a new program. In the history of the computer, this offered new quality compared with the machines controlled by tapes or cards, and it gives an indication of why the stored program concept may have been such an immediate success.

Nevertheless, the possibilities of the stored program concept mentioned above were not implemented equally in early machines. The EDVAC, for example, was only able to modify the address field of its instructions.

Furthermore, the activities processed both by humans and machines can be discussed with regard to the respective levels of the classification scheme. The ABC seems to be on the second level of the scheme. The output can be used as input. The output of the machine had no effect on the calculation process. It is usually said that the ABC is a computer capable of solving systems of linear algebraic equations. A closer look reveals that the machine can only eliminate one variable from a system of two equations with up to 29 unknowns. Human intervention is required to solve systems of equations

Table 2: Summary of the classification and its relation to the stored program concept. The arrows on the right indicate the advantage of the stored program concept for the practical implementation of using the output to influence control and of using output as a control sequence itself. Furthermore, stored program allows fast access to the instructions, and the same memory can be used to store data and program

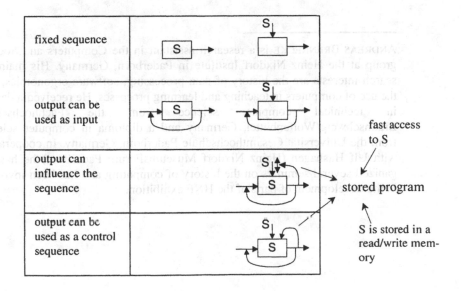

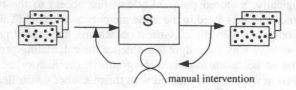

Figure 9: When the ABC was used to solve systems of linear algebraic equations, the output could not automatically be reused for input. The cards containing the intermediate results had to be fed manually.

completely (Fig. 9). The output cards for the intermediate results must be fed manually. However, the ABC could be used for other problems if they could be reduced to systems of linear equations, e.g. special numerical solutions of differential equations. The ABC also introduced very popular concepts like electronic calculation by logical operations and binary number representation

Similar examination of other machines can be helpful for the stated objective of this volume, i.e. a fruitful comparison of the architectures of the first computers of the world. However, the proposed classification is only one aspect and it is not sufficient to capture the historical importance of the machines. Therefore, other factors, such as technological conditions or social circumstances, must be also taken into consideration.

ANDREAS BRENNECKE is a research assistant in the Computers and Society group at the Heinz Nixdorf Institute in Paderborn, Germany. His main research interests are the history of data processing, software ergonomics, and the use of computers in teaching and learning processes. He received a degree in   technical   computer   science   from   the   Fachhochschule Braunschweig/Wolfenbüttel, Germany and a diploma in computer science from the Universität Gesamthochschule Paderborn, Germany. In cooperation with Ulf Hashagen (Heinz Nixdorf MuseumsForum Paderborn) he has organized several seminars on the history of computing and has been involved in the development of parts of the HNF exhibition.

Hardware Components and Computer Design

Harry D. Huskey

Abstract. This paper explains how available components affected the design of electronic computers. Why were the early computers serial? Why were limited instruction sets used? What kind of input/output equipment was used and why?
In presenting these ideas we shall take time to look in detail at the development of the ACE computers at the National Physical Laboratory (NPL) under Alan Turing.

1. Components available around 1940

Prior to 1940, "relays" were available to switch electricity in circuits. These involved mechanical motion which took substantial amounts of time – milliseconds.

The first electron tube with grid control (Fig. 1) was developed by Lee De Forest in about 1906. Later, interest in radio communication, stimulated by World War I (1914–1918), led to mass production of vacuum tubes in the late 1930s (100 million tubes per year). Multiple control grids were introduced in the early 1930s. These tubes promised switching times hundreds or thousands of times faster than relays. By 1940 radio broadcasting had led to tremendous production of electron tubes. World War II further increased the tempo of tube production, leading to the use of pulse techniques, for example in radar.

The simplest electron tube, called a triode, is shown in Fig. 1. It has a cathode which emits electrons when hot – hence, a heater is needed. Next to the cathode is a structure called a grid. It controls the electric field near the cathode. If it is sufficiently negative relative to the cathode, the electrons cannot leave the cathode, but if the grid has the same potential as the cathode, then electrons do leave; some land on the grid but most pass through it. The grid must be able to maintain an electric field near the cathode but must not be such as to impede the flow of electrons. Thus, it is fragile. If the grid becomes positive relative to the cathode, it attracts the electrons, causing the metal to heat and melt.

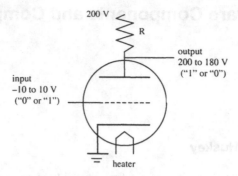

Figure 1: A triode

Consider Fig. 1 again. At the top is the "plate," which is connected outside the tube through a resistor R to a steady high voltage, say 200 volts. Being positive, the plate attracts any electrons which pass through the grid. They flow through the resistor to the 200 volts. However, as the electrons flow through the resistor R the voltage at the output drops. Just how much, depends upon Ohm's law, $E=IR$, where E is the potential across the resistor and I is the current produced by the electron flow in the tube. Typical values in a computer might be 0.01 amperes (10 milliamperes) through a resistor of 2000 ohms (2 K); therefore the product IR is 20 volts. Thus, the output drops from 200 to 180 V.

Another important point about Fig. 1, or any circuit, is that although there is no explicit condenser between the output and ground, there is capacity between the output wires and everything else. This capacity depends upon the surface area of the signal wires, tube and socket components, etc. Thus, there is considerable incentive to use small wires and place elements close together.

Fig. 2 shows the resistor-capacitor equivalent circuit. Assume the input voltage starts at 0 volts and instantaneously changes to +10 volts. The current

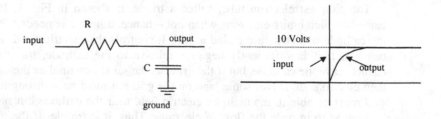

Figure 2: RC circuit

through the resistor $I(t)$ and the output voltage $E(t)$ are functions of time and obey the equation

$$E(t) = 10 \times (1 - e^{-t/RC}).$$

When $t=RC$ the voltage reaches about 63 per cent of the final value. The product RC is called the time-constant of the circuit; it takes about three time-constants before the output is near its final value (~95 per cent). This means that any time a signal passes through an electron tube there will be a delay of up to 3 time-constants. Thus, to attain speed, the circuit designer will wish to minimize the number of tubes through which the signal passes. Note that the input is the output of some other circuit and cannot change instantaneously, further contributing to the delay.

As a logical device the triode has the following properties: i) it inverts (negates) the input, ii) with appropriate choice of resistance there is no loss in signal size (amplitude), iii) the output is nearly 200 volts above the level of the input! To cope with this change in level the ENIAC (Electronic Numerical Integrator and Computer, University of Pennsylvania, 1946) used many voltage levels ranging from large negative to high positive. Another solution to this problem is to use an output network as shown in Fig. 3. As current flows from the top of R_1 to the –200 volts at the bottom, Ohm's law states that the output voltage changes in proportion to the change in the plate voltage. For example, if $R_1=R_2$ and the plate is at +200 V then the output is 0 volts; if the plate is at +180 V then the output will be –10 V.

What purpose does the condenser C (in Fig. 3) serve? Because of the time required to charge the condenser, the output voltage will initially move in step with the plate voltage. Thus, a 20 V decrease in the plate potential will immediately cause a drop of 20 V in the output potential, which in three time-constants will come close to the value determined by the resistors R_1 and R_2.

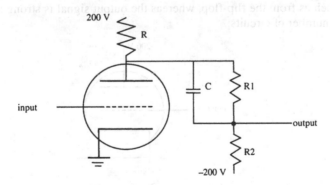

Figure 3: Output network

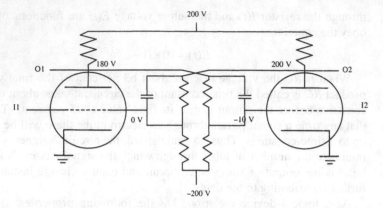

Figure 4: A flip-flop

Flip-flop memory

Two circuits like those in Fig. 3 can be connected as shown in Fig. 4. The circuit (called a flip-flop) has two stable states; if one tube is conducting, the other is off and tends to stay that way. An input to the non-conducting tube will cause it to conduct, turning the other tube off. This flip-flop stores one bit. Example voltages are shown for the left tube conducting.

If we wish the flip-flop to change its state quickly, then we must minimize the stray capacities; thus, the output may be via a cathode follower and the input may use extra tubes. The output capacitance load can be reduced by using a cathode follower (see Fig. 5). The output (cathode) is perhaps 5 to 10 volts above the input. The advantage is that the input may be a weak signal such as from the flip-flop, whereas the output signal is strong and could drive a number of circuits.

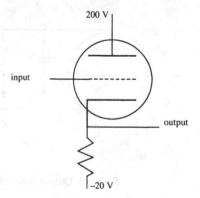

Figure 5: Cathode follower

The dual grid tube or gate

If the triode of Fig. 1 has a screen and a second grid added, then the output (negative) occurs only if both grids are at cathode potential. The first grid (*input 1* of Fig. 6) determines if electrons leave the cathode. If they leave, then the second grid (*input 2*) determines if they go to the plate or the screen. Since the screen is less robust than the plate, designers favored supplying control signals to the second grid and pulses to the first grid – the arrangement was called a gate.

We now move on to consider how these components can be used in computer design.

The ENIAC design

When the ENIAC was started in 1944, the components available to build a computer were the ones described above. The basic unit of the ENIAC was the accumulator, of which there were twenty. Each accumulator was to receive, store, and transmit ten-digit signed numbers. The received number could replace, add to, or subtract from the contents of the accumulator.

Each decimal digit was stored in a ring of ten interconnected flip-flops. Circuitry was arranged so that each ring was stable with one flip-flop "on" and the others "off." Thus, storing a ten-digit decimal number required 100 flip-flops. Numbers were transmitted as strings of 0 to 9 pulses over each of ten lines. The sign of the number required more flip-flops and another transmission line. With the control circuitry the accumulator had about 550 tubes.

There was a multiplier, a divider, and a square rooter. Each used several accumulators to accomplish their tasks. Certain accumulators communicated with punched card input and output equipment. A master-programmer initiated sequences of operations. In all there were about 18,000 tubes.[1]

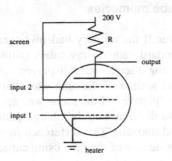

Figure 6: Dual grid NAND gate

---

[1] See in this volume: Jan Van der Spiegel et al., "The ENIAC: History, Operation and Reconstruction in VLSI."

2. New components – 1940s

Mercury delay line

The ENIAC had a very limited memory – twenty 10-digit numbers. Before it was completed, several ideas for better memory components were being explored. Some members of the ENIAC team were experimenting with storing numbers as pulse trains in mercury delay lines (mercury in a cylindrical tube).[2] Mercury lines had been used for precise measurement of time intervals in radar applications. Pulse trains would be applied to a piezoelectric crystal at the end of a tube of mercury. The generated sound waves travel at about 5 feet per millisecond. Another crystal at the receiving end picked up the signal, which was amplified, standardized and sent back to the input crystal. Operating at one megacycle, there would be about a thousand pulses in the line. A 32-pulse sub-train could represent a binary number which would have approximately the same range of values as 10-digit decimal numbers. Less than ten electronic tubes provided the amplification and gating needed. Thus, a mercury line with crystals at the end, and a few tubes would store thirty-two 32-bit binary numbers.

In the ENIAC, units were operating in parallel, transmitting and receiving numbers over multiple lines, whereas the serial nature of the mercury lines led to a quite different structure. Furthermore, the complexity of arithmetic circuits using the components described above led to a single arithmetic unit shared by all the memory units.

In the summer of 1946 the ENIAC team gave a series of lectures on the ENIAC and on their plans for the EDVAC, their delay line computer.[3]

Cathode ray tube memories

During World War II the military had investigated the detection of moving targets using standard cathode ray tubes (similar to TV tubes). The radar return from scanning a scene was "displayed" on the tube. Due to secondary emission the inner surface of the tube was charged by varying amounts, depending upon the "picture." If the scene was again scanned a short time later, there would be no difference in the charge pattern in the tube unless something had changed (moved). This difference in charge could be detected by a screen on the outside face of the tube using capacity coupling.

---

[2] See in this volume: Michael R. Williams, "A Preview of Things to Come," Fig. 1.

[3] See: *The Moore School Lectures, Theory and Techniques for Design of Electronic Digital Computers*, edited by Martin Campbell-Kelly and Michael R. Williams, MIT Press, Cambridge, 1985.

Cathode ray tube memories were studied by the ENIAC team and by F. C. Williams at Manchester University in England. Williams developed a system using scan lines on the face of the tube. This led to a computer design in which numbers (words) were represented by serial trains of pulses, but access to words in memory was random. This was in contrast to delay lines, where one had to wait until the number appeared at the end of the line. Williams was able to have an operating computer in 1949.

At MIT special memory tubes were developed which were similar to the cathode ray tubes; these were used in Whirlwind I.

Magnetic drums

Although magnetic recording had been invented in 1898, it was not until 1947 that magnetic memory for computers received attention. One of the first instances was a drum memory developed at the University of California, Berkeley. The drum consisted of an aluminum tube about eight inches in diameter by 20 inches long. The surface was sprayed with 3 mil coating of iron oxide, and the unit was vertically mounted and directly coupled to a 3600 rpm motor. Fifty heads were mounted along the cylinder so that their magnetic gaps were about 1.5 mils above the surface. There were 10 tracks to the inch and each track stored 90 bits per inch, giving a memory of about 500,000 bits.

This memory held its information during power shut-off, and was much larger than the other proposed alternatives; it was equivalent to about 500 acoustic delay lines.

Input/output equipment

The ENIAC used punched card equipment developed for business data processing for input and output, and the punched cards were used for shelf storage. Other computer projects used teletype equipment and punched paper tape.

Magnetic wire was developed by NBS (National Bureau of Standards) for the SEAC (Standards Eastern Automatic Computer). Magnetic tape was developed by Raytheon for a Navy computer and by Eckert and Mauchly for their UNIVACs.

Logic components

Crystal diodes had been known since the 1920s. They were used as the detector in crystal radio sets. They consisted of a crystal and a "cat's whisker" wire touching the surface. The junction had a low forward resistance and a

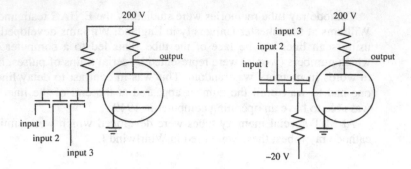

Figure 7: Diode NAND gate   Figure 8: Diode NOR gate

high back resistance. It converted (rectified) high frequency radio waves into low frequency sound waves.

The commercial version appeared in the late 1940s, using a germanium crystal, the same "cat's whisker," in a package smaller than a pencil and less than an inch long. It was delicate. If dropped two feet it was probably ruined. Heat sinks had to be used when soldering or the contact would be damaged. Forward resistance was perhaps 200 ohms and back resistance 50K (50,000 ohms). Ideally, one would like near zero forward resistance and infinite back resistance.

The crystal diode made possible more complex logic circuits. Two examples are shown in Figs. 7 and 8. If inputs swing from –10 V to 0 V, then the grid in Fig. 7 is high only if all inputs are high – an AND circuit. The tube inverts the signal giving NOT an AND, i.e. a NAND circuit. In Fig. 8 the grid is high if any or all of the inputs are high – an OR circuit. The plate output gives a NOR signal. The decrease in physical size and complexity reduced stray capacity giving faster circuitry.

3. Computers based on these components

EDVAC

The team that built the ENIAC lost some of its senior people. Eckert and Mauchly started their own company to build UNIVACs. Goldstine and Burks went to Princeton to work on a computer for von Neumann. I went to the National Physical Laboratory (NPL) in England on a one year appointment to work on the Automatic Computing Engine (ACE) under Turing. Sharpless

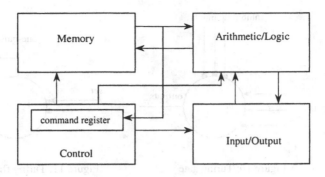

Figure 9: Eckert-von Neumann computer

became the leader and continued the work on the EDVAC using mercury delay lines for memory. Then he went into business for himself, obtained a lucrative patent on magnetic disks and built mercury delay-line memories for sale.

The EDVAC was divided into i) a memory, ii) an arithmetic unit, iii) a control unit, and iv) an input/output unit (see Fig. 9). Von Neumann wrote the "First Draft of a Report on the EDVAC" (June 1945, University of Pennsylvania) describing this structure. Thus, it is now called the "von Neumann computer"; perhaps a better name would be "Eckert-von Neumann Computer."

The delay lines were 384 microseconds long and operated at a clock frequency of one MHz. Words were 44 bits long and an instruction consisted of a 4 bit opcode with four 10 bit addresses. Sixteen instructions kept the control simple and the 384 microsecond lines reduced the delay waiting for operands or instructions.

EDSAC

M. V. Wilkes of Cambridge University attended lectures at the University of Pennsylvania given by the ENIAC team. Following their EDVAC plans he returned to England and built the EDSAC. Using half of the clock frequency of the proposed EDVAC, 500 KHz, he was able to have his computer operational by May, 1949.

Turing and the ACE

Turing used a general logical symbol shown in Fig. 10. There was an output "1" if at least m inputs were "1" and no inhibiting input was "1," otherwise

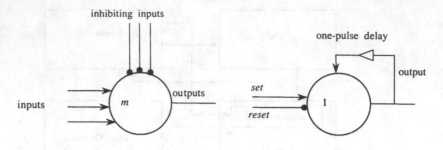

Figure 10: Turing gate   Figure 11: Turing flip-flop

the output was "0." The motivation for this symbol comes from the study of neurons in animal nervous systems. It is interesting to note that von Neumann used similar symbols and ideas in his 1945 EDVAC report.[4] Turing's notation had little relation to possible hardware realizations of the circuits. It is a challenge to use the components described above to accomplish the effect of this symbol. Compare the flip-flop of Fig. 4 with that shown in Fig. 11.

We describe the work at NPL under Turing in greater detail, since it represents a major effort to cope with the delays caused by waiting for operands or instructions in cyclic memories.

D. W. Davies reports in a 1972 reprint (NPL, Com. Sci. 57) entitled "A. M. Turing's original proposal for the development of an electronic computer" that after the end of World War II, J. R. Womersley became head of the new Mathematics Division at NPL and, having met Turing, encouraged him to propose to Sir Charles Darwin (grandson of the famous Charles Darwin), director of NPL, the building of a stored program computer. His proposal consisted of 51 pages of single spaced text and 52 figures, and was presented to the NPL Executive Committee in February, 1946.[5]

He proposed an "electronic calculator" with: i) a memory of 50 to 500 mercury tanks holding about 1000 binary digits each, ii) 50 quick reference storage units each holding about 32 binary digits, iii) an input organ capable of transferring "instructions and other material" into the computer from the outside world, iv) an output organ to transfer results out of the calculator, v) a logical control to interpret the instructions, and vi) a central arithmetic part. He also mentioned switching "trees," a megacycle clock, temperature control for the mercury lines, binary-decimal converters, a starting device and a power supply. The input and output was to be to and from punched cards.

Problems suitable for the computer might involve 5000 real numbers, would not take more than 100,000 times that of a human to solve, and the list

---

[4] John von Neumann, *First Draft of a Report on the EDVAC*, Moore School of Electrical Engineering, University of Pennsylvania, 1945.

[5] *A.M. Turing's ACE Report of 1946 and other Papers*, edited by B.E. Carpenter and R.W. Doran, MIT Press, Cambridge, 1986.

of instructions would be comparable to an ordinary novel. He included ten sample problems, one of which was to compute "winning combinations" in chess to a depth of three moves on either side.

In January of 1947 there were three people working on the project: Turing, J. R. Wilkinson and Michael Woodger. I joined them on a one year visiting appointment.

The group was working on the logical design of a delay line computer. The distinctive feature of the design was the placing of instructions and data in the delay lines so as to minimize wait times (latency). In contrast, other computers with cyclic memories usually obtained their instructions from consecutive addresses, which meant that any instruction took at least one memory cycle. Another distinctive feature of the ACE design was the dispersion of arithmetic and logical operations in the memory switching (this means that the operation units were mapped to the address space).

There was a source switch which connected any memory tank to a "highway" or "bus," which connected to a destination switch connecting, in turn, to the same array of memory tanks. As we shall see, some of the sources were logical combinations of two memory lines, and some destinations performed arithmetical operations. Also, some source destination combinations controlled input and output actions. This made control simple: set up the connections and decide when to transfer. In 1953 using the ACE principles, I designed a magnetic drum computer. The result was manufactured by Bendix and was called the G15. Delay lines were replaced by recirculating tracks on a drum. Each track held more than 100 words so the speed improvement by optimal location of data and instructions was more significant.

There are three ACE computers and the Bendix G15 that we shall consider. The three ACE computers are:

1. The ACE Test Assembly (ACE-TA), which I proposed that the group design and build to acquire experience and, perhaps, to have a usable computer. The project was approved in April, 1947.
2. In November, 1947, the computer design and construction activity was moved from the Mathematics Division to the Radio Division and the Test Assembly was re-designed to become the ACE Pilot Model (ACE-PM).
3. The ACE was the full scale computer that Turing wished to have built. The Mathematics group was working on this in January, 1947. It was to have two source switches supplying data to a function box, which, in turn, connected to a destination switch. The function box was to have 64 functions including logical functions, addition/subtraction, shifting, and discrimination. After the success of the ACE-PM and its commercial cousin the DEUCE, NPL went on to build the ACE.

The instructions for the four computers were coded using the following bits in a word:

ACE-TA		ACE-PM		ACE		Bendix G15	
Bits	*Function*	*Bits*	*Function*	*Bits*	*Function*	*Bits*	*Function*
1-2	Spare	1	Spare	1-5	Wait time W	1	Single/dbl
3-7	Source S	2-4	NI Source	6-11	Source A	2-6	Destination
8-12	Destination D	5-9	Source	12-17	Source B	7-11	Source
13-15	NI Source N	10-14	Destination	18-23	Function	12,13	Characteristic
16-18	Spare	15	Characteristic	24-29	Destination D	14-20	NI Time N
19-24	Trans time T	16	Spare	30	Stop bit	21	Breakpoint
25	Spare	17-21	Wait	31-35	NI Source	22-28	Trans time T
26-31	Wait time W	22-24	Spare	36-40	No effect J	29	Block/item
32	Spare	25-29	Trans time	41-45	NI Time		
		30-31	Spare	46-47	Characteristic		
		32	GO digit	48	Spare		

Trans: Transfer, NI: Next Instruction

The Test Assembly control circuitry was quite simple, consisting of a means to set the source S, destination D, and N switches, and a one-word line with a half-adder which controlled the timing.

The diagram in Fig. 12 illustrates the way the Pilot ACE and the other machines worked. The mercury delay lines are numbered DL1 to DL11. Each mercury line is read sequentially and the information cycles "on place." Information can be moved from one delay line to another by setting the source and destination switches. Closing S1 and changing the position of D2 for the appropriate number of cycles, for example, would transfer the number stored in DL1 to DL2. The TS units are used as registers. TS16 is used as an accumulator which can be loaded with a number (by switching D16). If D17 is accessed, the number arriving from another unit is added to the contents of TS16. If D18 is activated, a subtraction is computed. As can be seen, the main command in this machine is the "MOVE" instruction, which transfers numbers from a source to a destination. Even unconditional jumps can be programmed in this way by using the switches N1, N2, etc. If N1 is closed and the switch in the control unit too, the instruction register is loaded with a new address (previously incremented by one). Conditional jumps are implemented by using the destination 25. If the number transmitted is zero, then the first of the two following instructions is selected.

Other computers

Other machines developed at the time included the following:

• Williams at Manchester University was developing a computer using cathode ray tubes (CRT). His work lead to the Ferranti I.

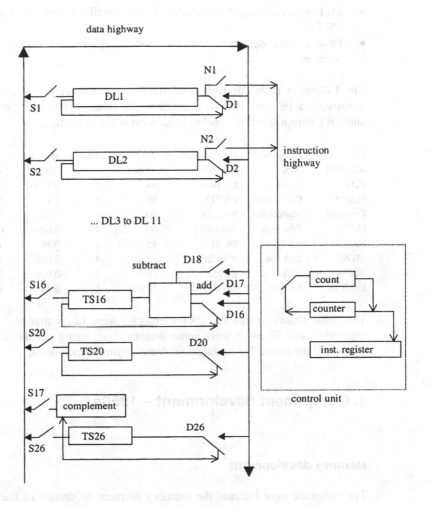

Figure 12: Simplified diagram of the Pilot ACE

- The Institute for Advanced Study (IAS) under von Neumann was designing a parallel computer expecting to use a memory tube developed by RCA.
- The National Bureau of Standards was building the SEAC (similar to the EDVAC) in Washington and the SWAC, a parallel CRT computer, at the Institute for Numerical Analysis (INA) on the campus at UCLA. MIT was building Whirlwind, a 16 bit parallel computer using special CRTs.

- MIT was building Whirlwind, a 16 bit parallel computer using special CRTs
- Others were designing and building computers with magnetic drum memory.

The following table (derived from data in the *Encyclopedia of Computer Science*, 3rd Edition, 1993) summarizes the characteristics of the delayline and CRT computers (DL = delay line, Wms = Williams tube).

Computer	Location	Memory(wds)	Wd Length	Address	Max Access	No. Diodes
ACE-PM	NPL	DL 360	32 bits	-	1.0 ms	-
EDVAC	Army	DL 1024	44	4	0.38 ms	10K
EDSAC	Cambridge	DL 512	35	1	1.1	0
Ferranti I	Manchester	Wms 256	40	1	0.64	0
IAS	Princeton	Wms1024	40	1	0.025	0
SEAC	NBS	DL 512	45	3	0.38	15.8K
SWAC	INA-LA	Wms 256	36	4	0.016	3K
Whirlwind	MIT	ES 256	16	1	0.016	22K
UNIVAC	E&M	DL 1000	12 char	1	0.40	18K

This table shows: i) that the CRT memories were faster than the delay line memories, and ii) there were some doubts about using germanium diodes. The small memories led to magnetic drum auxiliary memories.

4. Component development – 1950s

Memory development

The magnetic core became the memory element of choice in the 1950s. It was a very small ring of ferrite which had a hysteresis curve similar to that shown in Fig. 13. If a current I flows through the horizontal and vertical select wires, the remanence moves to H (Fig. 13) and to F when the current ceases. Similarly, if both currents are $-I$ then it moves to D then to B when current ceases. With pulse current I in one of the select wires the remanence moves from F to G and back to F, or from B to A and back to B, depending upon the initial state of the core. The point F may correspond to a "1" and point B to a "0," or conversely, F may be a "0" and B a "1."

The cores may be arranged in a square array, say 32 by 32, with such an array for each bit in the computer word. If a selected core receives current I in both select lines, it will move from F to H to F or from B-A-H-F. The movement F-H-F represents little change in magnetism so there is no induced

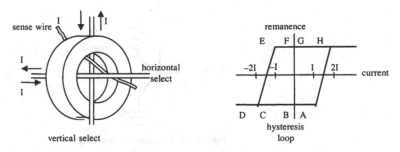

Figure 13: Magnetic core memory

signal in the sense wire; on the other hand, *B-A-H-F* gives an induced signal in the sense wire. Thus, sending current *I* through particular horizontal and vertical select wires in all the planes produces a representation on the sense wires of the binary number at that location. Unfortunately, it erases the word, so circuitry must be provided to restore it.

A nice feature: memory is not lost when the power is turned off. Initially, the hysteresis loops are not as square as one would like. This lead to schemes requiring more than the three wires through each core. Speed depended upon the core being small, but this increased fabrication difficulties. Materials were improved, the loops became more square, but fabrication remained a problem.

Logic components

Soon the germanium in diodes was replaced by silicon, the junction was made more rugged, back and forward resistances were improved. For example, in a junction diode the reverse current was less than 20 microamperes and the forward current at 0.8 volts bias was, perhaps, 50 milliamperes. All computers made in the 1950s used silicon diodes.

The first transistor, a point contact using germanium, was developed in 1947. As in diodes the technology soon moved to silicon and to junction devices. Originally, transistors more or less replaced electronic tubes in similar circuit configurations (see Fig. 14). This reduced voltage swings to, perhaps, five volts and there was no heater structure.

Thus, there was a tremendous reduction in power, components could be closer together reducing stray capacity and increasing speed, and reducing the need for air conditioning.

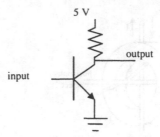

Figure 14: Transistor amplifier

Printed circuits

The cost of connecting circuit components (wiring) and the cost of checking the result led to the development of printed circuit boards. A circuit diagram was printed on a copper clad substrate. The board was then etched to remove the copper not protected by the printing. Components were placed on the board and it was dip soldered. The wiring on one side of the board had to be one-layer (no cross-overs). With through-the-board connections both sides of the board were used, giving two layers of wiring. If actual cross-overs were required then a wire jumper was used.

Magnetic tapes, magnetic drums and magnetic disks

The typical magnetic tape in the 1950s was 2400 feet long and 1/2 inch wide. Information was recorded 9 bits across the tape, perhaps using an 8 bit byte and a parity bit. At 1600 bits to the inch a tape could hold over 40 million bytes, in practice, inter-record gaps would substantially reduce this number. Locally, the tape could be started and stopped very quickly. There was a slack arrangement and servo system driving the reels to match (catch up) with the fast movement at the read/write station. Data transfer rates were about a Megabyte per second.

A drum might be 8 to 20 inches in diameter and up to 4 foot long. Data was recorded on a track around the drum, there was a read/write head for each track. At 3600 rpm the access time for an item might be as much as 17 milliseconds.

A typical disk had a number of platters, there was a pair of read/write heads for each platter, and the assembly of heads for all platters was moved in or out by a track seeking servo. Access might involve track seek time plus time until the data passed the heads. Furthermore, disks rotated slower than drums but had much larger capacity.

5. Effect on Computer Design – 1950s

The low cost of magnetic drum memories led to "small" computers whose cost was in the $50,000 range.

Magnetic cores were more reliable than CRTs or mercury delay lines. There was no waiting for data as in the delay lines. Physically, they occupied less space, were much less sensitive to temperature than delay lines, and more reliable. Cores quickly became the memory component of choice for large computers. Magnetic core memories and transistor logic brought significant improvement in reliability. People no longer talked about "mean time to failure."

Magnetic tape replaced punched cards for shelf storage of data. Punched cards remained the primary input medium. Tape drives became the primary auxiliary memory for computer systems.

Two classes of computers were designed, those oriented toward scientific computation and those toward data processing. Since the main computer was expensive, organizational systems (computer centers) were developed to keep it busy. The user was no longer allowed to laboriously debug his program one instruction at a time.

The user was separated from the computer, he submitted his program and data to a queue, computer center staff moved it to the computer and placed printed output in the output area. The user picked it up and, if he was lucky, had his results. If unlucky, he hoped he had enough information to correct any errors, so he could resubmit it.

So-called "assembly" languages were developed which made it easier for the user to write a program. Particularly, programs in assembly language were easier to correct. Labels for points in the program were introduced and more-or-less mnemonic names could be used for variables. The language processor translated the program into machine language. Also, user oriented languages such as FORTRAN were developed – the user was further removed from the hardware details.

6. Component development – 1960s

Impurities could be introduced (doping) into a sheet (substrate) of polycrystalline silicon so as to change its electrical characteristics. For example, arsenic introduced in the silicon matrix produces an excess of electrons, boron produces a deficiency of electrons (called holes). These are called, respectively, n-type and p-type silicon.

The next significant breakthrough was the field effect transistor (see Fig. 15). Usually the gate is made of metal and the insulator of silicon dioxide, which explains the name metal oxide semiconductor (MOS). Control of electron flow between the source and drain depended upon the field established

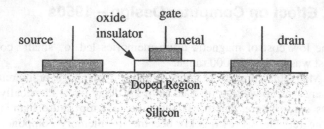

Figure 15: Field-effect transistor (Metal-Oxide-Semiconductor)

by the gate terminal. No control current was required beyond that needed to charge the condenser represented by the gate terminal. This reduced heat allowing components to be even closer together, giving more and faster logic units on a chip.

The multiple steps manufacturing process involves ion implantation, pattern transfer, etching, etc. Size of the individual elements depends upon purity of materials, cleanliness of environment, and stability of equipment.

Computer systems – 1960s

The late 1960s saw memory chips storing 1000 bits. This was the beginning of the end for magnetic cores. The repetitive pattern of a memory chip simplified the mask design, naturally leading to its popularity with chip manufacturers.

The availability of memory chips and transistor circuitry led to cathode ray tube (CRT) displays. These were a welcomed replacement for printers and their stacks of paper.

The new hardware capability opened the way for new operating systems. Instead of the computer center with its queues, several CRTs were connected to the central computer and the operating system gave time slots to the active CRTs in sequence. This allowed the user to process small problems quickly. On-line storage using magnetic tapes or disks allowed him to submit large problems and to debug them on line. Outputs could be sampled and printing ordered when results were thought to be correct.

The success of FORTRAN led to other languages. COBOL (Common Business Oriented Language) for data processing applications was developed. BASIC, a small highly portable (easily moved to new computers) language, was developed on a time-sharing system. ALGOL (Algorithmic Language) was defined by an international committee. It was not commercially successful but is significant for introducing important ideas, such as block structure and explicit declaration of variable types. Other languages with more specific areas of application were developed.

7. Computers 1970 and beyond

The 1970s brought the beginning of refinement: the number of tracks and the number of bits per inch on magnetic disks grew dramatically and the number of transistors on a chip doubled every two years.

The increasing number of transistors on a chip gave the possibility of building a CPU on a chip. The first of these was the 4004 microprocessor. It had 2300 MOS transistors on a tenth inch square chip. Since then there has been continuous improvement. Clock rates have gone up and up, main memory size has increased dramatically. Processor speeds, memory, and transmission rates have been increasing 60% per year (Moore's law). Dudley Buck in the 1950s talked of vacuum triode components produced by photolithography (like integrated circuits). This would make possible microcomputers like we have today. They would cost 25 cents. If they failed you threw them away! He was right about the result but wrong about the technology. Last year's $2000 personal computer costs less than $1000 this year.

Networking started in the 1970s. Now the majority of my correspondence is via e-mail and I buy my London theatre tickets via the World Wide Web (WWW). There are more computers in my car than there were in my state a generation ago!

HARRY D. HUSKEY was an Instructor in Mathematics at the University of Pennsylvania in 1944, when he started working part-time on the ENIAC project. He wrote a technical description of the ENIAC and participated in the early design of the EDVAC. In 1946 he took one-year visiting appointment at NPL (UK) working under Turing on the design of various ACE computers.

Prof. Huskey spent 1948 at the National Applied Mathematics Laboratories of the National Bureau of Standards in Washington, DC, monitoring US computer activity. In 1948 he moved to the Institute for Numerical Analysis, a field station of NBS located on the campus at UCLA. There he designed and led the team that built a parallel William's Tube computer, which was dedicated in August 1950. On leave from INA, he spent a year helping Wayne University (Detroit) set up a computer center, and designed a "small" computer using a cyclic magnetic drum memory which became the Bendix G15.

In 1954 he joined the faculty at UC Berkeley, teaching Numerical Analysis and Computer Design. He helped set up and direct the group that designed the Berkeley Time Sharing System, which was marketed as the SDS940. Prof. Huskey was active in both ACM and the IEEE Computer Society, being President of ACM for 1960-1962. He is a Fellow of ACM and the IEEE Computer Society, and has received Pioneer Awards from both societies.

Part II: The American Scene

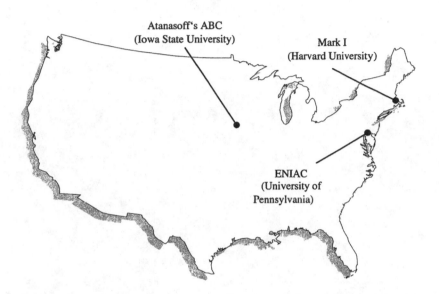

Reconstruction of the Atanasoff-Berry Computer

John Gustafson

Abstract. The Atanasoff-Berry Computer (ABC) introduced electronic binary logic in the late 1930s. It was also the first to use dynamically refreshed capacitors for storage, as in current RAM. Perhaps most astonishing is that it was *parallel,* supporting up to 30 simultaneous operations. Yet, it had far fewer parts than the serial computers that followed in the 1940s. Atanasoff and Berry completed the computer by 1942, but it was later dismantled. Only a few parts of the original computer remain. In 1994, a team of engineers, scientists, and students at Iowa State University/Ames Laboratory began rebuilding the ABC. We demonstrated the functioning replica on October 8, 1997.

In this paper, I describe the computer in modern computer architectural terms to facilitate technology comparison. While patent applications, purchase orders, and photographs gave much information for the reconstruction, a number of mysteries about the computer were solved only as a result of the effort to reproduce it. The answers to those mysteries are presented here for the first time.

1. Atanasoff's motivation

John Atanasoff was a physicist by training, whose interests included molecular spectra and crystallography. As a graduate student, he often found it necessary to solve systems of linear equations by hand, and as a professor he had advisees similarly engaged in the arduous process. He noted:

> Since an expert [human] computer takes about eight hours to solve a full set of eight equations in eight unknowns, k is about 1/64. To solve twenty equations in twenty unknowns should thus require 125 hours

... The solution of general systems of linear equations with a number of unknowns greater than ten is not often attempted.[1]

Alt made a strikingly similar remark eight years later:

> ...13 equations, solved as a two-computer problem, require about 8 hours of computing time. The time required for systems of higher order varies approximately as the cube of the order. This puts a practical limitation on the size of systems to be solved ... It is believed that this will limit the process used, even if used iteratively, to about 20 or 30 unknowns.[2]

In the evolution of automatic computing, solving a system of linear equations was the original "Grand Challenge." Trigonometric functions and logarithms could be tabled, and mechanical desktop calculators such as the 10-decimal Marchant could handle short sequences of the four basic arithmetic operations. However, Gaussian elimination requires of the order of n^3 operations for n unknowns, which makes this manual method unscalable. Atanasoff listed his target applications, which look much like the typical workload at a modern supercomputer center:

- Multiple correlation,
- Curve fitting,
- Method of least squares,
- Vibration problems including the vibrational Raman effect,
- Electrical circuit analysis,
- Analysis of elastic structures,
- Approximate solution of many problems of elasticity,
- Approximate solution of problems of quantum mechanics,
- Perturbation theories of mechanics, astronomy, and the quantum theory.

Atanasoff struggled for years to find a physical way to perform arithmetic that was digital instead of analog. He appears to have been the first to draw the distinction between the two, and to coin the term "computer" for a mechanical device. He thought about parallel processing much the way we do now, in that he considered connecting 30 commodity devices (mechanical Monroe calculators) to attain the necessary speed.[3] He discarded this ap-

---

[1] J. Atanasoff, "Computing Machine for the Solution of large Systems of Linear Algebraic Equations," reprinted in: B. Randell, *The Origins of Digital Computers*, First Edition, (Springer-Verlag, New York, 1973).

[2] F. Alt, "A Bell Telephone Laboratories Computing Machine," reprinted in Randell, n.1.

[3] J. V. Atanasoff, "Advent of Electronic Digital Computing," *Annals of the History of Computing*, 6 (1984).

proach as clumsy and error-prone. In late 1937,[4] he suddenly made the mental leap that provided him with what he was seeking while at a roadhouse near the Mississippi River. He jotted four principles on a napkin,[5] paraphrased here:

- Electricity and electronics, not mechanical methods,
- Binary numbers internally,
- Separate memory made with capacitors, refreshed to maintain 0 or 1 state,
- Direct 0-1 logic operations, not enumeration.

On the other side of the world, the designs of Konrad Zuse were independently paralleling those of Atanasoff, but the Zuse designs were mechanical or relay-based.[6] While the Zuse automatons were less advanced than the ABC in switching and storage technology, they were far ahead of their time in having full floating-point arithmetic and a real instruction set.

By 1940, Atanasoff and graduate student Clifford Berry had taken the above ideas to practice. I will present details of the design, including features not previously explained in the literature on the ABC.[7]

2. Block diagram

Fig. 1 shows an overview of the ABC. It uses terminology more like that for modern computers than like that of the original documentation. Terms like "keyboard abacus" have little meaning for present-day computer engineers.

Because the add-subtract modules could be used for base conversion (both directions) as well as vector addition and subtraction, the total vacuum tube count was very low: about 300 for the entire machine. Much of this economy is the result of operating on only one bit of each number at a time, keeping the carry/borrow bit in a capacitor for use in the next cycle. The 60 Hz line power served as the system clock; one 50-bit number could be added or subtracted in 5/6 of a second, with 1/6 of a second idle time.

The separation of memory from processing is one we now take for granted. On analog computers, there is no such separation. Atanasoff and Zuse independently made the same profound breakthrough in realizing the

[4] C. R. Mollenhoff, *Atanasoff: Forgotten Father of the Computer*, (Iowa State University Press, Ames, 1988).

[5] A. R. and A. W. Burks, *The First Electronic Computer: The Atanasoff Story*, (University of Michigan Press, Ann Arbor, 1989).

[6] See in this volume: R. Rojas, "The Architecture of Konrad Zuse's Early Computing Machines."

[7] A. R. Mackintosh, "Dr. Atanasoff's Computer," *Scientific American*, (1988). J. L. Gustafson, "First Electronic Digital Calculating Machine Forerunner to Cornell's FPS-164/MAX," *Forefronts*, (Cornell University Theory Center, October 1985).

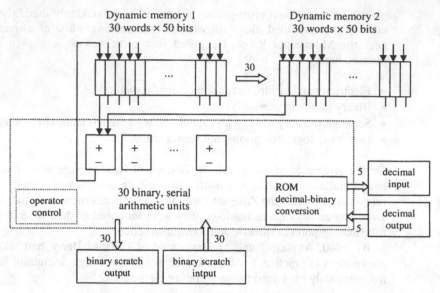

Figure 1. ABC Block Diagram

need for separate "memory," as Atanasoff anthropomorphically called it. One of the few news releases of the Iowa State College (now Iowa State University) about the ABC used the headline, "MACHINE 'REMEMBERS' ."

3. Physical design

The ABC is much smaller than the other early computers. The original dimensions were 1.5 m long, 0.91 m high, and 0.91 m wide. The seemingly minor decision about the width had much to do with the eventual destruction of the original device.

It was constructed in the basement of the physics building at ISU, which at the time was an open area interrupted only by support pillars. The basement was later finished with poured concrete walls and standard doors; the standard door width is 0.84 m. Hence, the computer was boxed in. After Atanasoff left ISU for Maryland, the ABC was seen only as an orphaned device taking up otherwise useful space. Since its frame was welded angle iron, the only way to remove it from the room was to cut it apart with a hacksaw. I feel we have most of the answer to the question: *Why was the ABC destroyed?* The answer is that it was 0.07 m too wide to go through the door. In reconstructing the ABC, we made one practical modification: we narrowed the frame enough so we would be able to go through a standard door.

Fig. 2 shows the reconstructed ABC, using a camera angle similar to that of the historical pictures of the original.[8]

The weight of the machine is about 750 pounds. It rolls on four heavy-duty casters, with the weight and maneuverability of an upright piano. Like a piano, the ABC must be "tuned" after moving to make sure that the timings of brush-triggered events are still within tolerance. The remarkable thing, of course, is that any such antique computer would be so portable. ABC successors such as the ENIAC and the Mark I were notorious for filling rooms with bulky, heavy equipment. The ABC replica has logged thousands of miles in its tours around the country, using a protective crate and a truck specialized for moving sensitive scientific equipment, and it remains functional.

We wondered about the power consumption of the system, but have found that the total power drawn does not exceed 1000 W. The heat generated by the tubes is barely perceptible if one stands near the computer. There is no need for fan-driven cooling; what heat is generated disappears by convection in the open design.

The ABC uses ordinary US line power, 117 VAC, 60 Hz. It was not designed with safety in mind; the two main bus voltages are +120 V and –120

Figure 2. The ABC Replica

---

[8] A web site with information about the ABC and the reconstruction effort can be found at http://www.scl.ameslab.gov/ABC.

V, and in many places on the computer large, unshielded surfaces at these voltages are separated from ground by a few centimeters. With protective covers removed, the rotating memory drums could easily snag loose clothing. We do not expect to operate the computer on a routine demonstration basis, but to rely on videotapes and simulations of its operation.

The physical design is true to that of the original, with the exception of the slightly narrower width and slightly more modern wire (plastic coated instead of cloth coated, coaxial shielding instead of twisted pairs). The contact brushes are the same IBM part used in 1939. The phenolic resin cylinders that hold the memory appear to be the same stock as used by Atanasoff.

The total amount of wire in the ABC is about 1600 m, and almost all connections are soldered. The manual effort required for the wiring (and correction of wiring errors) was probably the single largest part of the cost of the reconstruction itself. We discovered an advantage of the original's cloth-covered wire when we found that the soldering irons had melted some of the replica wire's plastic sheathing enough to cause short circuits in shielded cables.

4. Finding 1939 vintage parts

People often ask us, "Where did you get the vacuum tubes? Weren't they difficult to find?" Several suppliers stock vacuum tubes of the correct type. An Army-Navy warehouse in California supplied us with enough tubes for the entire ABC plus a few spares. About half the tubes tested were gassy, but the remainder worked; most of those that worked were still within design specifications. Finding vacuum tubes was by no means our biggest challenge.

It was much more difficult to find a proper synchronous motor to drive the rotating drums. Modern synchronous AC motors can synchronize on the positive *or* the negative part of the AC power cycle; in 1939, they were wound to synchronize on only the positive part. This does not affect any of the computational logic, but it does affect the base-2 card writer and base-2 card reader. Cards written while synchronized to one phase of the motor will not read properly if the machine becomes synchronized to the opposite phase (as can happen whenever the machine is turned off and back on).

In building add-subtract modules, we found the circuits very demanding of precise resistor values. We have evidence that Berry hand-selected resistors from bins until he found ones that worked, and we attempted the same tactic. In measuring a collection of ±10% resistors, we discovered the distribution about the nominal value shown in Fig. 3.

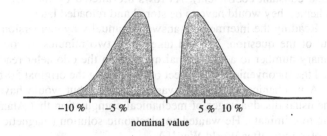

-10 %   -5 %   5 % 10 %
nominal value

Figure 3. Distribution of ±10% Resistor Values

Apparently, the manufacturer had already segregated the resistors close to nominal value. Hence, we found it necessary to use 1% tolerance resistors.

5. Arithmetic and rounding error

The ABC arithmetic is 50-bit binary, 2's complement integer arithmetic. There are no tests for overflow. There is also no rounding when it divides numbers by two via shifting. While this may initially seem to pull values toward zero with every operation, closer examination shows that the Gaussian elimination process causes the data to alternate from positive to negative values, and thus the truncation of bits is balanced on the average. After order n iterations of the subtract-shift process, the values will be off by about $\sqrt{n}$ bits in the least significant place.

The simple examples in Berry's thesis describing the operation of the ABC use linear equations with integer coefficients, avoiding discussion of rounding error. Now that the reconstruction has allowed us to experiment with the computer, it is obvious that Atanasoff and Berry intended to scale all input values so they would occupy the most significant bits in the 50-bit words. The $\sqrt{n}$ bits of error in the least significant place would be negligible for well-posed problems. With the exception of circuit analysis, the physical applications Atanasoff had in mind tend to produce positive definite matrices for which the ABC could easily generate answers correct to ten or more decimal digits.

6. A stumbling block: the ABC mass storage scheme

Atanasoff and Berry knew they would need to record intermediate results somehow. The refreshed-memory storage was only sufficient to hold two rows of the system of equations (up to 29 variable coefficients plus the right-hand constant coefficient). As rows are altered by the Gaussian elimination scheme, they would have to be stored and reloaded later.

Reading the intermediate answers manually by conversion to decimal was out of the question. It could take up to two minutes to convert each 50-bit binary number to a 15-decimal number on the odometer readout, which also had the inconvenient side effect of destroying the original 50-bit number.

A mechanical cardpunch was considered, but would have brought in all the usual disadvantages of mechanical computation that Atanasoff was seeking to eliminate. He wanted an electronic solution (magnetic storage did not evolve until after World War II.)

The solution he and Berry came up with was to use high-voltage arcs to burn holes in paper; a hole for a "1" and no hole for a "0." Berry refers to the paper as "dielectric material." It appears that paper was the only material ever used. The thyratron tubes visible on the left side of the ABC provided the high-voltage pulse to create the arcs. Reading the cards was done by passing the card between electrodes at a lower voltage than that used to burn the card, and with blunt electrodes instead of pointed ones to allow for small differences in alignment between reading and writing. The idea is that the arc will form in the card reader if there is a hole in the card, but the dielectric strength of the paper will prevent this if there is no hole.

Note that there are at least five variables to adjust for this method to work:

1. The voltage used to write
2. The voltage used to read
3. The thickness of the paper
4. The spacing of the electrodes used to write
5. The spacing of the electrodes used to read.

From telephone discussions with Clifford Berry's widow, we learned that Berry found the optimum paper to be Strathmore No. 2. This paper is no longer manufactured, but we know it was thicker than ordinary bond paper and not as thick as IBM card stock. A too-thin paper would buckle in the mechanism and would not prevent arbitrary arcs in the reading of the card; a too-thick paper might prevent arcs in the writing of the card, or snag on the electrodes. Unlike the rest of the ABC, this technology was not prescient ... unless viewed as a precursor to paper tape punch or magnetic tape recording.

Berry's M.S. thesis was to find a combination of these competing design variables that worked. He found it. The write voltage was 3,000 V, and the read voltage 2,000 V. With it, the ABC is able to record the entire contents of one memory drum (1500 bits) in one second. It would be many years before there was another method capable of this I/O rate. The scratch paper cards were about 12 cm by 18 cm and were loaded individually. As they ejected from the mechanism, they were apparently caught by hand; there is no record of any kind of tray.

The Burks[9] cite the reliability of the arcing scheme as roughly one error in every 10^4 or 10^5 bits. This is high, but probably not high enough for the solution of 29 equations in 29 unknowns. The likelihood of a one-bit error increases rapidly beyond about five equations in five unknowns, and this may be the source of the debate, "Did the Atanasoff-Berry Computer ever work?" From hands-on experience, we can now give the answer: yes, but reliability problems prevented it from solving systems large enough to fill its memory. It was still much faster and more reliable than hand calculation, which is what Atanasoff had hoped to achieve.

7. The ABC "Instruction Set"

The operator invokes "instructions" via the buttons and switches on the control panel. A *vector* is a set of the 30 numbers in either memory drum. (The vector in memory 1 is called CA in earlier descriptions of the ABC, and the vector in memory 2 is called KA.) A *short vector* is five numbers, aligned to end on a multiple of 5 within the vector; a decimal input card held one short vector, whereas the binary scratch cards held an entire vector. The possible instructions, with approximate execution times, are as follows:

1 s	Set short vector to coefficients 1 to 5.
1 s	Set short vector to coefficients 6 to 10.
1 s	Set short vector to coefficients 11 to 15.
1 s	Set short vector to coefficients 16 to 20.
1 s	Set short vector to coefficients 21 to 25.
1 s	Set short vector to coefficients 26 to 30.
1 s	Vector clear memory 1.
1 s	Vector copy memory 1 to memory 2.
16 s	Read a short vector from the base-10 card reader into part of the memory 1 vector, converting to binary using table look-up and the add-subtract modules.
1 s	Read a short vector from the base-2 card reader into part of the memory 1 vector.
1 s	Read a short vector from the base-2 card reader into part of the memory 2 vector.
1 s	Select a coefficient in memory 1 (for elimination or decimal output).
100 s	Add or subtract (chosen automatically) one row from another to eliminate the leading bit of the chosen coefficient, shifting right when successful and stopping when the chosen coefficient is zero.
100 s	Write a value to the decimal readout, using the lookup tables and add-subtract modules to convert the base-2 value to base-10.

---

[9] A. R. and A. W. Burks, (n. 5 above).

The timings of more than 1 second are data-dependent, and the values given represent averages for 15-decimal numbers. By using small integers instead of the full dynamic range, the time for those long operations drops by an order of magnitude.

A striking omission in the design of the ABC is the concept of *addressing*. Binary data is not used anywhere to select data locations. The operator performed the selection, which is why the design has so many sizes that are not powers of two (such as 5, 30, 50, 1500).

8. Base-10 output

Descriptions were frustratingly sketchy when describing where the answers finally appeared in human-readable form. We knew that Berry had attempted to use a car odometer to record the output, but eventually custom-made what he needed. The black-and-white photographs of the machine did not reveal anything obviously intended for the output, and we had to solve this mystery before reconstruction could begin.

The man who took those photographs was also the man who did most of the wiring of the original ABC: Dr. Robert Mather. A physicist residing in Oakland, California, he is perhaps the only person still living that saw the ABC in operation. (Atanasoff was alive when the reconstruction project began, but had suffered several strokes and was unable to communicate with us). We contacted Mather, who pointed out the cylinder next to the base-10 card reader in the photograph, and all became clear. The same mechanism that moved the card reader could be used to move a solenoid past odometer-type wheels, poking them by one decimal every time a value on the drum with the conversion table subtracted that decimal. Unlike a car odometer, there is no "carry" when a wheel passes 9; the wheels are independent. The display is small, since it has the same spacing as columns on an IBM punch card.

Because the algorithm alternates between adding and subtracting as the solenoid moves across columns of the number, the wheels are numbered in alternating forward and reverse order. This simplifies the mechanical aspects of the conversion, since the solenoid always moves the same way but the electronics changes state.

One does not find clear answers to some questions in the literature on the ABC. I will attempt to answer those now, based on our experience with the replica and the additional sources of information we found in our quest for details about the original.

9. Exactly when was it completed?

Unlike an invention such as the Wright Brothers' airplane, we do not have a precise date in history for the first successful electronic computation. There does not seem to be a precise date when one can say the ABC worked for the first time. The binary logic certainly was working by the summer of 1940, but the base-2 scratch storage method described above became reliable enough to use in a gradual process and not a dramatic one. By the time World War II had taken everyone away from the project in June 1942, the ABC was in the state that we have reproduced, and that reproduction is a working computer.

Had the war not interfered, Atanasoff was planning to make the instruction sequencing automatic instead of entered manually from the control panel, and to make the computer more general-purpose. With the exception of Zuse's paper-tape mechanism for instruction sequences, stored instruction sequencing had to wait until the late 1940s.

10. Was the ABC electronic or electromechanical?

The fractional-horsepower motor and gear trains suggest that the ABC was an electromechanical computer and not an electronic one. This is not the case. The mechanical function was similar to the motor that turns the hard disk or CD-ROM drive inside a modern computer; the gears and mechanical parts were not used for computing or to record data in any way.

A small number of relays were used, but only for control. They more closely resemble the on-off switch on a modern PC than the gate elements of the Zuse Z3.

Modern electronic computers have many moving parts in the input keyboards and output printers, and so did the ABC. It is true that the *clocking* of the system was mechanical and not electronic. With an oscillatory circuit used to set the system clock instead of a rotating cylinder making contact with brushes, the ABC logic could have been made ten thousand times faster. Note, however, that this would have been a gross mismatch to the I/O limitations of the system. Even the later ENIAC, which used electronic clocking, experienced its bottleneck in the punch card input and output.

The ABC was fully electronic in its calculation and in its storage of data. For that reason, I argue that its mechanical aspects are no different from those of any modern computer; a motor to rotate the storage medium, and mechanical switches for the human interface.

11. Was the ABC ever actually used?

Some have claimed that no one ever used the original ABC for production computing. We have found evidence to the contrary. The first three applications listed by Atanasoff are all statistical, and Atanasoff collaborated with the well-known applied statistician Snedecor at ISU. Publicity that resulted from our reconstruction effort led Clara Smith, a secretary in the Mathematics Department now living in rural Iowa, to contact us. She said one of her tasks was to hand-verify solutions to problems that Snedecor was sending to Atanasoff. It appears Snedecor sent a steady stream of small linear systems to the ABC for solution, and it would have been very well suited to regression, least squares, curve-fitting problems. Clara Smith verified some of the results for Snedecor to establish his confidence that the ABC was producing correct answers.

Robert Mather also says the original machine solved problems up to size five, but more typically size three during testing and debugging. The experience we have had with the replica makes this recollection very plausible. Five is the size of a short vector that fits on one input card, and does not require any switches to be changed on the short vector location. It also involved about 3×10^4 bits to be sent through the base-2 mass storage system, which is about where the reliability of that system becomes limiting. A linear system size of three is very useful in testing and debugging, since one can solve a 2 by 2 system plus right-hand-side vector without any use of scratch storage. Since we have done nothing to improve on the technology of the original, I feel we have settled the question of whether the original ABC was ever operational: It was. Moreover, it was probably used to solve real statistical problems.

12. How fast was the ABC?

As in all computer performance measurement, it is better to take into account the time to solve an entire problem and not excerpt the time to do a single operation as a measure of speed. The latter is usually much more flattering, but seldom reflects true performance. For example, one could cite the fact that a 30-element vector addition on the ABC takes only one second, implying 30 arithmetic operations per second. Perhaps this is the "peak performance" rating for the ABC. I instead consider "sustained performance." To measure that, there is no substitute for a working replica.

Because of the parallelism in the architecture, the sustained performance is maximized if all 30 add-subtract units are used; that is, if one solves a system of 29 equations in 29 unknowns. Burks has estimated that this would take about 25 hours, including human operator time. With a LINPACK-type operation count of $2/3\ n^3 + 2n^2$, the Gaussian elimination of a system of size 29 requires about 18,000 operations. This implies 0.2 operations per second. Because the ABC had parallelism greater than n, the time complexity grows as n^2 and not n^3.

I have noted that the ABC probably never was used for such a large system because of the very high reliability requirements for the scratch storage. We can look at the other extreme, two equations in two unknowns. The usual floating-point operation count for a 2 by 2 system (counting reciprocation as three operations) is 19. Our experience is that such a problem can be solved in about five minutes, somewhat less than the estimate in the Burks' book. This implies 0.06 operations per second. The 2 by 2 system need not involve scratch storage, which saves time beyond what would expect from problem scaling.

13. How "special-purpose" was the ABC?

Some refer to the ABC as a "special-purpose" computer, perhaps to diminish its place in computer history. "Special-purpose" and "general-purpose" are not scientifically defined adjectives. Most computers are designed with a certain range of applications in mind, and the list that Atanasoff mentions above is broad. We already know that the ABC did not have an automatic instruction stream like the Harvard Mark I; its only capacity for branching was in testing zero crossings and zero results during elimination. It relied on the human operator to deliver commands and make decisions about what to do next.

After the ABC reconstruction began to be operational, I realized that the ABC could in fact be employed the way one would use a pocket calculator, and that it could be "programmed" in a sense by the choice of the coefficients matrix. To see this, consider what happens when one solves the system

$$ax + by = u$$
$$cx + dy = v$$

The result of one application of the ABC row-elimination step results in

$$ax + by = u$$
$$0 + (d - bc/a)y = v - uc/a$$

The quantity $(d - bc / a)$ is easily read out on the decimal display. Quantities u and v need not even be entered. If we want to obtain the four basic operations $+, -, \times, \div$, then

$$b = -1, \quad u = +1 \quad \text{gives } c + d.$$
$$b = +1, \quad a = +1 \quad \text{gives } d - c.$$
$$d = 0, \quad a = -1 \quad \text{gives } b \times c.$$
$$d = 0, \quad c = -1 \quad \text{gives } b \div a.$$

Although this may seem clumsy, it was certainly easier and less error-prone than doing 15-digits decimal arithmetic by hand.

14. Perspective: computers now and computers then

The ABC illustrates two remarkable things about the history of computers: First, that Moore's law seems to work if extrapolated all the way to 1939, and second, that a surprising number of things have *not* changed much from the ABC.

Moore's law was first posited in 1970, using only three data points. It primarily applied to chip density, and by implication the cost per bit ... but because speed tends to scale linearly with memory, it has been found to be a good guideline for processor speed as well: *Performance doubles every 18 months*. So if we extrapolate back to 1942 from say, late 1996, we should have doubled performance about 36 times. 2^{36} is about 70,000 millions. Are current supercomputers 70 billion times more capacious and faster than the ABC? Does Moore's law hold even before the invention of integrated circuits?

The ABC had 0.3 kilobytes of main storage. The Intel Teraflops computer delivered to Sandia National Laboratories last year now has 0.3 terabytes of main memory and a terabyte of disk storage. It isn't clear which one we should use, because the ABC memory used refreshed capacitors like DRAM yet spun mechanically like a disk. The speed of the ABC was, translated to modern terms, about 0.06 "Flops" where we politely ignore the lack of exponent management in the ABC and look at the 50-bit precision as similar to a modern 52-bit IEEE mantissa. The Intel Teraflops computer, true to its name, has demonstrated a trillion Flops with that precision, and running the same application. That represents a factor of about 20 trillion.

With this improved baseline, we can recalibrate Moore's law – but it doesn't need much modification. It looks like DRAM technology doubles every 20 months, and processor speed doubles every 28 months. It's a little like recalculating the Hubble Constant when a telescope finds another more ancient and more distant quasar.

15. Cost

If the reader will forgive my use of American units, someone once noted that computers cost about $400 per pound – give or take $100. This amusing statistic is surprisingly good at predicting the cost of everything from a pocket calculator (0.1 pound, $40) to a Cray vector mainframe (30,000 pounds including motor-generators, $12,000,000). While we'd like to think that the cost of a computer stems from its intellectual content and not its mass, the heuristic seems to work. The ABC weighs 750 pounds, and it cost about $300,000 to reconstruct. This fits the $400/pound estimator despite the use of vacuum tube technology. This ABC price includes the engineering labor cost; some quoted prices for the original ABC list $5000, and include only the parts in 1939 dollars. If one adjusts for inflation and estimates the cost of Atanasoff, Berry, and the several students that helped with the original, the cost of the reconstruction is very close to the price of the original.

16. Summary

Scientific computer architectures have long been designed with linear algebra in mind. The vector computers of the late 1970s and early 1980s (Cray-1, Cyber 205, etc.) and array processors of the same era were strongly optimized for the kernel operations of matrix factoring and matrix multiplication. The ABC was the first linear algebra computer, and its 1940 performance is very close to what Moore's law predicts. World War II prevented its innovations from being publicized and credited to Atanasoff and Berry via patents or published papers. However, the ideas of fully electronic digital logic and dynamic refresh capacitor storage were communicated to other early designers and were thereby added to the body of knowledge of computer design.

Acknowledgments

The reconstruction of the ABC was driven initially by two men: Delwyn Bluhm, Manager of Engineering Services at Ames Laboratory, and George Strawn, former director of the Iowa State University Computation Center. They obtained initial funding from Charles Durham, a successful executive who had been a student of John Atanasoff. Without the enthusiasm and vision of Bluhm and Strawn, the ABC reconstruction would not have happened. Engineers Gary Sleege, Dave Burlingmair, John Erickson, and many others worked long and hard to reproduce what Atanasoff and Berry had created over 50 years earlier. The entire project was privately funded; no government money was used.

The person who took the untested prototype to a functioning computer was Charles Shorb, a dedicated graduate student and as close to a modern-day Clifford Berry as one could ask for. Shorb had worked on the Intel Teraflops computer just prior to his work finishing the ABC, so he went from the world's fastest computer to the world's slowest one (a ratio of about 10^{13}:1). When the press release went out from Intel that its parallel processor had demonstrated over a trillion operations per second, the problem that it solved was one Atanasoff would have appreciated had he lived to see it: It solved a linear system of 125,000 equations in 125,000 unknowns.

JOHN GUSTAFSON is a computational scientist at Ames Laboratory in Ames, Iowa, where he is working on various issues in high-performance computing. He has won three R&D 100 awards for work on parallel computing and scalable computer benchmark methods, and both the inaugural Gordon Bell Award and the Karp Challenge for pioneering research using a 1024-processor hypercube. Dr. Gustafson received his B.S. degree with honors from Caltech and M.S. and Ph.D. degrees from Iowa State, all in applied mathematics. Before joining Ames Laboratory, he was a software engineer for the Jet Propulsion Laboratory in Pasadena, Product Development Manager for Floating Point Systems, staff scientist at nCUBE, and member of the technical staff at Sandia National Laboratories.

Howard Aiken and the Dawn of the Computer Age[1]

I. Bernard Cohen

Abstract. Howard Aiken, one of the pioneers who introduced the computer age, earned his place in the historical record by several different sets of achievements. One was his design and completion of four giant calculators (or computers) at the dawn and during the first stages of the computer age, another his pioneering program in what we know today as computer science. He also was one of the very first explorers of the application of the new machines to business purposes – problems of the life insurance industry and computer billing in the utilities industries. He also contributed to the computer age in sponsoring new areas of application for computers, including machine translation of foreign languages, the use of the new machines in textual and historical analysis, and the application of computers to economics. He was much in demand as a speaker in America and in Europe, and he constantly urged the introduction of the computer into new areas of research and action. His contributions to the new age also include two symposiums he organized (in 1947 and 1949) to bring together those who were pioneering the designs and construction of new computers or planning new applications.

1. Aiken's background

Aiken's primary claim to a notable place in history is usually said to be his first machine, converted from his specifications into engineering reality by IBM engineers. This machine, known as Mark I (or the Harvard Mark I), and originally named the IBM ASCC, gave the world of scientists and engineers a visible proof that a complex machine could solve complicated mathematical

---

[1] This talk is based on the writer's recently published book, *Howard Aiken, Portrait of a Computer Pioneer* (Cambridge: MIT Press, 1999); also a companion volume of essays, edited by the writer together with Gregory M. Welch and Robert V. D. Campbell, *"Makin' Numbers": Howard Aiken and the Computer* (Cambridge: MIT Press, 1999).

problems by being programmed to execute a series of controlled operations in a predetermined sequence – and do so without error.

A strong man in both his quality of mind and his physical being, Aiken had all the strengths and weaknesses of successful pioneers. He forced Harvard, an essentially humanistic university, into adopting a leading role in the new art and science of computing. Some of his visions of the future so outran the course of events that his predictions were often not validated until decades later. He was strongly opposed to the new concept of the stored program, primarily because he feared that the mixing of program or instructions with data in the same store would jeopardize the integrity of the program, and he had a basic dislike of binary number systems.

Aiken's early life set him in a personality mold which affected his relations with students and colleagues and which determined his character. A daring and bold pioneer, he was self-willed, and combative. Aiken was a giant of a man – in his physical stature, his force of will, his originality of mind, and his achievement. Standing erect at six feet and some inches tall, he towered over most of his students and colleagues. Graced by nature with a huge dome of a head, he had piercing eyes crowned with huge beetling and somewhat satanic eyebrows.

Aiken related to people in extremes. When he met you, you were almost at once graded, placed at the top of his scale or the bottom – there was never a middle ground. On a scale from one to ten, Aiken would rate you as a zero or an eleven. People reacted to Aiken in the same way. His students and associates either admired him and established a friendly relationship or found him to be "impossible." Friends and colleagues and former students on the "plus" side remained loyal and devoted for the rest of their lives and Aiken himself cherished long-term relations. Those on the "minus" side tend to remember only those occasions when he was intransigent and difficult.

Howard Hathaway Aiken was born in Hoboken, New Jersey, in 1901 and was educated in the public school system of Indianapolis. Throughout his four years of high school, he held a full-time job, working a twelve-hour shift at night, in order to support his mother and grandmother who had no other source of income. On graduation, he enrolled as an undergraduate in the University of Wisconsin, where he continued to work at nights in order to support himself and his family. During an oral-history interview which Henry Tropp and I conducted with Aiken shortly before his death, on 24 February 1973, he explained that Wisconsin had adopted "the eight hour day" and so he "had to work from four to midnight on that job" which "was much easier" than the twelve-hour shift at Indianapolis.

Aiken graduated from the University with a bachelor's degree in electrical engineering. After a decade as a successful electrical engineer, he found himself elevated to a managerial position without contact with daily problems of engineering practice. He resigned his business post and decided to go back to the university for higher training in the sciences. His first choice was the University of Chicago, but he didn't like Chicago and transferred to Harvard, where he entered the graduate program in physics in 1931. At that time, Aiken was thirty years of age, much older than his fellow graduate students.

At Harvard Aiken became a member of a small group of faculty members and students associated with the Physics Department and known by the subject name of "communication engineering." Basically, the members of this group were concerned with the physics of electromagnetic waves (their transmission, reflection, reception, and so on), the study of radio transmission problems, and the physics of the operating units of radio transmission. Thus, one member of the group investigated the way in which radio waves bounce off the ionospheric layer, another the design of antennas and antenna theory generally. The head of the group, the person with whom Aiken did his research, E. L. Chaffee, was primarily interested in the physics of vacuum tubes.

2. Computing machines

As was the custom of those days, Aiken was assigned a problem for his doctoral research and eventual dissertation. His problem was the conductivity of space charge, "a field where one runs into [partial] differential equations in cylindrical co-ordinates ... in nonlinear terms, of course." Before long his thesis research came to consist primarily of "solving nonlinear [differential] equations."

The only methods then available for numerical solutions of problems like his made use of electromechanical desk calculators, of about the size of today's cash registers, so that calculations like those he needed were "extremely time consuming." It became apparent – "at once," according to Aiken – that the labor of calculating "could be mechanized and programmed and that an individual didn't have to do this." He was also aware that a computing machine would also be of great use in solving pressing problems in many sciences and in engineering and even in the social sciences.

By April 1937, he had progressed sufficiently far in his general thinking and design to be ready to seek support from industry. In preparation, Aiken drew up a proposal stating the need for such a machine, together with the principal features of its mode of operation and its general method of solving problems. His philosophy was later expressed in a student's assignment, drawn up for one of his Harvard classes in computer science. "The 'design' of a computing machine," the students were informed, "is understood to consist in the outlining of its general specifications and the carrying through of a rational determination of its functions, but does not include the actual engineering design of component units."

In this clear statement, as was often the case for Aiken, the primary concerns were the logic of the machine, the mathematical operations, and the general architecture, while the actual technological specifications or the choice of components was secondary. To judge from all the information available, Aiken's design would not have specified what particular components (nor even what types of components – mechanical, electromechanical, electronic) would be used, nor how the various components of the machine would be linked. He would have specified the need for performing certain types of mathematical operations and a means of programming them so that they would be performed in a certain predetermined sequence. He also would have indicated the need for storing certain tables of numerical data. These specifications would have been definite but not necessarily confined to any particular type of functioning elements. Thus the design would apply equally to a machine that would be constructed of mechanical, electromechanical, or electronic components, or any combination of them.

Once Aiken had completed the general design of his proposed machine, his next step was to find some company willing to build it. During the course of our interview, Aiken explained that because of the size and complexity of his proposed machine, only a large manufacturer of calculators or business machines could possibly have been induced to produce it. Accordingly, he turned first to America's foremost manufacturer of calculators, the Monroe Calculating Machine Company. Armed with his document of specifications, Aiken obtained an interview, which took place on 22 April 1937, with George C. Chase, a distinguished inventor in the calculator field who was then Monroe's director of research.

Chase later reported how Aiken outlined his conception of the machine and "explained what it could accomplish in the fields of mathematics, science, and sociology." Aiken told Chase that "certain branches of science had reached a barrier that could not be passed until means could be found to solve mathematical problems too large to be undertaken with the then-known computing equipment." Although Aiken referred to "the construction of an electromechanical machine," he had not as yet specified what kind of actual components were to be used. Chase was quite emphatic on this point. The "plan he outlined," Chase wrote, "was not restricted to any specific type of mechanism." Rather, his design "embraced a broad coordination of components that could be resolved by various constructive mediums."

Aiken's attempt to elicit the support of Monroe came up rather early during the interview, when I pressed him to explain why he had chosen to build Mark I out of electromechanical parts. After all, his thesis was on vacuum tubes, on space charge, and his own graduate specialty was the field of electronics. Why, I wanted to know, did he even consider electromechanical systems rather than electronic systems? Why had he not contemplated using vacuum tubes? I must confess that I expected Aiken to frame his reply in terms of his great often-expressed ideal: reliability! I will even confess that I asked the question less as a means of obtaining information than as an opportunity to record on tape – direct from Aiken's mouth – his thundering condemnation of unreliable vacuum tubes and his preference for slower and more reliable relays.

It was only much later that, thanks primarily to a little tutorial given to me by Bob Campbell and to the insightful comments of Maurice Wilkes, I came to understand that Aiken's study of the physics of vacuum tubes was only indirectly related to the use of vacuum tubes in designing electronic circuits. In fact, in a statement written by Aiken toward the end of the war, in 1945, he reviewed the goals of education in the Cruft Laboratory, where the "plan for instruction" had been designed around "basic scientific material of communication engineering," together with "much of the allied branches of science." There was "no attempt to apply this material to specific engineering problems." Instead, the program had been directed exclusively to "the elucidation of fundamental scientific principles." For the purposes of computation, however, what was needed was not the scientific principles underlying circuitry, not a knowledge of the physics of space charge, but rather some experience in the design of high speed pulse circuitry. In this latter area Aiken had little or no experience.

In a talk given in Sweden and in Germany in 1956, Aiken recalled that, in his undergraduate days at the University of Wisconsin, in his senior year, when he was a student of electrical engineering, "there was offered for the first time a course called 'thermionic vacuum tubes.' " He didn't explore this new field, however, because his professors advised him that he "would do far better" if he "took the course in transformer design rather than this new and untried subject."

Once started, Aiken continued his recollections of Chase. He was "Chief Engineer at Monroe," he said, "and a very, very, scholarly gentleman. He took an almost immediate interest, and we kept up an association for quite a few years thereafter. He wanted, in the worst way, to build Mark I. He would supply me with the parts and we would collaborate and do it together, that's what he wanted to do."

Chase was enthusiastic about Aiken's project. According to Aiken, Chase "went to his management at Monroe and he did everything within his power to convince them that they should go ahead with this machine because, although it would be an expensive development." Chase had the vision and foresight to recognize that the proposed machine "would be invaluable in the company's business in later years." But, although "Chase could see this," his "management, however, after some months of discussion turned him down completely."

Aiken's remarks about his not having been wed to a single type of components for his dream machine was very revealing, but I was not completely satisfied by Aiken's presentation. I wanted him to discuss what he remembered about the relative advantages and disadvantages of mechanical systems, electromagnetic devices, and vacuum tube circuits. Accordingly, a little later in the interview, I returned to the subject of why Aiken had chosen to have his machine built of electromechanical components such as relays – why he had not made use of vacuum tubes. This time I stressed the fact that this choice of relays had always seemed astonishing to me in view of the fact that Aiken had been a student at Harvard of E. L. Chaffee, under whose direction he had written his doctoral dissertation; Chaffee's specialty was vacuum tubes and vacuum tube circuits. To be specific, I asked whether at one time there hadn't been some thought given to having quenching circuits in that first machine, using vacuum tubes. Aiken replied, "Yes. But your question really is: since I had grown up in 'space charge' in a laboratory like Cruft [at Harvard], why wasn't Mark I an electronic device? Again the answer is money. It was going to take a lot of money. Thousands and thousands of parts!"

Then he explained: "It was clear that this thing could be done with electronic parts, too, using the techniques of the digital counters that had been made with vacuum tubes, just a few years before I started, for counting cosmic rays." And then he concluded with the following dramatic assertion: "But what it comes down to is this: if Monroe had decided to pay the bill, this thing would have been made out of mechanical parts. If RCA had been interested, it might have been electronic. And it was made out of tabulating machine parts because IBM was willing to pay the bill."

Clearly, at this time, Aiken was not wedded to any particular technology, his top priority was not the choice of relays.

To most historians and computer specialists, it will seem just as astonishing as it was to us to learn that the choice of the kind of machine to be built was determined solely by financial considerations, by the willingness of one or another company to put up money for the machine. This disdain for the technological components was, I believe, a very significant part of Aiken's intellectual make-up. We shall see in a moment how this aspect of Aiken's system of values was a major factor in producing the eventual rift between him and IBM. Aiken never appreciated the degree to which the technology of IBM's product line may have made IBM the only company that at that time would have undertaken to build Aiken's machine. It is to be noted that when Eckert and Mauchly designed the ENIAC, constructed at the Moore School at the University of Pennsylvania, they did not base the machine on any company's off-the-shelf technology but rather developed new types of circuitry and design for the special purpose they had in mind.

3. The role of IBM

When Chase found that his company would not undertake to build Aiken's dream machine, he advised Aiken to try IBM. At IBM, Aiken's project won the immediate support of James Wares Bryce, IBM's chief engineer, then known affectionately within IBM as "the Father Engineer." Bryce was the holder of more than 400 patents, making an average of about one per month. In 1936, on the centenary of the U. S. Patent Office, Bryce was honored as one of the ten "greatest living inventors." Aiken's meetings with Bryce were the inaugural steps toward the construction of the Automatic Sequence Controlled Calculator.

As all histories of IBM make clear, no important decision was ever made at IBM without the explicit approval of IBM's president, Thomas J. Watson, senior. Watson was a powerful figure, a titan in his sphere and endowed with just as forceful a personality as Howard Aiken. Anyone who has read anything at all about these two figures will know that there would be an eventual collision, a terrible clash. And, after IBM built Aiken's dream machine, there was just such an inevitable conflict, also supported by Thomas J. Watson (Jr.), IBM's CEO.

Bryce wanted to be certain that Aiken would know about the different technologies used in IBM business machines, accumulators, sorters, and printers. And so Bryce arranged for Aiken to attend IBM's training school for technicians and then go out on the job of repair and maintenance of IBM machines. Only after Aiken had become familiar with the potentialities and limitations of IBM's product line did Bryce advance to the next step of getting the machine built.

At first it was envisaged that IBM would supply the parts and that, under Aiken's supervision and with some assistance from IBM engineers, the giant machine would be built at Harvard by mechanics in Harvard's own machine shops. Eventually, this proved to be impractical, and all of the work was done at IBM's facility in Endicott, New York. Bryce appointed Clair Lake to be the engineer in charge of the project. Francis ("Frank") Hamilton was the immediate supervisor, and Ben Durfee was actually in charge of day-to-day construction of the circuits and the design of the controls. These three engineers were skilled technicians of extremely high ability, but they were not trained in college level mathematics and really didn't understand the nature of the mathematical problems that the machine they were designing was to perform.

Aiken spent the whole summer of 1939 in Endicott with the IBM engineers. He set forth the mathematical requirements, listed the constants that had to be in the store, and helped design the circuits. During the next twelve months, while Aiken was busy teaching at Harvard, he nevertheless got to Endicott most weekends and again spent the whole summer (1940) working with the IBM engineers at Endicott. During the following autumn and spring, Aiken continued to visit Endicott. On 6 March 1941, Aiken sent Hamilton 36 logarithms and 21 sine values to be stored in the machine for use in calculating values of those two functions. He also made specific his requirements for interpolation. One month later, Aiken was called to active duty as an officer in the U.S. Naval Reserve. He had enlisted in the Reserves some time earlier but was not called to active duty until April 1941, eight months before Pearl Harbor. Resplendent in his new uniform, he visited the Endicott laboratory on 24 May and announced that he would have "very little time to spend on this machine from that time on."

In order to ensure that the machine would be completed, Aiken appointed as his deputy a Harvard graduate student in physics, Robert Campbell. Campbell not only visited Endicott in the last phases of design and construction, he also became the chief operator and programmer of the machine after it had been delivered to Harvard in February 1944. During the spring of 1944, Aiken and his supporters succeeded in getting the Navy to appreciate the importance of the new machine in the war effort, with the result that the operation of the calculator became a Navy project. In the spring of 1944 Aiken received orders transferring him to Cambridge to take charge of his computer. He once remarked that he was the first officer in the history of the U.S. Navy to be put in command of a computer.

In the summer of 1944, with the computer in full operation as a Navy installation, Harvard's President James B. Conant decided that the time had come to have a formal dedication of the new machine. With the consent of the Navy, a ceremony was planned at which Aiken and Watson would give talks, to be followed by an inspection of the giant machine. A distinguished company of guests – including admirals, government officials, leaders of business and technology, and members of the Harvard faculty would attend. The date was 7 August 1944.

Watson arrived in Boston the day before the ceremony and was shown a copy of the news release prepared by Harvard for the press. This story stressed the contributions of Aiken as THE inventor, down played the contribution of the IBM engineers and barely mentioned the role of IBM as the constructor. Watson was so angry – and with good reason – that he threatened to boycott the ceremony altogether. Harvard's President Conant, Aiken himself, and several other members of the faculty and of IBM rushed to Watson's hotel and succeeded in calming his rage, promising to issue an emended news release that would give credit to IBM and its engineers.

This episode reveals a difference in philosophies. Aiken's philosophy of what we would call "architecture" centered around the functions that the machine would perform, the sequencing of the operations and the way in which the sequence of operations would be programmed. From his point of view, he was THE inventor, or at least the primary inventor, for without him there would have been no machine. Watson and the staff of IBM thought in more practical terms. An inventor was the person (or group of persons) who designed and built an actual working machine. From IBM's point of view, the three engineers – Lake, Hamilton, and Durfee – who had built the machine were the inventors. Here was an intellectual impasse that could not be bridged.

In any event, Watson did attend the ceremony the next day and magnanimously made a gift to Harvard of $100,000. The equivalent in today's money of about a million dollars. The news coverage of the dedication was worldwide. What impressed scientists, engineers, philosophers, and ordinary mortals was the fact that this giant machine could perform a complex sequence of operations (or commands) automatically according to a program and do so without error. Pascal, Leibniz, Babbage and others had dreamed of such a machine that might emulate functions of the human brain, but Aiken and the IBM engineers had done it.

4. Mark I

The IBM Automatic Sequence Controlled Calculator or Harvard Mark I, as it was soon called, was so large and so complex that most accounts stress the huge size and the enormous number of parts rather than its functioning. The completed machine was the largest and most complex electromagnetic device ever constructed. Mark I was an imposing sight, sheathed in stainless steel.

The news spread rapidly throughout the world, carried by newspapers, magazines, radio, and word of mouth. The news was that a new world was dawning, heralded by Aiken's machine. Let me give you a single example to show how the news spread.

Konrad Zuse told me an amusing anecdote about how he first encountered the work of Aiken. The occasion of our conversation was a luncheon in Zuse's honor, hosted by Ralph Gomory at the Watson Research Laboratory of IBM before a lecture given by Zuse to the staff of the lab. When Zuse learned that I was gathering materials for a book on Aiken, he told me that he had come across Aiken and Mark I in an indirect manner, through the daughter of his bookkeeper. She was working for the German Secret Service (*Geheimdienst*) and knew through her father of Zuse's work on a large scale calculator. According to Zuse, the young woman never learned any details about his machine, which was shrouded in war-time secrecy. But she knew enough about Zuse's machine to recognize that the material filed in a certain drawer related to a device that seemed somewhat like Zuse's. She reported this event to her father, giving the file number of the drawer, and the father at once informed Zuse of her discovery.

Zuse, of course, could not go to the Secret Service and ask for the document since that would give away the illegal source of his information. Zuse was well connected, however, and was able to send two of his assistants to the Secret Service, armed with an official demand for information from the Air Ministry, requesting any information that might be in the files concerning a device or machine in any way similar to Zuse's.

Zuse's assistants were at first informed that no such material existed in the files, but they persisted and eventually got to the right drawer. There they found a newspaper clipping (most likely from a Swiss newspaper), containing a picture of Mark I and a brief description about Aiken and the new machine. But there was not enough technical information to enable Zuse to learn the machine's architecture.

The importance of Mark I is primarily its role in making known to the world at large that a machine could successfully perform a programmed sequence of operations and do so automatically without error. As mentioned above, the Mark I was thus the herald of the computer age. We have adequate testimony to this role in history. In a widely used and standard reference work, the *Encyclopedia of Computer Science and Engineering*, edited by Anthony Ralston & Edwin D. Reilly, Jr. (1983), Maurice Wilkes declares that "the digital computer age began when the Automatic Sequence Control Calculator started working in April 1944." In the same encyclopedia, another article (by E. L. Stoll) begins: "the Harvard Mark I, also called the IBM Automatic Sequence Control Calculator ... marked the beginning of the era of the modern computer." In a volume of *Perspectives on the Computer Revolution*, Aiken and his "Automatic Sequence

Controlled Calculator, or Harvard Mark I," are given credit by the editor for the "real dawn of the computer age," which occurred with "the construction of a machine" which "could control the entire sequence of its calculations, reading in data and instructions at one point and printing results at another."

Aiken's commanding place in the unfolding world of the computer was recognized in 1964, when AFIPS (the American Federation of Information Processing Societies) established the Harry Goode Memorial Award to honor its second president, Harry H. Goode, by recognizing "outstanding achievement in the field of information processing." The inaugural award in 1964 went to the recognized pioneer, the inventor whose giant machine had inaugurated the computer age: Howard H. Aiken. I need not rehearse here the many honors and awards bestowed on Aiken during his life-time.

During the war years 1944–45, Mark I ran almost continuously, twenty-four hours a day and seven days a week. The war-time problems the machine was asked to solve included studies of magnetic fields associated with the protection of ships from the destructive action of magnetic mines, and mathematical aspects of the design and use of radar. No doubt the most important war-time problem was a set of calculations for implosions, brought from Los Alamos to the Harvard Navy installation by John Von Neumann. These were programmed for the machine by Dick Bloch. Only a year or more later did Bloch and the rest of the staff learned that these calculations had been made in connection with the design of the atomic bomb.

Mark I was gigantic, an imposing sight, standing 8 feet high and extending in length to 51 feet and almost 3 feet in depth. The portion of it on permanent exhibit in the main lobby of the Science Center gives only a partial notion of its original grandeur. It weighed 5 tons and used 530 miles of wire and was composed of 760,000 individual parts. Making use of IBM technology, as developed by IBM in its statistical and accounting (business) machines, it used traditional IBM types of parts such as electro-magnetic relays, counters, cam contacts, card punches, and electric typewriters (for the output); it did not make use of vacuum tubes and other elements of electronic circuits which were then foreign to IBM practice. But it is important to note that the new machine also incorporated functional elements of a new design, including – among other – new forms of relays and counters never before used in IBM machines. These were smaller in size and faster in operation than the ones then in use. The relays, the invention of Clair D. Lake, also had the advantage of being pluggable, so that they could easily be replaced as needed. The advantage of smaller elements was that the over-all size of the machine – gigantic as it was – was yet small enough to be functional or practical. With larger traditional (off the shelf) elements, a machine with the same computing power would have been impossibly large, perhaps too large to function effectively. Similarly, the availability of the new high-speed elements reduced the times required for such operations as multiplication or division. These were long enough, but they would have been impossibly long without the new high-speed elements.

The operation of the separate parts was powered by a long horizontal rotating shaft. This shaft rotated continuously, making a hum that has been described as like that of a gigantic sewing machine. There were 2200 counter wheels and 3300 relay components.

In later language, Mark I would be described as a parallel synchronous calculator. It had a word length of 23 decimal digits, with a 24th place reserved for algebraic sign. Calculations were done in decimal numbers with a fixed decimal point. There were 60 registers for the input of numerical data (the constants that appear in any algebraic or differential equation), each one containing 24 dial switches corresponding to 24 digits. For any problem, these had to be set by hand.

The location of each of these sixty registers was assigned a number, so that the instructions could use the location to identify a number being called up in the course of a calculation. The operative portion of the machine was composed of seventy-two registers, each of which was an "accumulator." Each such register was made up of twenty-four electromagnetic counter wheels – again providing the capacity for twenty-three digit numbers, with one place reserved for sign. This second set of panels comprised both the store or storage and the processing unit.

A typical line of coding in the program would instruct the machine to take the number in a given register (either a constant or a number in the store) and enter it in some designated register in the store. If there already was a number in that register, the new number would be added to it. The programmer had a code book, stating the designation of each location and each operation.

There were separate devices for multiplication and division and four tape readers. One was used to feed the instructions into the machine. The other three held tables of functions and could supply values as needed. There were internally stored programs (called "sub-routines") for interpolation and for sines, exponentials, logarithms, and for raising a number to some power. Programs were fed into the machine by punched paper tape. The programmer first reduced the problem to a sequence of mathematical steps and then used the "code book" to translate each step into the necessary coding or instructions for the machine. Mark I's instructions were essentially single-address instructions. Those who wrote programs for Mark I later recalled that the process was very much like programming later computers in machine language.

The chief programmer of Mark I, Richard M. Bloch, kept a notebook in which he wrote out pieces of code that had been checked out and were known to be correct. One of Bloch's routines computed sines for positive angles less that 45 degrees to only ten digits. Rather than use the slow sine unit built into the machine, Grace Hopper simply copied Dick's routine into her own program whenever she knew it would suit her requirements. This practice ultimately allowed the programmers to dispense with the sine, logarithm, and exponential units altogether. Both Bloch and Bob Campbell had notebooks full of such pieces of code. Years later, the programmers realized that they were pioneering the art of subroutines and actually developing the possibility of building compilers. A short time later, this approach was formalized in the book by Wheeler, Wilkes, and Gill.

Although Mark I produced results faster than conventional methods of computing, it was very slow compared to the machines to be constructed or unveiled soon afterwards, such as ENIAC or the Colossus machines. Addition or subtraction required one machine cycle or took 0.3 seconds. Multiplication required 20 cycles and took 6 seconds. But division could require as much as 51 cycles and take as long as 15.3 seconds. Accordingly, division was later performed on the Mark I by the method of reciprocals.

Although Mark I was slow, as compared to the ENIAC, completed and put into operation a few years later, Mark I was versatile. Not only was Mark I programmed, rather than being hard-wired for each problem, but it was an extremely versatile machine. Whereas ENIAC was restricted in its original design by the mission of computing ballistic tables, Mark I could be programmed to solve a large variety of different types of programs as well as being very efficient in producing tables of functions, such as Hankel or Bessel functions.

Mark I continued to function at Harvard for 14 years after the war, continuing to produce useful work until it was finally retired in 1959. During that time, Mark I also served generations of students at Harvard, where Aiken had established a pioneering program in what was later to be called computer science – with courses for undergraduates and graduate students leading to a master's degree or a Ph.D. Many important figures in the developing world of computers were introduced to the subject on the Harvard Mark I.

5. After Mark I

In the last months of World War I, Aiken was asked by the Navy to design and construct a second machine. The product was Mark II, similar in many ways to Mark I, also a relay machine, but based on improved types of components. Aiken and his staff went on to build two more machines. Mark III went to Dahlgren to join Mark II, but Mark IV was built for the Air Force and remained at Harvard. Mark I and Mark II were relay machines, but Mark III used some vacuum tubes and solid-state devices. Mark IV was all-electronic, using selenium solid-state devices, later replaced by ones made of germanium. Both Mark III and Mark IV introduced some important novel features, primarily the use of magnetic drum storage and (in Mark IV) the use of solid core magnetic memories. One of the truly innovative elements in Mark III was an automatic coding machine to simplify the work of the programmer and avoid human error in writing programs. Although Aiken's designs included some conditional branching, none of these machines used the stored program. Aiken's philosophy in these matters was narrow and strict: he insisted on maintaining the separate identity of data and instructions.

Technologically, Mark I can be considered an important breakthrough because it embodied a convincing demonstration of the possibility of large-scale error-free complex calculations in a programmed sequence. The public announcements concerning the new machine made known to the scientific and engineering worlds at large that the new computer age was dawning. At this time, 1944, ENIAC had not as yet been completed and tried, while war-time secrecy and ignorance veiled the Colossus machines and those designed in Germany by Konrad Zuse. But the degree of "state of the art" represented by Mark I did not extend to the later machines, even though Mark III and Mark IV had very innovative features. Although Mark II did useful work, it represented a relay technology being made obsolete by the advent of ENIAC, the all-electronic machine constructed at the Moore School of the University of Pennsylvania. By and large, Aiken's four machines did not greatly influence the on-rushing main stream of developing computer technology.

In retrospect, therefore, Aiken's most important and lasting contribution to the computer age may have been his pioneering applications of computers to data processing, notably, computer billing, and his development of an educational program in the area we know as computer science. Inaugurated in 1947, Aiken's program at Harvard was – so far as I know – the first full-scale program in this area, including courses for undergraduates and graduates and degrees at both the master and doctor level.[2] In retrospect, some old computer hands consider that his greatest contribution may have been the pupils he trained, who then went on to advance the art and science of the computer and to direct the new departments of computer science in different universities.

In 1961, Aiken took advantage of Harvard's policy of allowing faculty members to retire early – that is, to retire at age 60 with full benefits, without having to wait until he was 66. By then, in certain respects, Aiken had become a conservative figure in the world of computing. In the 1950s, at the age of fifty-plus, he was already "old" by the standards of this rapidly advancing science, art, and technology. Computer science and invention had become a young man's game. Even in the years just after the war, many of the major advances had come from young men, trained in the new electronics of radar rather than in classical electrical engineering, as was the case with Aiken. In Maurice Wilkes's words, the new computer innovators were young men with "green fingers for electronic circuits," many of whom in the early days had come from experience with radar and "were used to wide band widths and short pulses."

After retirement, Aiken moved to Fort Lauderdale, Florida, where he was given an appointment at the University of Miami. This did not require any

---

[2] Although Aiken's Harvard program seems to have been the first to offer full-scale graduate instruction leading to both a Master's and Doctor's degree, it should be noted that a year before the Harvard program was put into operation, Columbia was already offering instruction in this new area – courses being offered by Wallace Eckert, Herbert Grosch, and others.

teaching responsibilities but gave him an office. He then became a business entrepreneur, taking over ailing businesses and nursing them back to financial good health, whereupon they were sold. He also kept up his computer activity, serving as a consultant to Lockheed Missiles and Monsanto (who were exploring the potentialities of magnetic bubbles for computer technology). His final contribution in the computer domain was a means of encryption of data to provide security of information. He died in 1983 in St. Louis, while on a consulting trip to Monsanto.

I. BERNARD COHEN is the Victor S. Thomas Professor (emer.) of the History of Science at Harvard University. His specialties are the history of mathematics and the exact sciences and the development of the computer. For many years, he was IBM's principal historical consultant. He is the founding editor, and current co-editor with William Aspray, of the series of books on the history of computing published by the MIT Press. He is currently seeing two works through the press, his "portrait" of Howard Aiken (MIT Press) and a new translation of Newton's *Principia*, produced in collaboration with Anne Whitman, together with a *Guide to the Principia* (University of California Press). Dr. Cohen was closely associated with Howard Aiken during Aiken's decades at Harvard and (together with Henry Tropp) conducted an extensive oral-history interview with Aiken shortly before the latter's death.

The ENIAC: History, Operation and Reconstruction in VLSI

Jan Van der Spiegel, James F. Tau, Titiimaea F. Ala'ilima,
and Lin Ping Ang

Abstract. This contribution gives a brief historical overview of the ENIAC and continues with a description of its architecture. The 40 units of the ENIAC are grouped in five broad categories: arithmetic, control, memory, I/O and interconnections (busses). The overall operation of the ENIAC and of the individual modules is described next in order to give the reader an appreciation of the capabilities and limitations of the machine, including conditional branching. The last part of the paper deals with the reconstruction of the ENIAC in silicon using CMOS technology. A description of the key building blocks of the ENIAC-On-A-Chip is given. The reconstruction resulted in a 7.4 × 5.3 square mm silicon chip that contains over 174 thousand transistors. The paper concludes with a discussion of the relative computational power of the ENIAC.

1. Introduction: Rediscovering the ENIAC

The ENIAC (Electronic Numerical Integrator and Computer) was unveiled to the public on February 14, 1946, at the Moore School of Electrical Engineering at the University of Pennsylvania. Half a century later, a team of students and faculty started the reconstruction of the ENIAC as part of its 50th anniversary celebration. The goal of the project was to re-create the ENIAC using state-of-the-art solid-state CMOS technology. The project was a journey back into the history of computing. It illustrated, in a rather dramatic way the evolution of computers in terms of architecture, technology, size, power and performance. The journey was at times tedious but it was also exciting and rewarding. The end result is a 7.4 × 5.3 square mm sliver of silicon that houses the components of the 18,000-vacuum-tubes, 30-ton ENIAC.

In order to give full tribute to the ENIAC, the design team decided to re-implement the machine using a full-custom design approach. Rather than using standard cells and pre-designed logic and functional units to design the ENIAC-On-A-Chip, the team wanted to recreate the experience of building the ENIAC from its basic and primitive building blocks. For the ENIAC, these were vacuum tubes, whereas for the chip, they are transistors. This approach has ensured the most faithful reproduction and enhanced the educational experience for each member involved.

The full-custom design strategy adopted for the project required a detailed understanding of the original ENIAC, well beyond the behavioral and functional level. Substantial effort was put into reading the reports and blueprints, which contained circuit schematics of each unit of the ENIAC. The archives of the University of Pennsylvania and the Smithsonian museum in Washington DC, were consulted for original manuscripts and blueprints. Vacuum tube circuits were analyzed and several of them were reconstructed in order to gain a better understanding of their operation.

2. Designing the ENIAC-On-A-Chip

After gaining sufficient insight into the operation of the ENIAC, the team proceeded with the actual design process. We generally followed a top-down, hierarchical design methodology, using Cadence VLSI tools. This resulted in a relatively small number of different basic cells that were used as building blocks in the larger units, an approach already used in the original machine.

We decided to preserve as much as was possible of the architecture, the functional and logic blocks of the original ENIAC, in the current technology. This implies that each of the original vacuum tube circuits have direct counterparts in the ENIAC-On-A-Chip implementation. The realization of these circuits is somewhat different from the original ones, mainly as a result of the differences between the vacuum tube and the MOS transistor and the lack of availability of on-chip capacitors and resistors.

The design process started, in many cases, with the creation of a behavioral/functional description of the major units of the ENIAC in order to verify their operation. Verilog-XL[†] was used as the hardware description language. This was followed by the design of the logic and the transistor circuits of the individual cells. These cells were simulated with the Hspice[‡] circuit simulator. The results of the simulation were used to annotate the behavioral and logic models in order to approximate the actual operation of the larger units of the ENIAC more accurately. Next, the layout of each cell was handcrafted. Although this design method was time-consuming, it was an inherent part of the recreation process and also ensured optimal utilization of chip area. Each of these cells was checked for electrical and design rule errors and re-simulated with Hspice. The cells were then connected together into larger functional units, which were simulated once again to verify their operation.

In the end, all the different blocks were put together onto the final chip. Extensive verification was carried out before the project was submitted for

---

[†] Verilog-XL is a trademark of Cadence Design Systems.
[‡] Hspice is a trademark of Avant! Corp.

fabrication. The ENIAC-On-A-Chip was fabricated in a 0.5 μm single poly-silicon, triple metal, nwell CMOS process. It measures 7.44 mm by 5.29 mm and contains 174,569 transistors. The difference in the number of transistors and vacuum tubes is mainly due to the fact that transistors are not only used to replace the 17,468 vacuum tubes, many of which are dual tubes, but also to implement the 70,000 resistors, 6000 switches, 7200 diodes and 10,000 capacitors. Fig. 1 shows a photograph of the ENIAC chip mounted in a 132-pin grid array package.

3. A brief historical overview

The ENIAC was designed and built between July 1943 and November 1945 at the Moore School of Electrical Engineering at the University of Pennsylvania. The project was carried out for the U.S. Ordnance Department of the War Department under contract No. W-670-ORD-4926 and cost approximately $486,000.[1] Mr. J. Presper Eckert was the chief engineer, Dr. John W. Mauchly the consulting engineer, Dr. John G. Brainerd the administrative supervisor and Dr. Herman H. Goldstine the representative of the Ballistic Research Laboratory. The project's primary objective was to build a machine

Figure 1: Photo of the ENIAC-On-A-Chip mounted in a 132 PGA. The chip measures 7.4 mm × 5.3 mm and contains 174, 569 transistors (courtesy Univ. of Pennsylvania).

---

[1] H. H. Goldstine, *The Computer from Pascal to von Neumann*, Princeton University Press (Princeton, 1972).

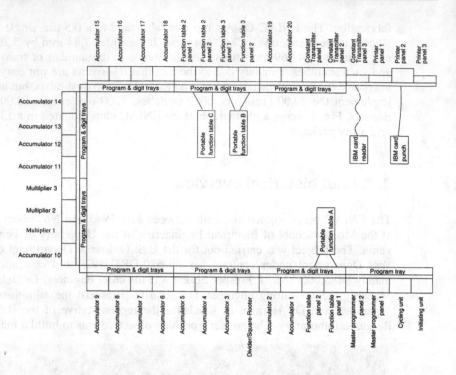

Figure 2: Floor plan of the ENIAC. The 40 panels, each 0.6 m wide, 2.7 m high and 0.7 m deep, are arranged in U shape occupying an area of about 10 m by 17 m.

that would *speed up* the calculations for the Ballistic Research Laboratory.

However, the inventors wanted to make the ENIAC as *flexible* as possible, so that it could serve as a general purpose machine. As its name implies, the ENIAC performs not only numerical integration, but is capable of solving a wide range of problems that involve various numerical operations, as well as storing and retrieving intermediate results. In addition, the ENIAC was designed "to perform these operations consecutively or concurrently, with automatic transfer of data from one step to the next."[2]

---

[2] "The ENIAC – Vol. I, A Report Covering Work until December 1943," University of Pennsylvania, Moore School of Electrical Engineering (Philadelphia, 1943).

The ENIAC's architecture was, to a large extent, shaped by the earlier calculating machines, the technology available, advances made in numerical
analysis methods, and the circumstances under which the ENIAC was developed. The inventors, Eckert and Mauchly, were familiar with desktop calculators (such as the Friden, Marchant, and Monroe type machines), punched
card and punched tape machines (from IBM and BTL – Bell Telephone
Labs), and the differential analyzer. Although the differential analyzer was
particularly well suited to solving ballistic equations, the goal of the inventors, i.e. to develop a more general and accurate device, meant that the differential analyzer was not a suitable candidate on which they could model their
machine. In order to achieve high speed, accuracy and flexibility, it is more
likely that the ENIAC was conceived in the tradition of the mechanical adding, multiplying and dividing machines of that time.[3] In addition, a considerable amount of work on electronic ring counters and scalers for experimental
physics had been done by tube manufacturers and several research institutions. These developments were known to the ENIAC engineers. It is also

Figure 3: View of the U-shaped ENIAC at the Moore School of Electrical Engineering in 1946, showing J. P. Eckert (left) and J. Mauchly (right) in the foreground
(courtesy Univ. of Pennsylvania).

---

[3] A. W. Burks, "From ENIAC to the Stored-Program Computer: Two Revolutions in
Computers," in *A History of Computing in the Twentieth Century*, Academic Press
(1980). M. Marcus and A. Akera, "Exploring the Architecture of an Early Machine:
The Historical Relevance of the ENIAC Machine Architecture," *IEEE Annals of
the History of Computing*, 18 (1996): 17–24.

said that Mauchly's thinking was stimulated by Atanasoff's work on digital computation. In 1941, Mauchly visited Atanasoff, who had built a small prototype of a special-purpose digital machine (for solving a set of linear equations through Gaussian elimination) that made use of vacuum tubes.[4] To what extent the ENIAC's architecture was influenced by Atanasoff's work has been the topic of considerable controversy.

Although present day computers dwarf the ENIAC in computational power, it was indisputably the fastest and largest machine of its time. It consisted of 40 panels, 3 portable function tables, a card reader and card punch. Each panel was about 0.6 m wide and 2.7 m high, organized in a U-shape occupying a 10×17 m room, shown schematically in Fig. 2. A photograph of the ENIAC, as it was set up in the Moore School of Electrical Engineering, is shown in Fig. 3.

Building such a machine required several innovations in construction and design methods. The machine consisted of a relatively small number of basic electronic elements organized as interchangeable modules, which could be easily plugged into the backside of the panels, similar to plugging a daughter card into a slot on a motherboard in today's computers. Reliability was always a major concern for the engineers. They took several measures to reduce the risk of breakdown or faulty operation, such as designing circuits that were insensitive to component variations, running-in the vacuum tubes and using carefully selected tubes well below their ratings.[5] The end results surprised even the most adamant of skeptics: the completed ENIAC failed only two or three times per week. Special test procedures were in place to identify the failed unit within a matter of minutes, which resulted in a down time of only a few hours per week.[6] This was an extraordinary accomplishment, considering that the machine was one of the most complex ever built under the constraint of operations with such a high degree of reliability.[7]

When the ENIAC was unveiled in February 1946, less than three years after its inception, it stunned the scientific, military and industrial community. The ENIAC captured the imagination of the public, not only because of its sheer size, but, more importantly, because of its lightning *speed*. Addition (or subtraction) of two 10-digit numbers was accomplished at an unprecedented rate of 5000 per second. This was about 1000 times faster than any other computing machine was capable of up to that point, with similar accuracy.

The ENIAC was a much more *flexible* and powerful machine than the individual mechanical adding machines on which it was originally modeled. The ENIAC could not only perform a programmed sequence of additions,

---

[4] Goldstine, n. 1 above.

[5] N. Stern, *From ENIAC to UNIVAC – An Appraisal of the Eckert-Mauchly Computers*, Digital Press (Boston, 1981).

[6] A. W. Burks, "Electronic Computing Circuits of the ENIAC," Proc. I.R.E., (August 1947): 756–767.

[7] Goldstine, n.1 above.

subtractions, multiplication, divisions and square-roots, but also had the capability to store intermediate results, and to communicate them among various units. Furthermore, it was possible to execute nested loops and conditional branching, as well as reading in and printing out numbers. The end product was a general purpose, highly parallel, digital electronic computer that allowed the calculations of solutions to a large class of numerical problems.

Despite similarities to modern computers, the ENIAC differed from them in one fundamental aspect: it was not a stored-program computer. As such, programming was done locally on the individual units by setting program switches and connecting the units to each other via digit and program trunks. A program pulse then stimulated the action of those units receiving it, and they emitted subsequent program pulses to activate other units. In this way, a sequence of operations could be carried out. Set-up was done manually and was highly time-consuming. The inventors were aware of this downside from the outset of the project, but it was thought to be acceptable because the ENIAC was intended to perform highly repetitive computations that used the same set-up. Ultimately, it was the time constraint facing the inventors that determined the architecture of the ENIAC, not allowing them to carry out research on more programmer-friendly architectures.[8]

4. Architectural and operational overview of the ENIAC

The goal of this section is to give the reader an understanding of the overall operation of the ENIAC in order to gain a better appreciation of the scope of its silicon reconstruction. A description of each unit is given in section 5, or can be found in the references below.[9]

---

[8] J. P. Eckert, J. W. Mauchly, H. H. Goldstine, J. G. Brainerd, "Description of the ENIAC and Comments on Electronic Digital Computing Machines," Moore School of Electrical Engineering, University of Pennsylvania (Philadelphia, Nov. 30, 1945).

[9] H. D. Huskey, "A Report on the ENIAC, Part II, Technical Description of the ENIAC," Moore School of Electrical Engineering, University of Pennsylvania (Philadelphia, 1946). J. F. Tau, "ENIAC-On-A-Chip: The Monolithic ENIAC," Masters Thesis, Department of Electrical Engineering, Moore School of Electrical Engineering, University of Pennsylvania (Philadelphia, 1996). T. F. Ala'ilima, "Recreation of the ENIAC using CMOS Technology," Masters Thesis, Department of Electrical Engineering, Moore School of Electrical Engineering, University of Pennsylvania (Philadelphia, 1996).

4.1 Architectural overview

The units of the ENIAC can be loosely grouped into five categories: arithmetic (general purpose and dedicated units), global control units, memory, I/O units and busses (trunks). Fig. 4 shows a functional organization diagram of the ENIAC. Of the 40 panels, 20 are accumulators, considered the main computational components around which the ENIAC is built. Other arithmetic units include a high-speed multiplier, and a combination divider/square-rooter. As multiplication is the second most frequently used operation after addition/subtraction, dedicated hardware (multiplication tables) is used to speed up the process. The master programmer is used for coordinating the operation of the accumulators and the execution of a sequence of operations and nested loops. Fast programmable, read-only memory is provided by 3 function tables. The constant transmitter in conjunction with a card reader constitutes the external input device. Finally, global control units include the Initiating and Cycling units that govern the overall operations of the ENIAC and take care of initiating computations, by providing digit and program, as well as reset pulses.

Various units of the ENIAC communicate with each other over the data, program, and synchronization busses (also called trunks). Digit trunks are carried in trays that are stacked on top of each other, allowing for multiple connections. Digit trays can also be used over again in the course of a program. Only one accumulator can transmit data on a digit trunk at any one time, but multiple accumulators can listen in. In addition to the regular transmission of digits over digit cables/trunks, adapters can be used to change the digit place between the transmitting and receiving accumulator. As an example, a shifter adapter is used to multiply a number by a power of 10, while a delete adapter is used to eliminate the pulses of one or more places of

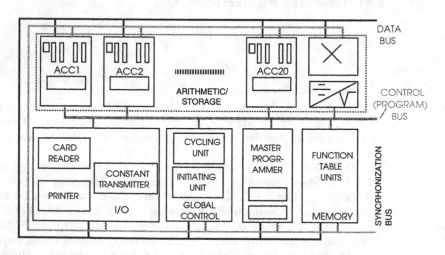

the transmitting number.

Program pulses are transmitted over program trunks, carried in programming trays. A third bus is the synchronizing bus (trunk), which carries the fundamental pulse train from the cycling unit to all other units and ensures that all units operate properly and in synchrony with each other. A description of the fundamental pulse train is given in the section on the Cycling Unit. The availability of multiple digit and programming trunks as well as the synchronizing pulses allow the execution of parallel operations. However, as was pointed out by Marcus and Akera, the ENIAC lacked an explicit mechanism to resynchronize parallel branches of a program, making programming for parallel operations tricky.[10]

The ENIAC is an accumulator-based computer. As such, the main arithmetic and data storage units are accumulators. A simplified diagram of an accumulator is shown in Fig. 5. It consists of arithmetic, local control and I/O circuits. The arithmetic unit receives a signed 10-digit number and adds this number to the one already stored. Whenever a decade counter overflows, a carry-over digit is generated and given off to the decade of the next signifi-

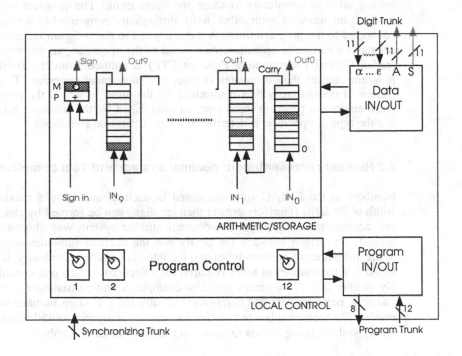

Figure 5. Simplified functional block diagram of an accumulator

---

[10] Cf. Marcus and Akera, n. 3 above.

cant digit (on its left), as is schematically shown in Fig. 5. A binary counter (Plus/Minus) on the far left of the most significant digit is used for the sign information.

The control unit of each accumulator determines which operation the accumulator performs (receive or transmit, additively or subtractively). From the user's point of view, the controls are simply settable switches (called Program Control Switches). There are 12 such program controls per accumulator, allowing each accumulator to perform up to twelve separate operations during the course of a program. Eight of these are capable of repeating their operation up to 9 times. Fig. 6 shows a photograph and a corresponding schematic representation of an accumulator's front panel.

Each accumulator has two Input/Output blocks. One is the *data*-I/O which transmits or receives a decimal number over the digit trunk (an 11-lead bus, 10 leads for digits and 1 lead for the sign). The accumulator has five input ports, labeled α through ε. The data outputs have two terminals, one called the A-port for transmitting the number as stored in the accumulator, and another called the S-port for transmitting the 10's complement of the stored number. The output port is tri-stated when the accumulator is inactive, allowing other accumulators to share the same trunk. The program control block communicates with other units through its *program*-I/O terminals, connected to the program trunk. A pulse applied to the program input terminal starts a particular operation. At the end of the operation, an output pulse (called Central Programming Pulse or CPP) is emitted from the finishing program control that stimulates (triggers) a subsequent program. The sequence of operations is thus determined by the order in which the program pulse enters the program input port, as established by the interconnections, and the type of operation is determined by the Operation switches.

4.2 Number representation: decimal system and 10's complement

Numbers in the ENIAC are represented in decimal and have a maximum width of 20 digits (numbers greater than ten digits can be formed by chaining two accumulators together). The decimal number system was chosen after careful comparison between the binary and the decimal implementation in terms of number of vacuum tubes and the interconnection complexity. It was found that the number of tubes required for a decimal system was considerably smaller than for a binary one. For example, a unit consisting of decade counters, pulse shapers and carry-over circuitry for a 10-digit number would require 280 vacuum tubes in a decimal system as compared to 450 tubes in a binary system (using 30 bits to represent the same range of numbers).[11]

---

[11] "The ENIAC," n. 2 above.

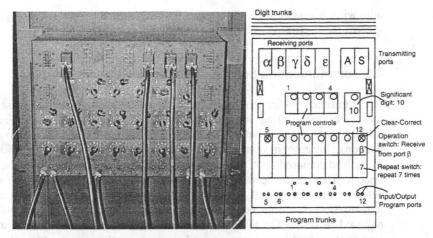

Figure 6: (a) Photograph of the front panel of an accumulator showing the Program Control and Repeat switches; (b) schematic representation.

The *complement* number system is used to represent negative numbers. Both the 9's and 10's complements were considered, but the designers found that the 10's complement would cause fewer problems regarding rounding off and deleting insignificant figures. Also the 10's complement system simplified the structure of the multiplier.[12] Whether a number is positive or negative is indicated by the state of the PM (Plus/Minus) unit. The PM unit is simply a binary ring-counter. An alternative method to indicate the sign would have been to use an additional decade to the left of the others, which would give a zero for a positive number and a nine for a negative number. This is the method used in modern digital systems working with binary numbers. However, using a full decade would be wasteful as only two states are possible (P and M). The 10's complement can be easily obtained by first subtracting each digit from 9 and then adding a 1 to the result, as illustrated for the complement of the number N=124 (where P means positive and M negative),

$$-N = 10^{10} - {}^{P}N = [(10^{10} - 1) - {}^{P}N] + 1$$

$$-124 = {}^{M}9\ 999\ 999\ 876.$$

The ENIAC makes it possible to use fewer than 10 digits by setting a Significant Figure switch, located on the accumulator front panel. Every time the accumulator is cleared to zero, the place below the last significant digit is set to 5. For example, for seven significant digits, the accumulator clears to P 0 000 000 500. When a number is then added to the accumulator of which the 8th digit is greater than or equal to 5, the 7th digit will be increased by one;

---

[12] Ibid.

otherwise it remains the same. The ENIAC uses the first seven digits and
ignores the remaining ones during subsequent operations.

4.3 Communication between units: pulse transmission

One additional choice that had to be made early on in the design phase is the
method of transmitting numbers: statically (i.e. using steady-state signals) or
serially in the form of pulses. The latter was chosen for general purpose connections, because it was believed that the pulse system was considerably
faster and required less vacuum tubes and interconnections than the static
system. The choice between the two systems was also related to the choice
between the binary and decimal number system. In the pulse system, the
transmission of a digit needed only one wire by sending as many pulses as
required in series. On the other hand, in the static decimal system, at least
four wires would have been required to represent the ten possible values of a
digit. However, the static outputs (outputs of each flip-flop of a decade
counter) were used for dedicated connections between specific accumulators
and special units, namely the multiplier, the divider/square-rooter, the printer,
and the function tables.

To illustrate how the transmission of numbers is done, let us consider a
simple example. We will transmit a number N consisting of a single digit
(e.g. "4") stored in a decade circuit of one accumulator to a decade counter in
another accumulator.[13] Fig. 7 gives a simplified block diagram of a decade
circuit in an accumulator consisting of a 10-stage counter and control circuitry. Each stage of the counter corresponds to one of the digits, from 0 to 9.
In our example the digit "4" is stored in the decade counter and is represented
by a "1" in the fifth flip-flop and "0" in all others (see Fig. 7).

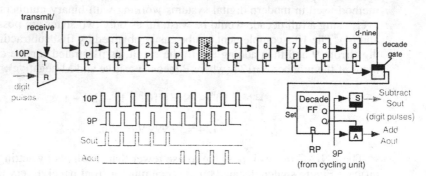

Figure 7: Simplified schematic of a decade counter illustrating the transmission of a
digit

---

[13] Burks, n. 6 above.

The data ("1" or "0") in a flip-flop will be shifted from one stage to the next every time a pulse is applied at its P input. The output of the flip-flop changes at the falling edge of the input pulse. During the *transmission* of a digit, 10 consecutive pulses, called 10P, are applied to completely shift the decade counter around, ending where it started. The 10P pulses, shown in Fig. 7, are supplied by the cycling unit. When the "1" in the counter reaches the last stage (i.e., 9) the *decade gate* will open at the falling edge of the P pulse. The next input pulse will thus pass through this gate and set the decade flip-flop (i.e., when counting from 9 to 0). In our example, this will happen after 5 pulses. The decade flip-flop controls two gates, i.e., a "Subtract" gate and an "Add" gate. The Subtract gate is normally open, and will be transmitting a number of 9P pulses on the S-port. However, as soon as the decade flip-flop is set, the Add gate opens and the remaining 9P pulses will be transmitted on the A-port. In the example discussed here, 5 pulses are transmitted through the subtract output and 4 through the add output. In this way, the digit N stored in the counter or the complement (i.e., 9 − N) can be transmitted. Which of the two is transmitted will be determined by the setting of the Program Control switch (A, S or AS). On the *receiving* accumulator, the program control is set to receive, leading the counter to ignore the 10P pulses. The counter will now be clocked by the digit pulses (i.e. gated 9P) coming from the transmitting accumulator.

It should be noted that in order to obtain the 10's complement a "1" needs to be added to the 9's complement. This is done by an additional pulse, called 1'P, supplied by the cycling unit on the wire that carries the least significant digit. The decade and PM counters of an accumulator are connected in such a fashion that proper carry-over occurs from one decade to the next as well as to the PM counter. The carry-over circuit has been omitted in Fig. 7 for clarity. A more detailed circuit will be given in section 6.2 of the chip implementation. The transmission of the sign information is somewhat different from that for a digit. For a positive sign, no pulses are transmitted to the PM counter and for a negative sign nine pulses are sent. The reception of an odd number of pulses cycles the PM counter to the opposite sign.

It is interesting to note that pulses are not transmitted directly by the decade counter. Instead a set of pulses is provided by the cycling unit to the accumulator over the synchronization trunk at a rate of 100 kHz. The accumulator then emits (through gating) a group of pulses on each of its digit lines, equal in number to the value it represents. This method of transmission not only prevents the pulses from being constantly degenerated, but also makes the operation of the ENIAC synchronous. Both the 10 digits and the sign information are transmitted simultaneously over the 11 lines of the digit trunk.

4.4 Instructions and number of cycles

The ENIAC can execute a variety of instructions. Some of these operations, such as an addition (transmit-receive), are built into the hardware, while others require more elaborate steps, involving different units. What and when an instruction is to be executed depends on the program control switch setting and the connections between the program input and output terminals, as was mentioned earlier.

The basic operation in the ENIAC is an addition/subtraction (or transmit-receive operation). The time period needed for this operation is called "Addition Time" and requires a total of 20 pulse times (200 µs). We will express the time that other operations require in units of the "Addition Times." Table 1 gives a list of the various operations that can be programmed in the ENIAC together with the number of required "Addition Time" cycles.

Table 1: Operations of the ENIAC and times required in terms of the Addition Time (0.2 ms).

Type	Operation	Description	Cycles (0.2 ms)
Arithmetic	Add		1
	Subtract		1
	Multiply	10-digit by p-digit	$p+4$
	Divide	Quotient of p digits	$13(p+1)^*$
	Square Root	Result of p digits	$13(p+1)^*$
Memory	Write to register (normal or complement)	Store in accumulator	1
	Read from register	Load from accumulator	1
	Read from table (up to r times)	Normal or complement number	$r+4$
Control	For loop	Nested loops possibly involving Master Programmer	Depending on operation
	If...then...else	Based on digit discrimination, involving several units	
I/O	Read from external memory	Punch card reader and constant transmitter	62 ms per 10-digit number (or 120 80-digits cards/min)
	Print result	Printer	75 ms per 10-digit

* On average. The exact time depends on the number of places required in the answers, as well as on the digits in each place of the answer.

It should be mentioned that some of the operations listed in the table were not called by the same name originally. Often there is no dedicated hardware available to execute some of these operations. As an example, the execution of the "*if ... then ... else*" statement requires an elaborate set of operations and connections among several units.

4.5 Programming

Programming the ENIAC is very different from what we consider programming on a modern-day, stored-program computer. The data-flow architecture of the ENIAC requires setting switches on and making connections between units. Programming consists of the following steps. First, the problem to be solved needs to be described by a set of mathematical equations, such as total or partial differential equations. Then, the equations are broken down into basic mathematical operations that the ENIAC is capable of executing. Also, one needs to plan for the storage of the numerical data. For each arithmetical operation one needs to set up a program control and make connections between the program control I/Os. Finally, the individual programs are tied together into a program sequence, so that a collection of programs is automatically stimulated upon completion of another set of programs.[14]

To illustrate the operation and setting up of the ENIAC, we discuss a simple program that involves subtracting and adding two numbers, using three accumulators. To start with, accumulator 4 stores some number a, accumulator 5 stores the number b, and accumulator 6 stores the number c. We will set up a program that calculates $(a-b)$ and stores the result in accumulator 4 and also calculates $(c+2b+359)$ and stores it in accumulator 6. In addition, the contents of accumulator 5 need to be increased by 359. Let us assume that the numbers a, b and c are already present in the accumulators (as results of a previous calculation).

Fig. 8 shows the three accumulators and their settings. Accumulator 4 needs to be instructed to receive the number b from accumulator 5. This is done on Program control 1 (non-repeat control) by setting the Operation switch to α to indicate that the number will be received on input terminal α. A digit cable connects the input terminal α to one of the digit trunks that we shall call trunk I. Accumulator 5 has to transmit its digits in two ways: once as the complement (for subtraction) and twice as they are stored. This is done by using a Repeat program control, e.g. control 5. The Operation switch is set to AS (indicating that both the positive and negative numbers will be transmitted), while the Repeat switch is set to 2. The "S" output port needs to be connected to digit trunk I to establish the desired connection between accumulators 4 and 5. Next, accumulator 6 has to receive the number b stored in

---

[14] A. K. Goldstine, "A Report on the ENIAC, Part I, Vol. I and II, Technical Description of the ENIAC," Moore School of Electrical Engineering, University of Pennsylvania (Philadelphia, 1946).

accumulator 5 twice. This is done by setting the Repeat switch of program control 5 to 2 on accumulator 6, and the Operation switch to α (or any of the other input terminals one wishes to use). Connect input terminal α of accumulator 6 to the output port "A" of accumulator 5 through digit trunk II, as shown in Fig. 8. In the final step of the operation, the number "359" (or any other number, e.g. received from the card reader) needs to be transmitted from the Constant Transmitter and added to the contents of accumulators 5 and 6. We shall use program control 6 and 1 on accumulators 5 and 6, respectively, as is shown in Fig. 8. Connect the output of the constant transmitter to the input terminals α and β of accumulators 5 and 6, respectively, over digit trunk III.

Finally, one must specify the start as well as the sequence of the operations. This is done by connecting the input and output program terminals in the proper way. The start pulse comes from the initiating unit's output terminal which will emit the "Initiating Pulse." Let us use program line A-1 of program trunk A to connect the initiating pulse to the input terminal of program control 1 of accumulator 4, and to the program input terminal 5 of both accumulator 5 and 6. When accumulator 5 has transmitted its number twice, it will generate a program output pulse on program control 5. This pulse needs to be connected to the program input terminals of control 1 of accumulator 6, and of control 6 of accumulator 5, as well as to the program input terminal of the constant transmitter. We have used program line A-2 for this purpose.

Thus, upon receiving the initiating pulse, the three accumulators start to work in parallel. After one addition cycle, accumulator 4 will store the number $(a–b)$ and stop, while accumulators 5 and 6 continue sending and receiv-

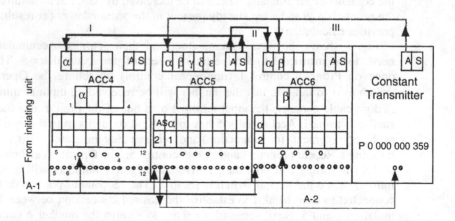

Figure 8: Example of the accumulator set-up. Accumulators 4, 5 and 6 initially store the numbers a, b and c, respectively. After executing the program they will store the numbers $(a–b)$, $(b+359)$ and $(a+2b+359)$.

ing. After the second cycle, accumulator 6 will store the number $(c+2b)$, and accumulator 5 will generate a program pulse that will initiate the transmission of the number between the constant transmitter and accumulators 5 and 6. At this point, the output terminal of program control 6 of accumulator 5 will generate a program pulse. This can be used to initiate another operation, or can be left unconnected so that the ENIAC will come to a halt until another initiation pulse is generated.

5. Units of the ENIAC

This section is intended to make the reader familiar with the main features and operation of the different units of the ENIAC. Details of the circuits will not be discussed. It should be mentioned that the descriptions are not intended to be complete, as space limitation does not allow us to cover every aspect of the original ENIAC. For a more complete description of the ENIAC, the reader should consult the original reports.[15]

5.1 Control Units: initiating and cycling units

The division between arithmetic and control units may be somewhat deceiving, as each accumulator itself consists of arithmetic circuits and control circuits. However, the main function of the two units described here, the initiating and cycling units, is to govern the operations of all other units.

The *initiating unit's* task is turning the power on and off, initiating a computation, initial clearing, selective clearing of a group of accumulators and generating control signals for the reader and printer. Pressing the "Initiating Pulse" button starts off a programmed sequence of operations as determined by the accumulators and master programmer controls. Also, the operator can clear the contents of all accumulators before starting a computation by pressing the "Initial Clear" button.

The *cycling unit* provides the fundamental signals (pulses) to all other units which act upon them for the transmission of numbers. This unit is connected to other ones over the synchronizing trunk. The cycling unit allows the operator to choose among two debugging modes by pressing the proper buttons: "Addition Mode" and "Pulse Mode." When the Addition Mode is pressed, the machine cycles through one full addition cycle, and upon pressing the Pulse Mode button, it progresses one pulse at a time.

---

[15] A.K. Goldstine, n. 14 above, H. D. Huskey, n. 9 above.

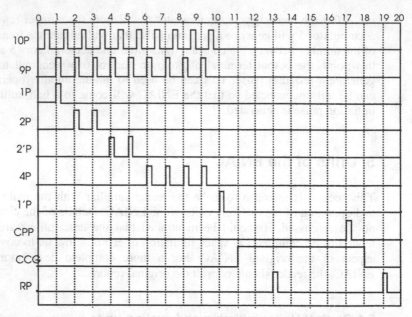

Figure 9: Fundamental pulse train (one Addition cycle) supplied by the cycling unit. 10P is for digit transmission while 9P is for both digit and sign transmission. 1P-4P are used for digit transmission by the Constant Transmitter and the Multiplication Tables. 1'P is used for complementary number correction. CPP synchronizes the program flow while CCG and RP take care of the carry-over and resetting the accumulator.

The Cycling Unit generates 10 types of pulse trains that govern the transmission and generation of numbers, program control (or data flow), number correction, and decade flip-flop reset. These pulse trains occur once every Addition Time, which is equal to 200 µs, based on the 100 kHz clock rate of the ENIAC. Each Addition Time is made up of 20 smaller time slices, in which the pulses are about 2 µs in duration. The first half of the pulse train is used for digit pulses, while the second half is used for program control. These pulse trains (Fig. 9) and their purposes are explained below:

- **10P**: Cycles the decade counter of the Accumulator during digit transmission (see also section 4.3, Fig. 7).
- **9P**: Pulses that are propagated on to receiving units (see also Fig. 7). Which of the 9P pulses are transmitted depends on whether the number is going through the A or S port, and are gated appropriately by the decade flip-flop. At the same time, the 9P pulse train cycles the Plus/Minus circuit of the Accumulator.

- **1P, 2P, 2'P, 4P**: Digit pulses gated by the Constant Transmitter to initialize Accumulators. The multiplication tables also use these pulses. By combining various pulse trains, any 0-9 number can be formed, as illustrated in Fig. 10.

- **1'P**: Used as a correction factor in subtractive transmissions to change from 9's complement to 10's complement.

- **CPP**: (Central Programming Pulse) Used to synchronize program flow. It demarcates the beginning of the next Addition Cycle and the end of the current Addition Cycle. An inactive program control in an Accumulator is induced to perform its programmed function when a CPP enters it, while an active program control ends its operation by clearing a flip-flop upon receiving the CPP.

- **CCG**: (Carry Clear Gate) Allows carry-over to occur by turning on the appropriate gates in the carry block of the decade counter. This pulse is called a gate because of its length, which is 70 μs, and function, i.e., to gate the Reset Pulse (RP). This pulse is also used for clearing the accumulators, provided the clear correct switch is set.

- **RP**: (Reset Pulse) Has two functions: first, whenever the decade counter counts beyond 10 to indicate that a carry-over should take place, the RP will be passed on to the next decade counter as the carry-out pulse; second, the RP will clear the decade flip-flop in the decade counter as soon as the carry-out pulse has been transmitted.

The Cycling Unit is implemented as a 20-stage ring-counter. By appropriately gating the pulses coming from the ring-counter, all the pulse trains described above can be created.

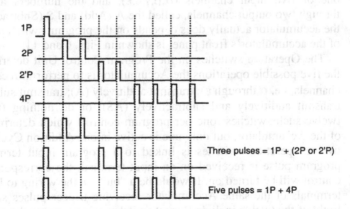

Figure 10: Combining the 1P, 2P, 2'P and 4P pulses to form any number from 0-9

5.2 Accumulator

The Accumulator of the ENIAC is analogous to an ALU plus a register in a modern microprocessor. As its name suggests the Accumulator "accumulates" by keeping its addend in memory, i.e., host Accumulator, and incrementing by the addend it receives from another transmitting unit, such as a Function Table or another Accumulator. As illustrated earlier, increments are carried out on a per digit basis, with each pulse adding one to the existing figure during the addition cycle. Subtraction is carried out like that in a modern microprocessor, by adding the complement of a number to the addend, but with an additional signal indicating the sign of the number transmitted, rather than using the instruction to determine how the number should be interpreted. This is a direct consequence of the data-flow architecture of the ENIAC.

As is shown in Fig. 5, the Accumulator can be divided into two functionally, as well as structurally, distinct sub-components: the arithmetic/storage unit and the program control unit. Physically, the arithmetic or storage unit is located directly above the control unit, and consists of ten of what look like long slabs of wood. Each slab's chief function is to represent a digit, which is indicated by the position of the lit neon bulb in one of the ten serially connected stages. Functionally, it is essentially a walking-one ring-counter, and since it counts to 10, it is called the decade counter by the inventors. In transistor technology this behavior is perfectly modeled by a closed loop ten-stage shift-register, in which all but one stage is initialized to a "0." The decade counter is meant to be easily removable from the whole component and replaced by a substitute. In this way, the inventors made the Accumulator modular so that when vacuum tubes, which are concentrated in the decade counters, failed in one decade counter, simply replacing it would quickly re-enable the Accumulator to work. Decade counters receive their inputs from one of five input channels ($\alpha,\beta,\gamma,\delta,\varepsilon$), and the numbers are transmitted through two output channels, called the A (Add) and S (Subtract) ports. What the accumulator actually does depends on the program switch setting. A view of the accumulator's front panel is shown in Figs. 6 and 11.

The Operation switches on the Program Control Unit determine which of the five possible operations the Accumulator is to perform: receive (on 1 of 5 channels, i.e. α through ε); transmit additively (A); transmit subtractively (S); transmit additively and subtractively (AS); or do nothing (O). There are twelve such switches (one per program control) which determine the action of the Accumulator, but only one is active in one Addition Cycle. Each of the twelve operation switches is linked to a program input terminal. When a program pulse is received on an input terminal, the corresponding Program control will be turned on. It would be a "bug" in the wiring to have two input terminals of the same Accumulator receiving trigger pulses simultaneously. Eight of the twelve include Repeat switches which enable the Accumulator to repeat a certain action a number of times (1–9) without receiving triggering pulses every time. Only one program output pulse is emitted at the end of r

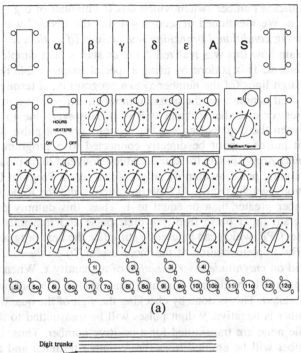

(a)

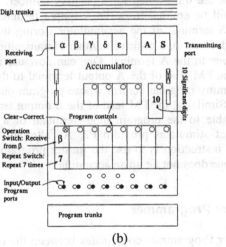

(b)

Figure 11: (a) View of an Accumulator's front panel. The operation switch's settings are α, β, γ, δ, ε, O, A, S and AS. The repeat switch's settings are 1 through 9. (b) Simplified front panel.

additions. A clear-correct switch corrects for the dropped unit pulse in complemented numbers during operations that require shifting such as in multiplication, division, or square-rooting. Moreover, each accumulator has one significant number switch which sets the number of digits used in the calculation. We mentioned in section 4.4 that the ENIAC was capable of implementing *conditional branching* instructions *(if ... then ... else)*. We will now explain how this was realized.[16] Let us start with a simple *"if x is zero halt"* statement. This is done by using a special adapter cable that connects one of the digit lines (of the number x) to a program input terminal. If the transmitted digit is non-zero, the program input circuit will be stimulated by the digit pulses, or else the program will stop. However, due to timing differences between the digit pulses and the program control pulse (CPP – see Fig. 9), digit pulses cannot be directly connected to a program input terminal. A *dummy program* has to be used which converts digit pulses into a true programming pulse. A dummy program is a repeat Program Control whose Operation Switch is set to O ("no operation") and the Repeat switch is set to a number greater than or equal to 1. Thus, this dummy program derives its program input pulse from the digit pulse and transmits a program output pulse, CPP, which can be used to stimulate the next program.

The implementation of the more useful *"if ... then ... else"* statement is based on *magnitude discrimination* of a quantity x. When for instance $x < b$, program P1 should be stimulated, and when $x \geq b$, program P2 needs to be stimulated. This is done by checking the sign of the quantity $x - b$. When the number is negative, 9 digit pulses will be transmitted to the sign lead (PM), while none are transmitted for a positive number. Thus, if $x \geq b$, a positive number will be emitted from the A output terminal and a negative number from the S terminal of the accumulator storing the quantity $x - b$. On the other hand, if $x < b$, a positive number is transmitted to the S terminal and a negative one to the A terminal. One can now use a special adapter cable to connect the PM lead of the A output terminal to the program input terminal of one dummy program control whose program output pulse stimulates program P1. Similarly, the PM lead of the S output terminal is connected by an adapter cable to the program input terminal of a second dummy program control that stimulates program P2. An alternative way to implement the branching instruction is to use the master programmer. This has the advantage that one does not tie up an accumulator.[17]

5.3 Master Programmer

The Master Programmer coordinates between the operations of 20 Accumulators and simplifies looping to include nested loops. It is basically a pulse counter which emits a program pulse every time it receives one (followed by one Addition Cycle delay), and at the same time increments its own counter. The program pulse it emits is used to trigger operations in Accumulators which receive the program pulse. When the number of input pulses it receives

---

[16] Markus and Akera, n. 3 above, A.K. Goldstine n. 14 above.

[17] A. K. Goldstine, n. 14 above.

matches the number set on the switches, no further program pulse will be emitted, thus stopping the program flow.

The Master Programmer consists of two panels that have 10 pulse counting channels, 5 on each panel. Each stepper is capable of receiving program pulses from one another, allowing the execution of nested loops. A schematic front panel diagram of the master programmer is shown in Fig. 12.

A unit of the Master Programmer has ten *decade counters* and five 6-stage *stepper counters*. Each stepper has a Stepper Clear switch and input channels. Each decade counter is grouped into separate "number-units" by the decade associator switch on the top of the Master Programmer, whereby those decade counters in the same group form an *n*-digit number belonging to a particular stepper (see Fig. 12). When the number of pulses going into a stepper input matches the number set by the decade switches in the same stepper group, the stepper counter will increment by one, at the same time clearing the content of the *decade counters*. There is a switch for each decade counter setting corresponding to each stepper stage. When the *stepper counter* reaches the stage indicated by the stepper clear switch, the stepper counter will be cleared, while emitting a final program pulse from that stage's program output port. Individual decade counters can be incremented directly, as can the stepper counters, through the decade direct input port and the stepper direct input port, respectively. The operation of the master programmer is summarized by the algorithmic state diagram of Fig. 13.

To illustrate the operation of the Master Programmer, let us discuss the following example, in which stepper C will be used to execute a sequence of two different operations. The first operation needs to be performed 21,509 times, while the second one will be executed two times. The stepper clear switch (C) will be set to 2 as illustrated in Fig. 12. One needs to use 5 of the 10 decade counters in order to set the number "21509," which requires that the 2nd and 3rd Decade Associator Switch be set in the position "C." The switches in the first stage will then be set to "21509" and the ones in the second stage to "00002." Upon receiving a pulse at the stepper input, the decade counter will increment by 1 and an output pulse will be emitted at the first output of stepper C one addition cycle later. This will continue until 21,509 pulses have been received. At this point, the decade counter is reset and the stepper counter C advances to stage 2. When the stepper has received 2 input pulses, both the stepper and the decade counters will be reset and the process halts.

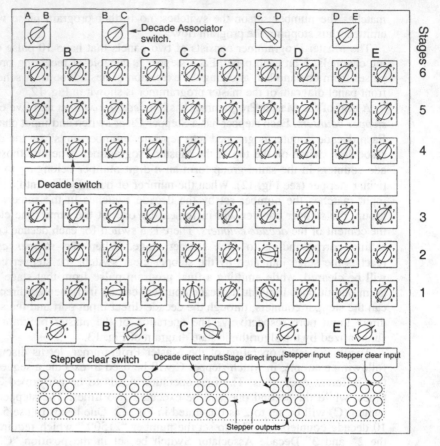

Figure 12: Front panel schematic of the master programmer. The Decade switch settings are 0 through 9, and the Stepper Clear switch settings are 1 through 6.

The Master Programmer can also be used to synchronize parallel operations and for digit discrimination. It should be mentioned that no special hardware was provided in the ENIAC to synchronize parallel branches, which made it hard to fully explore the ENIAC's parallel architecture. As mentioned by Marcus and Akera, one of the reasons why the inventors did not make use of the parallelism of the ENIAC, may have been their concern for reliability. By using as little of the ENIAC's hardware as possible at any one time, the chances of a tube breaking down are minimized.

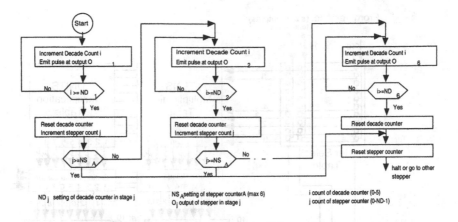

Figure 13: Algorithmic state diagram describing the operation of one of the 10 steppers of the master programmer. ND_j (max is function of the decade associator switch setting) is the setting on the decade counter in stage j; NS is the setting of the stepper counter (max 6); O_j is the output of the stepper stage j; i is the current count of the decade counter and j the count of the stepper counter. The above diagram can be looped inside another state diagram to implement nested loops.

5.4 High-Speed multiplier

The high-speed multiplier is used to multiply two signed, ten-digit numbers. It works in conjunction with four accumulators (or six in case twenty-digit products are required). Two of the accumulators are used for storing the multiplier ('Ier) and the multiplicand ('Icand). Multiplication is performed by multiplying the entire multiplicand by consecutive digits of the multiplier, and accumulating the partial products. The multiplication of a ten-digit multiplicand with a p-digit multiplier takes ($p+4$) addition times. The high-speed multiplier consists of large tables that map digits of the 'Ier against the digits of the 'Icand. Fig. 14 gives a schematic block diagram of the multiplier. The 'Ier select table is used to step through the various digits of the multiplier, one per addition time. This is followed by two multiplication tables, which output in pulse trains all the possible results of multiplying a digit by the selected 'Ier digit. One table is used for the tens results and one for the unit results. From this point on, the data path is split into two parts, for tens and units, including the accumulators which store and accumulate the partial products. The outputs of these tables are passed into the 'Icand select tables, which select the appropriate results from the possibilities according to the actual contents of the 'Icand accumulator.

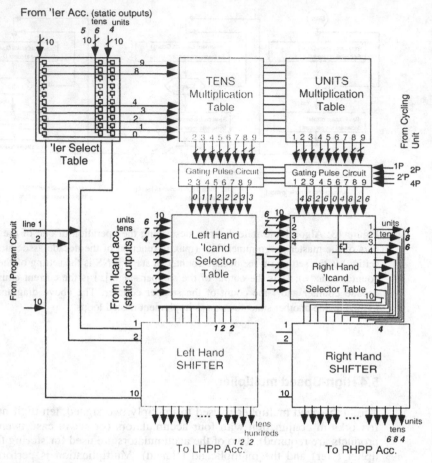

Figure 14: Block diagram of the high speed multiplier. The numbers in italics correspond to the example of the multiplication of 476 ('Icand) by the number 4 ('Ier).

The last pair of tables are called shift tables, and are responsible for making sure the partial products are transmitted to the correct places of the result accumulators. The tens digit goes to the left-hand partial product (LHPP) accumulator and the unit digits to the right-hand partial product (RHPP) accumulator. The tens results are transmitted to the LHPP accumulator shifted one place to the left of the corresponding units of the RHPP accumulator. Once all the p digits of the 'Ier are taken care of, the two partial product result accumulators are combined into the final result.

If one or both of the operands are negative, a correction factor for both the sign and the product needs to be applied. After the correction factor is applied, the two partial products are added together to produce the final product that is stored in one of the two partial product accumulators. So far, the multiplication requires $(p+2)$ addition times. The remaining two are needed for the reception of the arguments and for setting up the selector tables and round-off in the LHPP accumulator. It should be noted that when twenty-digit results are used, the upper and lower digit of the left- and right-hand partial products are stored in two accumulators, denoted LHPP I&II and RHPP I&II.

The use of multiplication tables allows the ENIAC to multiply two 10-digit numbers at high speed by eliminating the need for successive additions. These tables are like ROM storing the 0-9 multiplication matrix. The tables consist of an array of resistors in which each column functions as a logic OR gate whose inputs are supplied by the 'Ier select table, as shown in Fig. 15. The output of the resistive network feeds into the gating circuit which generates a pulse train corresponding to the actual number. The control of the gating circuit is done in an inhibitory fashion: the gates pass a pulse, unless explicitly directed not to by the table. To illustrate the operation, assume the number 4 is presented to the multiplication tables by activating the horizontal line, (labeled 4 in Fig. 15). The first column (labeled 1 on top) in the Units Table (right side matrix in Fig. 15) will generate control signals which produce the number 4 by combining the pulses 2P and 2'P. Similarly, column two will produce 8 (= 2×4) or (= 2P+2'P+4P) in the Units Table and 0 in the Tens Table (see left side matrix in Fig. 15); line 3 will give 2 (= 2P) and 1 (= 1P) in the Units and Tens Table, respectively, to produce the number 12 (= 3×4).

The 'Icand select table will then choose the actual (partial) product among all the potential partial products issued by the multiplication table. Both the 'Ier and 'Icand select tables use static connections to all ten states of all ten decade counters of their respective accumulators, as well as the PM (Sign) units. This requires 101 lines between each accumulator ('Ier and 'Icand) and the select tables of the high-speed multiplier.

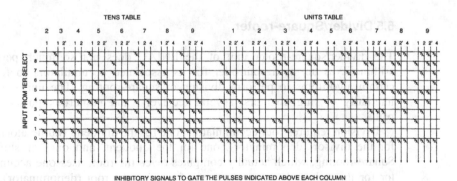

Figure 15: Multiplication table. The left matrix is the tens and the right matrix is the units table.

Program controls

The Multiplier has 24 Program Controls, allowing one to perform up to 24 different types of multiplications during one program. One program control can be used several times, but each time the settings will be the same. The only difference is the actual operands. Each program control consists of a transceiver, with input and output terminals for program pulses, 'Ier and 'Icand accumulator receive switches, 'Ier and 'Icand accumulator clear switches, a significant figures switch, a places switch, and a product disposal switch.

- Receive Switches: Used to allow the program control to stimulate the accumulators to receive their respective arguments. The switches have six positions, five corresponding to the five digit inputs of the accumulators (α through ε) and one off position. The first addition time is devoted to this process of receiving arguments.
- Clear Switches: Determine whether or not the argument accumulators are cleared at the end of the multiplication. This takes place during the $p+4^{th}$ addition time.
- Significant Figures Switch: Used for rounding off the final product.
- Places Switch: Determines how many places of the 'Ier are multiplied by the 'Icand. The possible settings are 2 through 10.
- Product Disposal Switch: Provides options for transmitting the final product at the end of the multiplication. If set to Off, the final product will be retrieved from the accumulator by explicitly stimulating a program control on the accumulator. The product can be transmitted additively (A), subtractively (S) or both (AS). If set to On, the product accumulator is instructed to clear as soon as the product is transmitted, again with the same three possible modes of transmission.

5.5 Divider/Square-rooter

The divider/square-rooter unit of the ENIAC does not so much perform arithmetic as act as a controller for its associated accumulators to follow certain algorithms to perform division or square-rooting. The unit consists internally of ring counters, several receivers, and numerous pulse-gating paths.

Division requires four accumulators, one for the dividend (numerator), one for the divisor (denominator), one for the quotient, and one for shifting. Square-rooting, which actually computes *twice* the root, uses one accumulator for the radicand (numerator), one for twice the root (denominator), and one for shifting. Operations can be divided into four periods: I, for initial set-up; II, for the main calculation; III, for rounding off; IV, for interlocking and clearing. The algorithms used for division and square-root will be explained next.

The division algorithm

Division is performed by repeated subtractions and additions. When the numerator and the denominator have the same sign, the denominator is subtracted from the numerator until the sign of the numerator changes. When the signs of the numerator and the denominator are different, the denominator is added to the numerator until the sign changes. Every subtraction causes an increment in a decade of the quotient accumulator, and every addition causes a decrement. The place (decade) in which this increment or decrement occurs is moved one place further to the right after every sign change.

Assume, at first, that both numerator and denominator are positive. The first set of subtractions leaves in the leftmost place of the quotient accumulator the integral number of times the contents of the denominator accumulator divides into the numerator, plus one. This is essentially a very rough estimate of the quotient. The sign change indicates that the division has overdrafted by, at most, one in this decade. After shifting the numerator to the left (thus multiplying it by 10), the magnified overdraft is then divided by the denominator. Thus, the estimate can be further refined to one more place of accuracy, until another sign change takes place. This process can be repeated for as much precision as is desired.

Rounding off is accomplished by shifting the numerator one more place to the left and subtracting or adding the denominator (depending on whether or not the signs of the two are matched) five times. If this does not cause an overdraft, then the quotient is decremented or incremented by one in the last place calculated.

In order to prevent loss of a significant figure during the shift of the overdraft accumulator, the denominator has to contain a 0 in its tenth place (or 9 for a negative number). Also, the first non-zero denominator digit can not be more than one decade to the right of the first non-zero digit of the numerator accumulator to prevent a left-side overflow in the quotient accumulator.

The time required to complete a division (and similarly a square-root) depends on the number of places required in the answers, as well as on the digits in each place of the answer. If one assumes that the average digit in the answer is 5 and if p is the number of digits, the time required for a division (and square-root) operation is approximately $13p$. This time is much greater than that required for a multiplication which made use of the multiplication tables.

The square-root algorithm

Similar to division, the process of square-rooting involves repeated subtractions, additions and shifts. The square-root algorithm is based on the fact that the square of the integer a is the sum of the first a odd integers:

$$\sum_{i=1}^{a}(2i-1) = a^2. \tag{1}$$

This method was often used in electric or manual desk computing machines at the time of the ENIAC. To find a rough estimate of the square root of M, one can subtract progressively larger odd numbers until a negative result is produced. The last odd number used is $(2a - 1)$ where a^2 is the first perfect square greater than M. Adding one and dividing by two provides an estimate of $\sqrt{M}$ that is no more than one greater than the actual value. However, for a ten-digit number, the process of subtracting larger and larger odd numbers is prohibitively time-consuming. Moreover, it is useful to refine this estimate using decimal places rather than just integers since more often than not, square-root calculations do not involve perfect squares.

Using the above equation, it is apparent that 100 is the sum of the first ten odd integers, 1 through 19. Adding 20 to each of these provides the next ten odd integers, i.e. 21 through 39. The sum of these 10 odd integers is $100 + 10(20) = 300$. Continuing in this way, it becomes obvious that the sum of the $(10n - 9)^{th}$ through $10n^{th}$ odd integers is $(2n - 1) \times 100$, and the values of these integers are $20(n - 1)+1$ through $20 \times (n - 1) + 19$, or $20n - 19$ through the value $20n - 1$.

This leads to a method of making a very rough estimate of the square root. First, subtract successively increasing odd hundreds (100, 300, 500, ... $(2n-1) \times 100$...) until a sign change is achieved. At this point, it has been determined that M is greater than (or equal to) the sum of the first $10 \times (n- 1)$ odd integers, but less than the sum of the first $10n$ odd integers (the last of which is $20n - 1$). Hence, $10 \times (n - 1) \leq \sqrt{M} < 10n$.

Since the last number used to subtract is $N = 200n - 100$, and the last of the $10n$ integers is $l =(20n-1)$, l can be calculated to be, $l = N/10 + 9$. The rough estimate can be refined by "un-subtracting" (i.e., adding) successively smaller odd numbers from the overdraft starting with l until a sign change. If the last odd integer added is b, it is now known from Eq. (1) that $b-1 < 2\sqrt{M} \leq b+1$. A good estimate is, therefore, $b \approx 2\sqrt{M}$. The same algorithm can be expanded by noting that 10^4 is the sum of the first 10^2 odd integers, 10^6 is the sum of the first 10^3 odd integers, and so on.

The ENIAC begins with 10^8 in the denominator ("two-root") accumulator (i.e., a 1 in the ninth decade) with the radicand stored in the numerator accumulator. The contents of the denominator accumulator are subtracted from the numerator accumulator, after which the denominator is incremented by 2 in the ninth place. This subtraction and increment is repeated until the sign changes. At this point, the denominator accumulator has already been incremented beyond the last number subtracted. The division by 10 is accomplished by shifting the numerator one place to the left. Then, instead of adding nine in the next place down in the denominator accumulator, an effective 20 has already been added, so 11 is subtracted.

The new number is added to the numerator, and the quotient is *decremented* by two in that next place down. After a sign change, the numerator is shifted, and the correction factor of 11 is added one more place down. Then, the process continues, alternating subtraction and incrementing with subtraction and decrementing at every sign-change, until the desired accuracy is achieved.

Rounding off in square-rooting is an approximate method. As in division, the residue of the numerator is shifted to the left again, and the denominator is subtracted from this five times. If no overdraft results, the last place of the doubled root will be incremented or decremented by two.

Again, because of shifting, significant figures in the numerator accumulator can potentially be lost. Since the denominator is initially incremented in the ninth place, any radicand greater than or equal to 25×10^8 will cause the tenth place of the denominator accumulator to be non-zero, and thus cause shift errors. To ensure that no such shift errors can occur, the general rule is to keep at least one zero preceding the radicand in the numerator accumulator.

Program controls

The divider/square-rooter has eight program controls. Each of these consists of a transceiver with input and output terminals, an interlock pulse input terminal, numerator and denominator accumulator receive switches, numerator and denominator accumulator clear switches, a divide/square-root and places switch, a round-off switch, an answer disposal switch, and an interlock switch. A brief description follows:

- Receive Switches: Used to stimulate the argument accumulators to receive their respective values. Each switch has three positions: α and β, to indicate on which input to receive, and Off.
- Clear Switches: Cause the argument accumulators to be cleared at the end of an operation.
- Divide/square-root and Places Switches: Specifies whether a division or square-root operation needs to be performed. It also controls the number of places the operation needs to be carried out.
- Round-off Switch: Used to determine whether or not a round-off operation is to be performed during period III.
- Answer Disposal Switch: Settings 1 and 2 are used for quotient disposal, and 3 and 4 are used for disposal of the doubled root. The interpretation

of these switch settings depends on how the interconnectors are wired to the accumulator terminals.

- Interlock Switch: Is used to synchronize the end of a division with any other operations, which have been occurring in parallel. Divisions and square roots take so long that other operations are normally scheduled in parallel. In order to keep the program from continuing before either operation is completed, each program control includes an interlock input. If the interlock switch is set, the divider/square-rooter can not progress beyond period III until an interlock pulse has been received.

5.6 Programmable Read Only Memory: Function Tables

In those days memory was sparse and expensive by today's standards. The ENIAC's memory consisted of internal and external storage. Internal memory is present in each accumulator's decade counter which stores the number during computation, and in the three function tables. External memory consists of punch cards that are used in conjunction with the constant transmitter for reading and writing numbers during the course of a computation.

The Function Table of the ENIAC finds its modern equivalent in the programmable ROM. It is required during the course of calculating the solution to difference equations, which require multiplying coefficients to variable values. These constants are looked up via the Function Table, which provides a quick way to read and transmit them (within 5 addition cycles per look-up). The values of the numbers are set on a panel of switches, which contains in total 104 entries. The position (argument) of the entry is referred to by a number ranging from −2 to 101. For each value of the argument there is a corresponding entry of 20 digits which can be divided into two groups of 10 digits each, called A and B. Each 10-digit number consists of 6 digits which are variable from entry to entry and are set by the switches on the portable function table. The remaining 4 digits are constant throughout the range of the table and are set by the Master Digit Switches on panel 2 of the Function Table. In addition, the sign can be variable or constant depending on the setting of the Master PM switch (see panel 2 below). Fig. 16 illustrates the composition of the 20-digit numbers available on terminals A and B on panel 2. The function table allows the 20 digit number to be used in a variety of ways: one signed 20-digit number or 2 signed numbers, one consisting of k digits and the other of $(20 - k)$ digits for 2 functions.

Using programming nomenclature, the Function Table can be said to be an array of signed 10-digit numbers of length 104. This is, indeed, miniscule compared to modern storage capacity, considering the size of the components necessary to realize it.

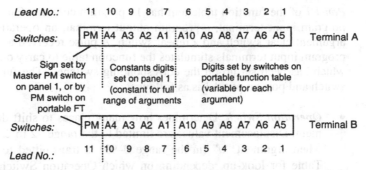

Figure 16: Formation of a 20-digit number and corresponding switches

The ENIAC contains three Function Table units, each independently accessible. Each of them consists of two stationary panels and a portable module that plugs into the stationary one. Figs. 17 and 18 show the panels of the Function Table. The Function Table is controlled by the output of two decades of an associated accumulator in which the argument is stored. The argument values –02, –01, 100 and 101 are included, even though the argument accumulator holds only the values 00 to 99 (see panel 1 below).

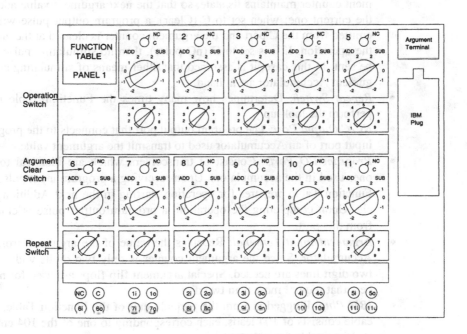

Figure 17: Panel 1 of the Function Table Module

Panel 1 of the function table unit has 11 program controls. Each consists of a program input terminal, a program output terminal, an operation switch, an argument clear switch and a repeat switch. A pulse received at one of the 11 program input terminals stimulates the function table to carry out the program which has been set up on the corresponding switches. The function of each switch and port on panel 1 is as follows:

- *Operation Switch*: Determines how many places to shift the input argument. Displacement values are limited to the range –2 to 2. For example, when argument "*a*" (in the range 0–99) is transmitted to the Function Table for look-up, depending on which Operation Switch is activated, the effective look-up value is "*a* + *d*," where *d* is in [–2,2]. Furthermore, depending on which side the dial points, the number transmitted is either the actual number itself, or its 9's complement. The 1'P which completes the 10's complement is picked up in the digit set in the Subtract Pulse Switch in panel 2.

- *Argument Clear Switch*: Used to determine from which of the two outputs a program pulse is to be emitted to stimulate the transmission from an argument Accumulator, and whether the argument counter is to be cleared at the end of the table look-up. When set to NC (No Clear), a program output pulse will be emitted from the NC port, and the argument counter maintains its state, so that the next argument value adds to the current one; when set to C (Clear), a program output pulse will be emitted from the C port and the argument counter is cleared at the end of the operation. If it is set to O (nO pulse), then no stimulating pulse will be sent. In this case, it is assumed that other means of stimulating an argument Accumulator are at hand.

- *Repeat Switch*: Determines how many times the Function Table is to perform the look-up.

- *NC/C Output Port*: The program output port that connects to the program input port of an Accumulator used to transmit the argument value.

- *Program I/O Ports*: Consist of transceivers which are identical to the ones used in the accumulators. Upon receiving a program input pulse the Function Table prepares for operation starting in the next Addition Cycle. Upon completing the operation, a program output pulse is emitted from this port.

- *Argument Input Terminal*: Receives the value of the argument from an Accumulator. Since the argument counter is only two digits wide, only two digit lines are needed. Special argument flip-flops are used for numbers that require more than two digits.

- *IBM Plug*: Plugged in from the portable unit of the Function Table. The cable consists of 104 leads, each corresponding to one of the 104 entries in the portable function table.

Panel 2 is used to specify the 4 digits which remain unchanged throughout the range of the table. It has the following switches:

- *Master PM Switch*: Sets the sign value to P, M, or to be read from the portable Function Table module. If the sign to be emitted is constant throughout the range of the table, the PM switch is set to "P" or "M," depending on the desired sign. If the sign varies from entry to entry, the PM switch is set to "Table."
- *Constant Digit Switch*: Sets the value of the constant digits to be transmitted either as a fixed value or dependent on the sign of the variable digits: zeros (P) or nines (M). The Constant Digit Switches form the upper 4 digits of the number transmitted (see Fig. 16). PM1 indicates that Master PM Switch 1 is in effect, and PM2 indicates that Master PM Switch 2 is in effect.
- *Digit Delete Switch*: If set to delete, then the Constant Digit Switch is nullified and no digit pulses are transmitted.
- *Subtract Pulse Switch*: If the transmission is done subtractively, then the digit (A5, ..., or A10) that has this switch set to S will pick up the 1'P. In practice, only one of the A and B switches should be set to S.
- *Function Output Terminals (A/B)*: Two terminals from which the table values can be transmitted. Each terminal has 11 leads, one to carry the sign and ten for the 10 digits. Transmission is done in pulse form.

The portable module of the Function Table is the variable part of the unit. By variable it is meant that values in these digits change across the argument range independently of the sign and Constant Digit Switch. It consists of switches on both sides of the rectangular module which, on either side, contains a two dimensional array of switches, 26 rows by 28 columns, divided into two halves.

Fig. 19 shows one half of the arrangement of switches on this unit. These switches form the lower 6 digits of the numbers A and B, transmitted as is shown in Fig. 16. The pattern of arrangement repeats for the other half of the function table and the whole facade is repeated on the other side. Starting at the left top corner on one side is the value for argument –2 and at the bottom left corner is argument 23. The top of right half is argument 24, reaching argument 49 at the bottom right. The argument then continues on the other side with the top left corner at argument value 50, then 75 at the bottom left; and 76 and 101 at the top right and bottom right, respectively.

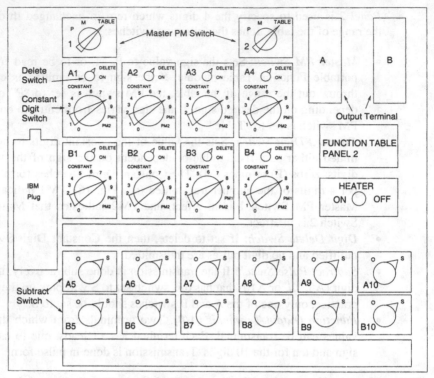

Figure 18: Panel 2 of the Function Table Module

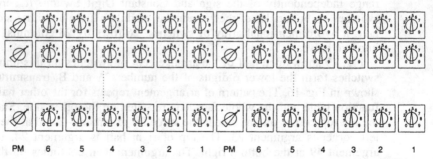

Figure 19: Switches on the portable function table module. The module contains 26 rows of 28 switches. Only one half of a row is shown. The switch settings for the number switches are 0 through 9.

When a program pulse enters one of eleven program input ports on panel 1 in Addition Cycle "t," the corresponding operation switch is armed. Table 2 below shows the procedure the Function Table follows to look-up an entry.

The transmission of the digit pulses via the function table is done by appropriately gating the pulses 1P, 2P, 2'P, and 4P from the cycling unit. The gates are set to allow digit pulse passage by mechanical switches on the portable function unit. Upon receiving a program input pulse, at Addition Cycle "t," which initiates Function Table action, a program output pulse is emitted from the appropriate port "N" or "NC," in the following Addition Cycle, and the receiving Accumulator is stimulated to transmit its value. Only the two lowest digits are used from the Accumulator. Then, this value is further incremented (the argument counter starts at minus 2 so that only increments need to be performed) and decoded so that only one of 104 lines is activated. The argument counter is implemented as a 10-stage ring counter, similar to that of the decade counter of the Accumulator and Master Programmer. Finally digit pulses are gated according to the switch settings on the portable Function Table and emitted from function output terminals A and B on panel number 2.

For example, if the values in lines "$x + 1$" and "$x + 2$" are to be looked up once, then the argument Accumulator should store the argument x, and an Operation and the corresponding Repeat switches are both set to 1. A second Operation Switch is also set to 1. The argument counter should not be cleared for the first Operation Switch, so the Argument Clear Switch is set to "NC." However, the counter should be cleared at the end of the second operation, so the corresponding Argument Clear Switch is set to "C." The program output for the first Operation Switch is connected to the program input of the argument Accumulator, while the program output of the second Operation Switch is not connected to anything. Since the first argument is not cleared, all that needs to be done to obtain the correct argument value is to add one to "$x + 1$." The switch settings and the numbers transmitted across Function Output Terminals A and B are shown in Table 3.

Table 2: Operations involved in reading a number from the Function Table Unit

Cycle	Operation
$t + 1$	Function table emits a program output pulse on C or NC output port on panel 1 to stimulate transmission of argument by the argument Accumulator.
$t + 2$	Function table (panel 1) receives 2 digit pulses for the argument from an argument Accumulator.
$t + 3$	Argument stored in the argument counters on panel 1 of the function table is adjusted to the value specified on the operation switch.
$t + 4$	One of the 104 lines corresponding to the appropriate argument row of the portable unit is activated, based on the value received at $t + 3$
$t + 5$	Function value is transmitted
$t + r$	If the repeat switch is set to a value r greater than 1, then function value is transmitted again. A program output pulse transmitted from one of the 11 program output terminals at the end of the repeat switch value.

Table 3: Example of settings and transmission of digits from the Function Table[18]

Line	Portable Function Table Switches
x	P 123 000 M 795 642
$x+1$	M 764 000 M 421 508

Example A:
Setting of Constant Digit Switches on Panel 2
All Digit Delete Switches are set to "On"
Master PM Switch 1 set to table
Master PM Switch 2 set to table

A4 at PM1	B4 at PM2
A3 at PM1	B3 at PM2
A2 at PM1	B2 at PM2
A1 at 3 B1 at PM2	

Subtract Pulse Switches:
A8 at S
B5 at S
All others at 0

Number Transmitted			
Transmit	Argument	Terminal A	Terminal B
Add	$x+1$	P 0 003 123 000	M 9 999 795 642
Add	$x+2$	M 9 993 764 000	M 9 999 421 508
Sub	$x+1$	M 9 996 877 999	P 0 000 204 358
Sub	$x+2$	P 0 006 236 999	P 0 000 578 492

Example B:
Setting of Constant Digit Switches on Panel 2
Digit DeleteSwitch of A4 is set to "DELETE." All others are set to "ON"

	B4 at 9
A3 at 0	B3 at 9
A2 at 0	B2 at 9
A1 at 0	B1 at 9

Subtract Pulse Switches:
All are at 0

Number Transmitted			
Transmit	Argument	Terminal A	Terminal B
Add	$x+1$	P 0 000 123 000	M 9 999 795 642
Add	$x+2$	M 0 000 764 000	M 9 999 421 508
Sub	$x+1$	M 0 999 876 999	P 0 000 204 357
Sub	$x+2$	P 0 999 235 999	P 0 000 578 491

---

[18] A. Goldstine, n. 14 above.

Keep in mind that the values stored in the function tables are transmitted in pulse form. Hence, the number stored in the function table could also be used as program pulses. In that case the pulse emitted by the function table would be directed to the program input terminals of target units.

5.7 Input/Output devices: constant transmitter, IBM card reader, neon lights

The Constant Transmitter, together with the IBM card reader, is the external input device of the ENIAC. Like modern external storage devices, the constant transmitter is a hybrid of mechanical and electronic components. It enables arbitrary numbers to be loaded into Accumulators for computation. The IBM card reader is the mechanical device that reads punch cards (IBM cards), which contain values that are used to set relay switches, in the Constant Transmitter. These switches gate digit pulses which are transmitted to Accumulators. An IBM card can store 80 digits, which are to be grouped into either 5- or 10-digit signed numbers. The Constant Transmitter itself has manual switches which allow 20 digits and 4 signs to be stored, in addition to the 80-digit read in from an IBM card. Like the digits on the IBM card, these digits are grouped into either 5- or 10-digit signed numbers. The rate at which IBM cards can be read ranges from 120 to 160 cards per minute. This rate is considerably slower than that at which the ENIAC performs calculations. This is not a serious problem as long as the number of calculations (iterations) per I/O is large, which is the case for most of the calculations for which the ENIAC is designed.

The IBM card contains 80 columns which span the length of the card, and the columns are divided into 8 groups of 10 digits or 16 groups of 5 digits (labeled A[LR] through H[LR]). In each column there are 12 positions, going from top to bottom, which indicate the sign and value of a number. Position 11 indicates the sign of the number, while positions 10 through 1 correspond to digits 0 through 9. Although there is a sign position in every column, only the sign in the most significant digit column matters. Negative numbers on the card are converted into 9's complement during the reading process, and into 10's complement during the transmission process.

The Constant Transmitter itself consists of two panels, shown in Figs. 20 and 21. *Panel 1* houses Constant Selector Switches, which are used to select the grouping of the digits on the IBM card. The letters on the switches, A-H, correspond to 8 groups of 10 digits stored on the IBM punch card, while letters J and K correspond to the 2 groups of 10 digits on Panel 2 of the Constant Transmitter. Within each letter, the 10-digit group can be further broken down into two groups of 5 digits, or be used as it is. There are 6 switches for a two-letter pair, each corresponding to one of the six program controls that can be activated during a computation. The first 24 switches are used to interpret the IBM card. In computations where only 5 digits are needed, setting the constant selector switch to L or R will limit the number of digits transmitted from the IBM card to those in the L or R group. When a constant selector switch, say A1, is set to L, one of the other 5 switches in the A group, say A2, is usually (but not necessarily) set to R. The redundancy in the switches allows the same group, L or R, to be transmitted to more than one destination. Organizing the numbers into 5-digit L and R groups effectively doubles the number of different values that can be stored on the IBM card. On the other hand, when the computation demands a 10-digit number, the constant selector switch is set to LR. Two rules on how the Constant Selector Switch can be set must be followed. In a group of six, if a switch is set to transmit 5 digits, no other switches in the group of six can be set to transmit 10 digits. Conversely, if a switch is set to transmit 10 digits, no other switch in the same group is allowed to be set to transmit 5-digits.

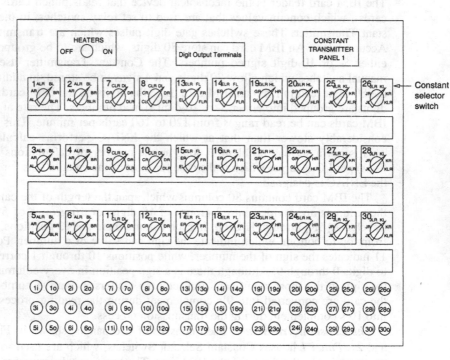

TERMINALs 1i, 2i, ..., 30i
Program input pulse terminals associated respectively with constant selector switches 1-30.

TERMINALS 1o, 2o, ..., 30o
Program output pulse terminals associated respectively with constant selector switches 1-30.

Figure 20: Panel 1 of the Constant Transmitter, used to select the grouping of the digits on the IBM card

The last 6 switches, letters J and K, are used to group the 20-digit switches settable directly on the Constant Transmitter. *Panel 2* (Fig. 21) of the constant transmitter contains two rows of 10 Constant Switches. The top row is lettered J, while the bottom row is lettered K. There are four PM switches (JL, JR, KL, KR), each corresponds to one of the 4 possible groups of 5 digits. If a 10-digit group is selected, only the L PM switch is used. Each switch is associated with a transceiver, which communicates with the rest of the ENIAC through a pair of input and output ports. A program pulse through the input port will start constant transmission, while one will be emitted from the corresponding output port when the operation is finished. The number transmitted is from any of the groups chosen from the 80 digits stored in the Constant Transmitter relays. The IBM card reader translates the card punches into relay settings which, ultimately, gate the passing of 1P, 2P, 2'P, and 4P from the Cycling Unit. Each of the 80 columns on the IBM card will set 4 relays that control the gates that will pass a combination of the pulse trains supplied by the Cycling Unit to produce the correct number. Therefore, there are 80 groups of 4 relays, or 320 in all, in the Constant Transmitter responsible for number storage.

The IBM card reader is a mechanical device attached to the Constant Transmitter and sets the Constant Transmitter's relay switches from information stored on the IBM card. Card punches are translated into relay switch settings through the wiring of the plug board, which was characteristic of the IBM card reader. The plug board is a detachable unit containing numerous single-hole terminals, called hubs, which allow wires to be plugged into

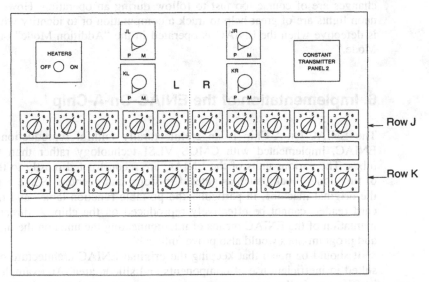

Figure 21: Panel 2 of the Constant Transmitter, used to set the constant part of the number

them. They are internally connected to read brushes in the card reader and coding relays in the Constant Transmitter. By connecting hubs to each other, connections are established among the card punch positions and relay switches. The IBM card reader's reading unit consists of 80 wire brushes that are in contact with the punch card as it is rolled in from top to bottom between two drums. As the card rolls between the read brushes, contact is made on the coding cams whenever a hole in the column is encountered. While the coding cams and the read brushes are in contact, a pulse is emitted to set the corresponding relay.

The card reader can be activated in two ways, via the Initiation Unit, which emits a pulse to the Ri input port on the card reader, or by pushing the emergency start button on the card reader itself. The latter method is intended to be used for testing purposes. Once started, the card reader will pick a card for reading from a stack of cards contained in the card tray. Because the card reading process is much slower than the transmitting process, the Constant Transmitter has an interlocking flip-flop which the card reader will set once it finishes reading a card. The Constant Transmitter will not begin operation until this flip-flop is set.

Neon lights as visual output

The numbers stored in each accumulator are visually displayed on neon tubes which are connected to the static outputs of the decade counter and the sign counter (binary PM counter). This makes it possible for the operator to see the numbers "rushing" through the accumulators during an operation. The changes are of course too fast to follow during an operation. However, the neon lights are of great help to track a computation or to identify which unit is defective when the ENIAC is operated in the "Addition Mode" or "Pulse Mode."

6. Implementation of the ENIAC-On-A-Chip

The ENIAC-On-A-Chip is an architecturally faithful reproduction of the ENIAC, implemented with CMOS VLSI technology rather than vacuum tubes. The various units of the ENIAC are reconstructed mostly to the level of functional blocks, such as gates, flip-flops, and counters. Some aspects of the original machine, in particular the portable function tables and the IBM card reader, cannot be effectively reproduced on the chip. A direct implementation of the ENIAC means of interconnecting the units on the digit trays and program lines would also prove unfeasible.

It should be noted that keeping the original ENIAC architecture often resulted in inefficient use of components and silicon area. An example of this inefficiency is that in order to represent numbers 0 through 9, the original ENIAC has ten stages of serially connected flip-flops forming the decade counter. This is equivalent to using ten bits to represent ten numbers! As there are ten decade counters per Accumulator, making a grand total of 100 bits, the range of numbers represented is 0 through $10^{10} - 1$. The same range of numbers can be represented by only 34 bits using the binary number system. In today's technology, there is an unwritten law that the binary number representation be used. However, as was explained in section 4.2 and 4.3, the choice of decimal number system and the use of pulse transmission resulted in fewer vacuum tubes and interconnections. The communication of a signed ten-digit number requires only eleven wires in the decimal pulsed system, while the static transmission of the same number in binary would have required 34 lines. Considering that communication was done over several long cables and trunks, all around the ENIAC, it is understandable that the inventors chose the pulsed decimal system. This allowed them to trade-off wiring complexity for speed of transmission (i.e. serial transmission on a per digit basis). The chip implementation adheres to the original design.

Another aspect of the ENIAC that presented an obstacle in the silicon version is its parallelism. Various units of the ENIAC could communicate with each other without restraint through manually programmable communication channels, and groups of such units could do so simultaneously. Not only did an Accumulator transmit pulses to other Accumulators via digit trunks, but it also statically transmitted its contents to the multiplier on dedicated static leads. Furthermore, the manual switches in the program control sub-units find their silicon incarnation in many 4-bit shift-registers. What these aspects of the ENIAC mean for the silicon version is that programmable data paths and a large number of wires are required, and these proved to be, as in all such designs, the most unwieldy parts of our project. How these were implemented on the chip will be described later.

The vacuum tube circuits of the ENIAC are fairly easily translated into CMOS transistor logic. However, there are a few differences that need to be pointed out. Vacuum tube circuits take advantage of the *wired-OR* logic. The output of gates and flip-flops could usually be simply wired together and the effect would be a logical OR. In such circumstances, the CMOS implementation requires an explicit OR-gate. In general, the logic gates are realized as traditional CMOS circuits composed of nMOS and pMOS transistors. Thus low static power consumption and strong logic levels are maintained. However, fully complementary NOR and AND-NOR gates with several inputs require large areas, due to the large pMOS transistors.

The following sections give a brief summary of some of the circuits used in the Silicon ENIAC. Not all circuit blocks will be described due to space limitations. The interested reader is referred to the thesis written by Tau and Ala'ilima.

6.1 Interconnection between units

As described earlier, the units of the ENIAC were connected to each other by a set of program and data lines. Replicating the complete flexibility with which the ENIAC could be interconnected and programmed proved nearly impossible. Reproducing even a minimal set of program lines on-chip would be too costly in terms of space, and the number of possible connections that would have to be somehow programmed would quickly become untenable. Therefore, a scheme was decided upon, which would allow a single physical line to implement all the program lines of the ENIAC. Likewise, two digit buses, each 11 lines wide, are used to implement all the digit lines. This limits the chip version to programs which only require two parallel data transmissions by way of the digit lines, but this was seen as an acceptable limitation, given the type of programs one would expect to run.

The scheme involves *rapidly reprogramming* the interconnections between the units every time a program pulse is passed. Since CMOS circuits can run much faster than the tube circuits of the ENIAC, large streams of setting data can be loaded into shift registers across the chip, within the scope of a single addition time. Each bit controls a connection between a unit's program controls and the program line, or its data ports and the digit lines. In order to keep connections intact while the new settings are being loaded, two parallel shift registers are used, one maintaining the current connections while the other is being re-loaded. Every time a program pulse is passed along the program line, the newly re-loaded shift register is activated and the re-load process for the other shift register is initiated.

6.2 Accumulator

Decade counter

The same decade counter design is used both in the Accumulator and the Master Programmer. As mentioned earlier, it is modeled by a ten-stage shift-register ring with one stage set to "1" at the start. Fig. 22 shows a simplified schematic diagram of the decade counter. The flip-flop uses a 12-transistor 2-phase master-slave design that can be compactly laid-out. The signal in the decade counter is propagated on the falling edge of phase 1 of the clock (generated by the d, Cin or 10P input pulses).

Each stage in the decade counter represents a number from 0-9. Digit pulses will cycle the decade counter a number of stages from its original place. When the decade counter moves from stage 10 to stage 1, or from 9 to 0, the decade flip-flop in the Carry circuit (see Fig. 23) is set. Thus, the decade flip-flop "registers" that a carry-over has occurred and opens the transmission gate T2. When the reset pulse (RP) arrives a few clock pulses later (at clock period 13 of the addition cycle – see Fig. 9), it will pass through

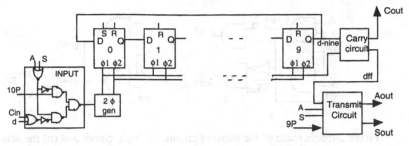

Figure 22: Simplified schematic of the decade counter. The signal in the decade counter is propagated on the falling edge of the input clock input.

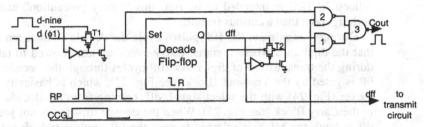

Figure 23: Carry circuit. The d-input is derived from the digit input (is connected to phase 1 of the clock, $\phi 1$), d-nine is the output of the 10th flip-flop (storing the number 9) of the decade counter.

gate T2 and generate a Carry-out pulse Cout, because the Carry Clear Gate (CCG) has opened the NAND-gates 1 and 3. At the same time, the RP will reset the decade flip-flop on its negative going edge, which occurs after the RP has passed through the circuit and become the carry-out pulse.

If a carry-out pulse from a previous decade-counter $(n-1)$ arrives (during period 13-17) at Cin, now the carry-in pulse of decade-counter (n), and if this decade-counter is at stage 10, the carry-in pulse will pass through transmission gate T1 and, since the CCG pulse is still on, continue through NAND-gates 2 and 3, thus generating another carry-out pulse feeding into the next decade-counter. Decade-counter (n) will also move to stage 0 by the carry-in pulse. The decade flip-flop of decade-counter (n) will also be set and the second RP will reset it (Fig. 9). It is possible that several such carry-overs take place, in the worst case, twenty. In the original ENIAC, the total propagation time through the twenty decade counters could be as long as 25 μs. To allow the safe rippling through of all possible carry-out pulses, it was deemed sufficiently safe to have CCG extend 40 μs beyond the initial carry-out pulse. However, in the silicon ENIAC this is not a problem as propagation delay from one stage to another, even in the 20-stage case, cannot be more than, at most, a few hundred nanoseconds. Compared to the 100 kHz at which the

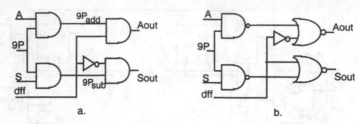

Figure 24: Schematic of the transmit circuit: (a) logic circuit and (b) the actual circuit implemented on the chip

Silicon ENIAC is intended to be run, this "safety precaution" amounts to nothing more than a curious feature.

During *transmission* the 10P pulses cycle the decade counter ten times so that the digit ends where it starts. Carry-overs are not allowed to take place during the transmission of digits. As 10P cycles through the decade counter, 9P is gated by the Transmit Block (see Fig. 22), which is basically a multiplexer (Fig. 24) with the select signal, dff, coming from the decade flip-flop in the Carry Block (see Fig. 23). When the decade flip-flop is not yet set (i.e. dff is low), the MUX is selected to pass the $9P_{sub}$ pulses (which is 9P when the program control is set to S, i.e. to transmit the complement of the content). When the decade flip-flop is set, the MUX will pass $9P_{add}$ (which is 9P when the program control is set to A, i.e. to transmit the number). The number that is passed through the S port will thus be 9-*n*, and the number through the A port will be *n*, where n is the original number in the decade counter.

Program control

The settings of the program control and repeat switches are implemented as a shift-register with 20 (twelve for program control setting, eight for repeat) four-bit registers. The switches are encoded in binary in order to minimize the number of bits needed to load in the program settings. The data in the registers are meant to be shifted-in serially at the start of the ENIAC operation.

Figs. 25(a) and 25(b) show the program control block. When a program pulse enters an Accumulator via one of the inputs (1-12), Fig. 25a, the corresponding SR latch is set. These latches are cleared at startup by the icg pulse. The output of the set latch will enter the "Program Control Switch MUX" block, which contains 12 "program switches'" of 4-bits each, and selects one of the 12 program switches. The output of the selected program switch (program select code stored in the 4-bit register) will then be decoded by the "Program Action Decode" block to perform one of the nine possible operations (α, β, γ, δ, ε, A, S, AS, O). If the latch set is in Group 1, then the Accumulator simply performs the operation once, as the latches in group 1 will be

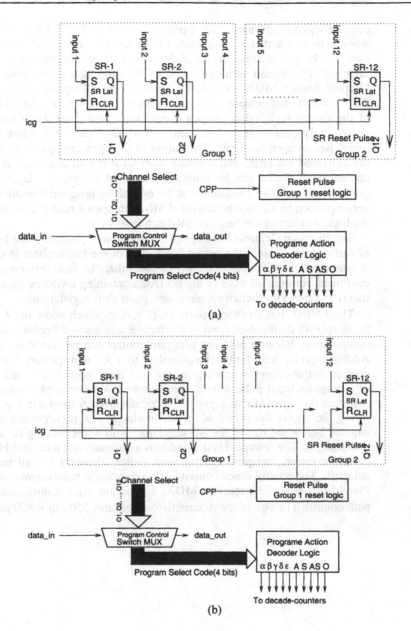

Figure 25: Simplified schematic diagram of the Program Controls in an Accumulator

reset in the following Addition Cycle when CPP arrives. On the other hand, if the program input pulse sets one of the latches in group 2, the repeat decade-counter will be activated. The repeat-counter (see Fig. 25(b)) counts CPPs when activated and is nothing more than a decade-counter whose static outputs will be passed through one of the 9 pass-gates corresponding to the SR latch set. The signals turning on one of the nine pass-gates come from the "Repeat Switch MUX," which is similar to the "Program Control Switch MUX" above, but contains only eight four-bit program switches. The output of the switch is decoded to open one of nine pass-gates. Each time a CPP enters the repeat-counter, it advances one stage until it reaches the stage which has an open pass-gate. The output of the open pass-gate will then reset the activated SR-latch, and together with CPP it will also reset the repeat-counter itself. Recall that for program controls 4-12 (group 2 latches), a program output pulse will be emitted at the end of the program execution. This is accomplished by having the activated SR-latch open a pass-gate which passes the logic product of CPP and the SR-latch reset pulse.

The Program Control- and Repeat- Switches are effectively one long series of shift-registers, since the output (data_out) of the last register in one block is connected to the input (data_in) of the other. In fact, this connection is continued between all units of the ENIAC containing switches (that is, all of them) and make the switch elements one giant shift-register unit.

The ENIAC has five input ports, α, β, γ, δ, ε, which allow the Accumulator to receive digit pulses from five separate sources at different times during computation. Because only one program control can be activated during one Addition cycle, these ports are equivalent to a 5:1 multiplexer. The program control settings determine which of the five input ports is selected. As soon as a program input pulse arrives, the switches in the multiplexer are enabled and steer the signals to the input of the decade counter (port d in Fig. 22).

Fig. 26 shows the layout of an accumulator. The registers are physically situated below the rest of the Accumulator, with wires running on top of the components. The floorplan is divided into a datapath and a control block. The decade counters, program control, and switch elements are all part of the datapath. The control block consists of random logic components such as the Plus/Minus unit, digit receiver MUX, significant digit control, and the 1'P path control. The size of the Accumulator measures 550 μm $\times$ 800 μm.

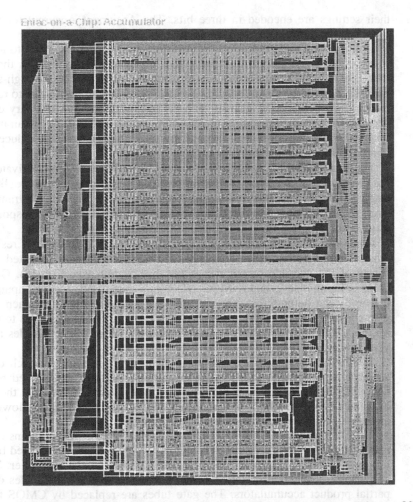

Figure 26: Layout of one of the twenty Accumulators. The size is 550μm × 800μm. Floor Plan and Layout of an Accumulator.

6.3 Multiplier

Similar to the Accumulator, the settings of the program controls and switches in the multiplier are stored in shift registers. The Significant Figures switch and Places Switch have ten and nine settings, respectively, so that their settings are encoded in four bits. The Argument Accumulator Receive switches and Product Disposal Switch have six and seven settings, respectively, and their settings are encoded in three bits. The Argument Accumulator Clear Switches are controlled with a single bit.

Both the 'Ier and 'Icand Select Tables require static connections to all ten states of all ten digits of their respective accumulators, as well as the PM units. Routing 101 lines from two separate accumulators to the high-speed multiplier on the chip would have proved too cumbersome. In order to reduce the number of lines, the state of each decade is transmitted in binary coded decimal. A four-bit encoder is included on each decade of the accumulator, with corresponding decoders on the high-speed multiplier. This reduces the number of lines needed to 41 for each accumulator.

The 'Ier select table of the ENIAC uses columns of gate tubes activated by the program ring. As in other parts of the chip, the ring counter is replicated using a ring of D flip-flops. The gate tubes are replaced by CMOS transmission gates, activated by the complementary outputs of the corresponding stage of the program ring.

The Multiplication Tables in the ENIAC consist of a network of resistors which formed column-wise OR-gates, the outputs of which inhibited some subset of the possible output pulses for that column (see Fig. 15). On the chip, column-wise pseudo-nMOS NOR-gates are used, whose inputs are driven by the outputs from the 'Ier Select Table. The output of each NOR gate controls the passage of a subset of the 1P,2P, 2'P and 4P pulses to make up the actual pulse train that is passed on to the 'Icand Select Tables as inverted pulses.

The ENIAC's 'Icand Select Tables are arrayed in columns, each corresponding to a decade of the multiplier and consisting of tubes gated by the decoded static connection to the 'Icand Accumulator. On the chip, they are organized by column into pseudo-nMOS AND-NOR gates, each followed by an inverter to drive the shift tables, leaving the pulses inverted.

The Shift Tables for the ENIAC are arrays of gates, with columns corresponding to the outputs of the 'Icand Select Tables, and rows selected by the program ring to correspond to the selected decade of the multiplier. Their outputs are connected in diagonal lines corresponding to the decades of the partial product accumulators. The gate tubes are replaced by CMOS transmission gates, with each diagonal line loaded by a weak pMOS pull-up to prevent drift in standby. One last inverter rectifies the pulses, producing the final output which is connected to the digit lines for transmission to the Product Accumulators. The entire chain, from Multiplier Select to Shifter, is shown in Fig. 27.

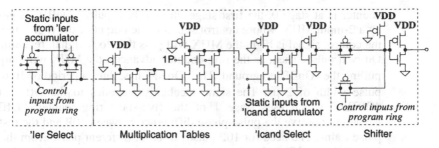

Figure 27: Cross-section of the data path from Multiplier Select through Shifter Tables

6.4 Divider/Square-rooter

The divider/square-rooter was converted fairly simply from ENIAC to the chip, given the standard receiver to flip-flop translation, and the use of gates to implement wired-OR logic. Pulse-gating in the divider/square-rooter on chip is accomplished variously, both with logic gates and transmission gates, depending on which requires the fewest transistors to accomplish the given task. In many cases, the pulses are passed inverted in order to minimize the number of inversions along the way.

The receivers were statically connected to the common programming circuits of the argument and result accumulators for a given operation, so that they could stimulate the necessary actions directly, without program controls. As in the case of program and digit line connections, the full flexibility of the original ENIAC cannot be reproduced on-chip. So these connections to the common programming circuits are hard-wired on the chip, allowing a limited number of possible configurations, but, at the same time, a selection which represents the majority of configurations that are needed.

6.5 Cycling unit

The silicon Cycling Unit is divided into two parts: a pulse generator and a 20-stage ring-counter. The pulse generator block divides an input clock of 500 kHz into five time periods of 10 µs each, which creates the desired 100 kHz clock that the ENIAC used. In the original Cycling Unit the 2 µs pulse in each Addition Time period was generated using delay lines operating on the 100 kHz input clock. For the silicon version we take the all-digital approach and divide down a 500 kHz master clock to create the 2 µs pulse for the 10 µs period. The pulse generator consists of a five-stage ring-counter, which appropriately lengthens the 2 µs period to 10 µs. The first stage in the ring-counter is set to 1, while the rest is set to 0 during reset. When the ring-

counter is at stage 0, the first stage, the On-beat pulse "OnBP" is generated
by a 2-input MUX, whose control signal is the output of the first stage. When
the control signal is high, the MUX will pass the clock signal, resulting in the
On-beat pulse. Then, as the ring-counter advances, the MUX will output a 0,
pulling the "OnBP" to ground. The "OnBP" has, thus, a period of 10 μs and a
pulse width of 2 μs. The same mechanism applies to the Off-beat pulse
"OffBP," which uses stage 1 of the five-stage ring-counter. "OffBP" is
needed to observe the pulse-phase difference between 10P and the rest of the
pulse trains. The need for 10P pulses to be of different phase from the others
is that since 10P cycles the decade-counters (see section VI-b) whose flip-
flops are used to gate the transmittance of 9P, the flip-flops must not change
until their outputs have allowed 9P to pass through properly. Besides the
normal, continuous mode, the pulse generator has two additional inputs
"Am" and "Pm," which stand for "Addition Mode" and "Pulse Mode," referring to the two debugging modes in the ENIAC. These signals are created
externally by a push during debug. "Addition Mode" is a signal intended to
allow only one Addition Cycle's worth of pulse trains to be generated, thus
stepping through the program one Addition Cycle at a time, and "Pulse
Mode" is intended to allow only one pulse to be generated at a time.

The 100 kHz, 2 μs "OnBP" pulse is gated by the outputs of a 20-stage
ring-counter block to generate all the pulse trains of the cycling unit (Fig. 9).
Fig. 28 shows the simplified schematic diagram of this block. The pulse train
1P is generated by enabling a 2-input MUX, "G1," which passes "OnBP"
while stage-1 of the ring-counter is asserted. 2P is generated via a similar
method by using the outputs of stage-2 and stage-3 to select the "OnBP"
input of the two-input MUX "G2." 2'P is generated similarly, using outputs
of stages 4 and 5. Pulse train 4P is generated by having the output of stage-6
setting an SR-latch, and that of stage-10 resetting it, with the output of the
latch controlling MUX "G4," which passes "OnBP." Furthermore, the output

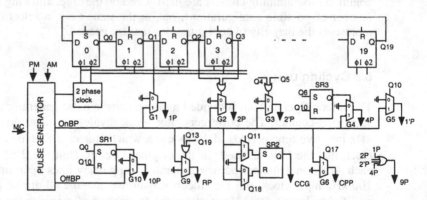

Figure 28: Simplified schematic diagram of the 20-stage ring counter that generates
the fundamental pulse sequence of Fig. 6

of stage-10 also controls MUX "G5" which creates 1'P. The 9P pulse train can be generated by combining 1P, 2P, 2'P and 4P with an "OR" gate. The Central Programming Pulse, CPP, is generated by taking the output of stage-17 to control MUX "G6," and RP is generated using "G9" which is switched on when the ring-counter is at stages 13 and 19. The Carry Clear Gate, CCG, is created by setting another SR latch, "SR-2" during stage-11, resetting it during stage-18 and taking the output of the latch as the desired signal. Finally, 10P is generated by setting latch "SR1" during stage-0, thus enabling MUX "G10" and passing "OffBP," and resetting it at stage 10. The output of "SR-1," will remain high until the latch is reset when the ring-counter reaches stage 10. In this way, between the time when "SR-1" is set and reset, "G10" is turned on to allow "OffBP" through exactly 10 times.

To summarize, the Cycling Unit takes a 500 kHz master clock and divides it down to two 100 kHz signals, "OnBP" and "OffBP." These two signals are then gated by the outputs of a 20 stage ring-counter in conjunction with SR latches and special MUXes. In this way, all of the Cycling Unit pulse trains are created as shown in Fig. 6.

6.6 Layout of the ENIAC-On-A-Chip

The layout of the ENIAC chip is shown in Fig. 29. The chip contains all the functional units of the original ENIAC and a limited number of programmable digit and program lines. It was designed and fabricated in a 0.5 mm single poly-silicon, triple metal, nwell CMOS process (HP CMOS14TB) through MOSIS.[+] The chip measures 7.44 mm by 5.29 mm and contains 174,569 transistors.

7. The Relative computational power of the ENIAC

By today's standards the ENIAC was a very slow computer. It ran at a snail's pace of 100 kHz, and took 20 cycles to perform one addition/subtraction operation, a period known as the Addition Cycle. To multiply, the ENIAC needed 14 Addition Cycles. Thus, the ENIAC could do 5000 additions/subtractions and 357 multiplications per second. A modern state-of-the-art microprocessor typically runs in the hundreds of megahertz range. It is interesting to note that, like the ENIAC, the clock rating on the modern microprocessor, depending on whether it is pipelined or not, does not necessarily mean that it is the rate at which a computation is carried out.

[+] MOSIS (Metal Oxide Semiconductor Implementation Service) is a low-cost and low-volume production service for VLSI circuits, supported by the National Science Foundation and DARPA.

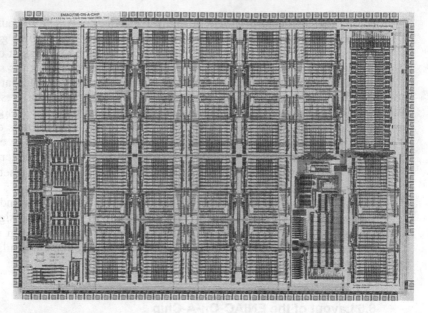

Figure 29: Layout of the ENIAC-On-A-Chip (7.44 mm × 5.29 mm)

The speed with which today's computers perform calculations is an indication of just how far computer technology has come since the ENIAC was first introduced. We will attempt to compare the relative speed of a modern microprocessor with the ENIAC, but at the risk of comparing apples to oranges. The comparison will, therefore, focus on the most basic of commonalties between the two: fixed-point addition/subtraction and multiplication.* While this restriction would render the modern microprocessor less than useful, this is what must be done in order to make the comparison meaningful. Moreover, let us further idealize the situation by making the following assumptions:

1. The microprocessor runs at a sustained rate one instruction per cycle. For a pipelined machine this means that latency has past and all instructions are such that there are no data hazards.
2. The microprocessor has a multiplier on the same die that can perform multiplication in one clock, with or without pipeline.

---

* Although both the ENIAC and the modern microprocessor are capable of performing divisions, this operation involves many addition, subtraction and multiplication operations as intermediary steps and is more algorithmic than purely computational, hence it will be ignored in the comparison.

3. The microprocessor is not running an operating system, and the only program it runs is our test program, so that it is never interrupted from execution of our test program.
4. Only one arithmetic operation is performed in one execution cycle.

It is well known that a machine's performance depends on the mix of instructions in the program. In this case, our performance calculation is quite simple since we only have 2 types of instructions. For a program with 20000 additions and 9500 multiplications, the ENIAC will run in roughly 30.61 seconds. On the other hand, a modern microprocessor running at, say 233 MHz, will only take 126 μs, or about 240 thousand times faster. Put another way, what takes the ENIAC a year to calculate will take a modern 233 MHz microprocessor only about 2 minutes.

It should be mentioned that programs could be parallelized to take advantage of the inherent multiplicity of components in the ENIAC that allows the simultaneous execution of many instructions, which will speed up computations. Modern superscalar microprocessors also have independent identical units which speed up program execution. It is not very meaningful to include this aspect in the comparison, since it does not increase the degree of contrast in the relative computational power more than we have already shown.

8. Conclusions

The reconstruction project of the ENIAC has been an interesting journey into the history of computing and technology. The project gave the design team a first-hand appreciation and understanding of the workings of the ENIAC, from the architectural level down to the functional, programming, and circuit levels. It is probably the first time that such a comprehensive study of the machine itself has been done since the ENIAC was constructed. It quickly became clear to the design team how complicated the ENIAC really was, and how creative and resourceful the ENIAC engineers were in putting such a large-scale machine together in such a short period of time. To be sure, it was the state of urgency created by World War II that provided the driving force for the ENIAC project. The emergency "fostered a spirit of cooperation and willingness on the part of the ENIAC engineers to subordinate their creative impulses."[19] Without this pressing sense of immediacy, the vision which emerged at the Moore School of Electrical Engineering to build the ENIAC would have had no chance of being funded, it would have been dismissed as too risky. Even so, the ENIAC project had its detractors from major research institutions. Convinced of the importance to the war effort of an electronic calculating machine, it was the engineers and the administration who were ultimately responsible for the success of the ENIAC project. At the same

---

[19] Stern, n. 5 above.

time, the time pressure of the war constrained the ENIAC engineers to use proven ways of architectural and circuit design. The sponsoring agency wanted a machine that was fast, accurate and that would be operational in the shortest possible time. This prevented the engineers from doing research on new methods of design and exploring architectures that would have eliminated some of the ENIAC's shortcomings, as pointed out in one of the reports.[20] The report addresses the desirability of automatic set-up of the program and preparation of the results. In addition, it is argued that in electronic machines "serial operation" is preferred, provided the components are fast enough. The advantages of a serial machine over a parallel one are the reduced hardware, increased reliability and ease of programming, as mentioned in the report.

Notwithstanding its shortcomings the ENIAC was an extraordinary accomplishment that ushered in the electronic computer age.[21] The success of the ENIAC convinced the industrial, scientific and military communities that digital electronic computing was not only feasible, but also desirable from the point of view of speed and accuracy. The fact that the ENIAC was a large-scale, general purpose machine meant that it played a key role in a variety of areas and provided solutions to problems which were, up to that point, beyond the scope of any known method. One of the first problems programmed on the ENIAC was a series of calculations of thermonuclear reactions associated with a project at the Los Alamos Research lab in 1945. According to a Newsweek article, the project ran 2 hours on the ENIAC (2 weeks including set-up time) and would have taken 100 man-years.[22]

Although the ENIAC was not a stored-program computer, it was more than a powerful calculator. It was capable of performing a "wired" (in effect programmed) sequence of instructions, storing intermediate results, reading and printing data, and, most importantly, of changing its course of computation based on previous results. It is this ability to take an alternative course of action that distinguishes the ENIAC from just a very powerful calculator.

The ENIAC implementation in VLSI highlighted some of the disadvantages associated with the ENIAC architecture, in particular, that of the data-flow architecture which requires many different connections and settings. Providing enough transistors for all possible connectivity patterns would have consumed prohibitively expensive silicon real-estate. It was hard to incorporate the same degree of flexibility of adding cables and using special adapters that the ENIAC had. In the chip implementation, we sacrificed some of the programming flexibility for reasons of economy and design practicability. Since parallel programming was not used extensively, the chip uses the sim-

[20] Eckert, Mauchly, Goldstine, n. 8 above.

[21] D. P. Winegrad, "The ENIAC – The Age of Information Begins..." in *Eniac a Tribute – 40th Anniversary Booklet,* University of Pennsylvania (Philadelphia, 1986).

[22] "Answers by Eny – All-Electronic Super Calculator Is a Whizz at Super Problems," *Newsweek,* Feb. 18, 1946, p. 76.

plification that only one programming line will be needed at any time. With the exception of the high-speed multiplier, a maximum of two 10-digit lines were needed at any one time. The chip tries to obtain a compromise between full programming flexibility and minimal chip area by using a rapid programming scheme. This reduces the number of physical lines on the chip and provides the number of "virtual" lines required by a program. Except for the implementation of the digit and program trunks, the portable function tables, the card reader and printer, the ENIAC-On-A-Chip contains all of the building blocks of the original ENIAC. Setting-up the ENIAC-On-A-Chip to run an application involves loading a configuration file into the chip's local memory that stores the setting of the switches and the interconnections. This is similar to what is currently done in Field-Programmable-Gate-Arrays (FPGA).

Comparing the original ENIAC with its chip implementation is hard to do, as it is not always clear what the frame of reference is in which to compare the two implementations. However, it is nevertheless instructive to compare some of the physical characteristics of the two implementations. The ENIAC contained 17,468 vacuum tubes, 70,000 resistors, 10,000 capacitors, 7,200 crystal diodes, and 6,000 switches, it had a footprint of about 33 m × 1 m, occupied a room of 170 square meters, dissipated about 140-174 kW and weighed 30 tons. In contrast, the chip realization contains 174,569 transistors, measures 7.4 mm × 5.3 mm (the PGA package measures 3.6 cm by 3.6 cm), dissipates a few Watts (depending on how many units run in parallel and the clock speed), and weighs a few grams. Also, in terms of power requirement the comparison is striking. In addition to the AC power for the heaters of the tubes, the card reader and the card punch, the ENIAC required 78 different DC voltage levels to power 10 different types of vacuum tubes. The power equipment was housed in 7 panels which were separate from the ENIAC's 40 panels. Special ventilating equipment consisted of an elaborate system of fans and blowers to keep the temperature inside the panels below 46° C. In contrast, the chip needs only one power supply of 5 V (or lower). The clock frequency used in the ENIAC was 100 kHz, while the one on the chip can easily run at 50 MHz or higher.

The above comparison illustrates the tremendous changes in technology which have occurred over the last 50 years. The replacement of the vacuum tubes by the transistor and the integrated circuit was the main reason for decrease in size, weight and power. Besides advances in technology there have also been fundamental changes in computer architecture. The ENIAC was a parallel, data-flow machine in which program memory was locally stored at each unit. Numbers were represented as fixed point decimal numbers. Current day computers are basically serial in nature and programs are stored in memory, separated from the processor. However, it is interesting to note that there has been an increased interest in parallel architectures as a way to further increase the operating speed. Also, the emergence of Time-Switched FPGA (this is the name given to FPGAs whose programmable cell-blocks can be altered), in which the configuration and interconnections between hardware units can easily and quickly be changed make a data-flow architecture an interesting alternative for specific types of applications.

Acknowledgments

The authors would like to thank Professor F. Ketterer and the many students who contributed to the ENIAC-On-A-Chip project, including R. Tong, D. J. Yoon, M. Feng, F. Chew, W. Wong, D. Seider, M. Zeno, B. Santos and C. Helfinstine. We would like to acknowledge the help of Mr. A. Akera and T. Rauenbush for getting us access to the ENIAC manuscripts, blueprints and other documentation. We are also grateful to Dean G. Farrington and Prof. S. Rabii for their continued support and enthusiasm. The chip was fabricated through the MOSIS Service with the support of the NSF.

JAN VAN DER SPIEGEL'S research interests are in biologically based sensory information processing systems, neural computers, micro-sensor technology, and low power, low voltage analogy integrated circuits. He has published over 120 papers and book chapters. He is the recipient of the UPS Foundation Distinguished Education Term Chair (1998), the Bicentennial Class of 1940 Term Chair, the Christian and Mary Lindback Foundation Award for Distinguished Teaching in 1990, the S. Reid Warren Award for Distinguished Teaching in 1987, and the Presidential Young Investigator Award in 1984. He is a senior member of the IEEE and editor of N&S America for Sensors and Actuators A. Van der Spiegel received his Masters and Ph.D. in Electrical Engineering from the University of Leuven, Belgium, in 1979. He is currently Professor and Chairman of the Department of Electrical Engineering at the University of Pennsylvania.

The Institute for Advanced Study Computer: A Case Study in the Application of Concepts from the History of Technology

William Aspray

Abstract. This paper uses the history of the Institute for Advanced Study computer built in Princeton, New Jersey, between 1946 and 1952 as a case study. The case study introduces computer practitioners who might be interested in writing about the history of early computers to some of the concepts that have been used by business and other kinds of historians of technology. These concepts include organizational mission, project objectives, organizational buy-in, organizational capabilities, technology transfer, value and impact, technological and other obstacles, first-mover advantages, and organizational continuity.

1. Introduction

The Institute for Advanced Study computer, built in Princeton, New Jersey, in the late 1940s and early 1950s, was one of the most important early computers. It deserves this recognition on account of the scientific work that was conducted on it, the people who were trained upon it, and to some extent for the computer design principles that it embodied. Several historical accounts of this machine already exist,[1] and it is not the purpose of this paper to duplicate this literature.

There is a vast body of literature, written during the past twenty years by computer practitioners, historians, and journalists, about the computers built in the 1940s and 1950s. Indeed, more has been written about this aspect of computing history than any other. Despite the considerable attention these

---

[1] See, for example, William Aspray, *John von Neumann and the Origins of Modern Computing*, Cambridge, MA: MIT Press, 1990; Julian Bigelow, "Computer Developments at the Institute for Advanced Study," In N. Metropolis, J. Howlett, and Gian-Carlo Rota, eds. *A History of Computing in the Twetieth Century*, pp. 291–310. New York: Academic Press, 1980; Herman Goldstine, *The Computer from Pascal to von Neumann*, Princeton: Princeton University Press, 1972.

early computers have received, our understanding of them is incomplete. Raúl Rojas, for example, has recently contributed the first detailed analysis of the working of Konrad Zuse's early computers.[2] Other papers in this volume have similar aspirations for other early computers. Rojas's book and these papers give what might be called a "technical reassessment" of early computing machines.

This paper has a somewhat different purpose. It intends to identify for computer practitioners interested in writing about the history of early computers some concepts that have been used by business and other kinds of historians of technology, and to suggest through one brief case study how these concepts can be applied to the study of early computers. The concepts that are discussed here include organizational mission, project objectives, organizational buy-in, organizational capabilities, technology transfer, value and impact, technological and other obstacles, first-mover advantages, and organizational continuity. Although these concepts were developed to understand how companies functioned, they can also be applied to other organizations that create technology, such as the Institute for Advanced Study.

Many existing accounts of early computers touch on one or more of the historical concepts addressed in this paper, but a deeper understanding of these computers might result from paying more explicit attention to these concepts. Thus this paper is as much about a conceptual vocabulary for rewriting the history of early computing machines as it is about the IAS computer. The length of this paper does not permit a more complete enumeration of the concepts being applied by historians of technology, nor a detailed treatment of any one of those concepts that are addressed. Thus this paper is intended to be merely suggestive of a way to think about computing machine history.

2. Background

Before turning to the discussion of these concepts, it is useful to give a brief history of the IAS computer project and a brief biography of the project director, John von Neumann. Von Neumann (1903–1957) grew up in Budapest, Hungary, in an upper middle class family. His formidable intellectual abilities were recognized at an early age, and he was given an excellent education in Hungary's best private school. His first love was mathematics, and he wrote his first publishable paper as a teenager. He completed a doctorate in mathematics at a university in Budapest, while simultaneously pursuing an undergraduate degree in chemical engineering at the technical university in Zurich so as to assure his family that he would learn a practical skill. Upon

---

[2] R. Rojas, ed., *Die Rechenmaschinen von Konrad Zuse*, Berlin: Springer-Verlag, 1998.

graduation, for several years he held postdoctoral positions in mathematics in Germany, where he came under the influence of the distinguished mathematician David Hilbert. During his few years in Germany he developed the mathematical foundations of quantum theory, wrote some early papers on game theory and economic theory, and worked on Hilbert's program to secure a logical foundation for mathematics.

With his international mathematical reputation already secure, a lack of professorships available in Germany, and looming political problems in Europe, von Neumann decided to emigrate to America in 1930. He first held a visiting position in the mathematics department at Princeton University. Then he became the most junior of the five faculty members originally appointed to the Institute for Advanced Study, a think tank established in Princeton in 1932, which rented space in the university's mathematics building until it built its own buildings across town in 1939.

Von Neumann continued his active research program in pure mathematics in Princeton, but beginning around 1937, through the influence of his colleague Oswald Veblen, he began to take an interest in applied mathematics, especially applications to war-related problems. During the war he worked for several different military organizations, most importantly for the national laboratory in Los Alamos, New Mexico. The computational problems associated with a triggering device for the atomic bombs being developed at Los Alamos led von Neumann to the computing projects at the University of Pennsylvania. By the time he joined the Penn group as a consultant in 1944, the ENIAC design had been frozen, development was advanced, and discussions had begun on the design of a successor computing device, the EDVAC. Von Neumann joined these group discussions about the design of the EDVAC and in 1945, based partly on these discussions, he wrote the Draft Report on EDVAC, which introduced the basic description of the stored-program computer.

By early 1945, even before the war was over, von Neumann had decided that the computer could be a breakthrough technology for scientific research and that he wanted to build a computer for these purposes. He secured permission to build a computer at the Institute for Advanced Study, with funding at first from the IAS, Army, and Navy (and later from the Air Force and the Atomic Energy Commission), with in-kind support provided by RCA and Princeton University. The project began in 1946, and the computer reached working order in 1952. The IAS continued to operate the computer until 1957, when it was sold to Princeton University. The university operated it for another three years, before donating it to the Smithsonian Institution for public display as an historic artifact.

3. Organizational mission

Some historical accounts of early computers unfortunately pay little or no attention to the organizations in which the computers were built. Computers at this time were large, capital items that consumed significant amounts of organizational resources (funding, space, staff). The decision to build or buy a computer was generally an important decision for an organization, made (or at least approved) at the highest levels of management. When good management practice was followed, the decision involved careful consideration of how the computer would fit with the mission of the organization. But no matter how good the management, the organizational mission typically shaped the kind of project and whether and how it succeeded.

The organizational mission of the Institute for Advanced Study was to pursue world-class research in a few selected areas of study. The areas selected for study in the 1930s were mathematics and theoretical physics. Faculty were free of all normal university duties such as committee work, administration, teaching, and student advising so that they could concentrate on their research. Subjects such as biology, although they offered promise of fundamental breakthroughs, were avoided at the Institute because they involved a large research support investment, such as laboratories and technicians, which were both costly and thought to detract from the protected environment in which the elite faculty did its scientific thinking. Thus it was not at all clear that a computer project, which required laboratories, technicians, programmers, and operators, and which led to the building of an artifact that could be used for business and military purposes, would be welcome at the Institute. Von Neumann politicked assiduously to convince the director and trustees of the institute that the computer would be a breakthrough technology for science, especially in the study of nonlinear physical phenomena – an area of scientific research that he noted had been largely stagnant since the nineteenth century.

4. Project objectives

It is useful to examine not only the objectives of the organization, but also the objectives of the computer project itself. In an entrepreneurial start-up, there are typically only a small number of projects, and they tend to have objectives that are closely aligned with the mission of the organization. But in larger, more established organizations a project might have many different possible objectives, not necessarily closely tied to the organization's mission. The commitment to a project and the resources assigned to it will depend on the project objectives. A computer manufacturer is likely, for example, to provide greater support to a project that results in a product that is a key offering than it is to satisfy the intellectual interests of a staff member. The objectives of the project also set the standards for success.

The main objective of the IAS computer project was to explore the value of the computer as a scientific instrument. This objective shaped both the machinery that was built and the way in which it was used. In the *Draft Report on EDVAC* von Neumann had advocated building synchronous computers with serial architecture and no floating-point. These choices were believed to simplify the overall design and construction of the hardware at a time when nobody had yet built a stored program computer.

But in the IAS computer project, von Neumann did not follow his own advice.[3] He returned to first principles to decide what kind of machine would be most suitable for scientific research. He chose to use parallel memory and parallel arithmetic, and to make the machine asynchronous. All of these choices were made to increase the speed of the machine. It proved to be a major task to work out the circuitry for this design, but once it was done, the machine was indeed fast (so long as it was supported by a fast random-access memory). The computing community was somewhat surprised that this design led to a physically small machine (by standards of those days), with fewer tubes and a simpler control mechanism than in the EDVAC. Speed of basic operations and size of the memory were determined both by the available components and by von Neumann's calculations about the capacity a computer would need to have in order to solve certain classes of scientific problems.[4]

5. Organizational buy-in

An organization might allow a project to be carried out without committing itself to the project fully – intellectually, emotionally, or in terms of the resources that are allocated to it. Such a project might be tolerated because it seemed as though it had potential to be important to the organization, or it was strongly desired by some sector or key individuals in the organization; but it was not yet seen as core to the mission of the organization. Projects without strong organizational buy-in often have difficulty in succeeding. Resources for carrying out the project are likely to be meager. Almost every project has some stage at which it experiences problems, and it is difficult to acquire the extra resources or time to overcome these problems if the management does not have faith in the proposal. Management is quick to terminate such projects, and the proponents are likely to be disappointed and may even become bitter.

---

[3] A good account of this is given in Michael R. Williams, *A History of Computing Technology*, Englewood Cliffs, NJ:Prentice Hall, 1985, pp. 353–359.
[4] This issue is discussed at length in Aspray (1990), chapter 3.

In order to succeed, von Neumann needed to have organizational buy-in at four places: from the IAS board of trustees, the IAS director, the other faculty, and the financial and technical partners. Von Neumann worked first on the director, Frank Aydelotte. Von Neumann was a productive and valued member of the faculty (of which there were then only six), and Aydelotte worked hard to see that the faculty was happy and well supported. There was a strong belief that, as some of the best scientists in the world, the faculty should be given great freedom in choosing the projects they wished to conduct. Thus, the culture was supportive of von Neumann's decision to pursue a study far removed from that typical in mathematics. However, von Neumann recognized that his computer project would seriously disturb the think-tank atmosphere of the Institute. He solicited support from his colleague Oswald Veblen, who had been instrumental in locating the Institute in Princeton and focusing it on scientific rather than medical research.

Von Neumann himself used a "carrot and stick" approach. Aydelotte and the trustees were seeking a breakthrough to more firmly establish the Institute as a world-class operation. While the Institute did have the world-renown Albert Einstein and other distinguished mathematicians on its faculty, it was still a young organization and Aydelotte believed that it needed some new achievements to secure its reputation. Von Neumann convinced Aydelotte that the computer might have this effect. He also made it clear that he was determined to build a computer, emphasizing this point by soliciting offers of faculty positions from Chicago, Columbia, and MIT. With Veblen and Aydelotte's encouragement, the Trustees voted to approve the project and provide $100,000 of initial funding.

During the course of the project, Robert Oppenheimer succeeded Aydelotte as director of the Institute. Oppenheimer was familiar with the importance of advanced scientific tools from both his university research as a physicist and his Los Alamos experience during the war. Not surprisingly, he seems to have had no major objections to the presence of a computer project at the Institute. His objection was instead to some of the uses of the computer. Having renounced at the end of the war any further personal role in the creation of weapons of mass destruction, he was distressed by the heavy use the Institute computer was receiving from weapons designers at Los Alamos, as well as by von Neumann's involvement in this work. But Oppenheimer adhered to the *laissez faire* practice that let Institute faculty decide for themselves what work they would pursue. Although Oppenheimer and von Neumann were not friends, and although they were diametrically opposed politically, they displayed strong mutual respect for one another as professionals. For example, von Neumann set aside his political beliefs in order to testify in Congress on Oppenheimer's behalf, when Oppenheimer was in danger of losing his security clearance.

Veblen and perhaps one or two of the other Institute faculty were supportive of the computer project. However, that support was mainly vested in support for von Neumann, not in the intellectual merits of the project itself.

This became clear in the later years of the project. Beginning around 1952, von Neumann began to spend more time in Washington, D.C. than in Princeton. He was working closely with the Atomic Energy Commission and eventually became one of its five commissioners. He was also doing active consulting for the Air Force on intercontinental ballistic missiles and for the predecessor organizations of the National Security Agency on cryptanalysis. He eventually decided to resign his faculty position at the Institute because he believed the intellectual climate was not conducive for his research on scientific computing and because he was eager to escape from a town he had always regarded as provincial. He signed a faculty contract with the University of California, Los Angeles, to become effective once his term at the AEC had ended. He believed that UCLA would be more accommodating to his research. This move would also locate him in his beloved southwestern United States, where the weather, cuisine, and geography were all to his liking. While in Washington, however, he became gravely ill with cancer. He died in 1957, before he could move to California.

As soon as it became clear that von Neumann was not going to return to the Institute, support within IAS for the computer project evaporated. Staff began to leave for other research organizations, and the computer was sold to Princeton University. This computer was rapidly becoming obsolete, but no serious consideration was given to building or buying another computer for the Institute or to continuing the research program in scientific computation.

Von Neumann also had to obtain institutional buy-in from his financial and intellectual partners in the project. The support from the military services and the Atomic Energy Commission was quite strong. It was not at all clear at the end of the war that the federal government would be willing to support a computer project at the Institute. The military was already supporting projects at the University of Pennsylvania, nobody knew how much of a need there was for high-powered computers, and one could question why the government should support a computer devoted to scientific research rather than to military objectives. However, the military leadership had great respect for von Neumann on account of his contributions during the war, his obvious intellectual abilities, his congenial hawkish political views, and his willingness to continue in peace-time to make his extraordinary talents available to the military. It should be remembered that von Neumann was virtually alone among the early computer designers in having already achieved a strong, international scientific reputation, and this no doubt helped in both giving scientific legitimacy to computers generally as well as in gaining support for his own project.

Von Neumann designed a plan whereby the IAS computer would receive substantial financial support from the armed services but would nevertheless be free from doing operational calculations for them. This was accomplished by positioning the IAS computer as the testbed for computer design for machines that the military and energy laboratories would build for their own use in the Cold War. This plan was fully endorsed by the military, and later by the Atomic Energy Commission, and the project was adequately funded throughout its lifetime, even during the early years when progress on the construction of the machine was slow.

The organizational buy-in from the other partners was less strong. This was certainly true of Princeton University. Several Princeton faculty members were supportive of the computer project, notably John Wheeler in physics and Martin Schwarzschild in astronomy. However, the physics department did not supply the electronics expertise that had been called for in the early negotiations. Other Princeton faculty members who had no particular standing at the Institute were not made particularly welcome when they tried to use the machine. When the Institute sold the computer to the university, the university believed that it had received less than full disclosure about the state of the computer and the costs of maintenance and operation. In fact, these costs proved to be so great that several years later, when an electrical storm temporarily disabled the computer, the university used this event as an excuse to decommission the machine.[5]

RCA's buy-in was also weak. The company was supposed to build the main memory device for the project. Work proceeded slowly, and RCA management was unwilling to take away resources from its core businesses to support the memory development effort. As a result, the Institute staff had to scramble to find an alternate memory, settling on the Williams tube that had been first developed at the University of Manchester. RCA's Selectron memory ended up being used in only one computer, the Johnniac, a copy of the IAS computer that was built at the Rand Corporation.

6. Organizational capabilities

Much of the literature on early computer developments does not give direct consideration to the capabilities of the organization in which the computer was built. Yet each organization has strengths and weaknesses, which have an effect on both what the organization tries to do and how well it succeeds. One common exercise I give to my students whenever I teach a course on the history of computing is a prosopographical analysis of the computer industry. Rather than looking at individual firms, we create a prosopography (a kind of group portrait) to uncover the strengths and weaknesses – marketing, customer base, technical knowledge, industry knowledge, technical management, capital, research facilities, etc. – of types of entrants to the computer industry – startups, defense contractors, electrical manufacturers, and business equipment manufacturers. While there are certainly important differences between NCR and Burroughs, for example, they share many similari-

---

[5] See Acton, Forman. Oral History Interview. Conducted by Richard Mertz, 21 January 1971, Princeton, NJ. Smithsonian Archives.

ties, such as strong customer bases but weak electronics knowledge, as business equipment manufacturers; and these similarities led them to have somewhat similar experiences as they entered the computer era.

Organizational capabilities to carry out a computer development project were weak at the Institute for Advanced Study. The Institute had to hire an entire staff of engineers, technicians, and programmers. Although there were some talented engineers available at the end of the war who had had experience with electronics, very few people experienced with building electronic or electromechanical computing equipment were available. The shortage of experienced engineers meant that the development phase proceeded more slowly and that all of the craft knowledge of building computers had to be learned as they went along.

There was also a problem with project management. Von Neumann had experience with the EDVAC, as did Herman Goldstine, who was hired as von Neumann's second-in-command to direct the daily operations. But they were both mathematicians, not engineers. Presper Eckert, the principal engineer on the ENIAC and EDVAC projects, was originally offered the job as IAS chief engineer. But the offer was rescinded because Eckert wanted to retain patent rights to the intellectual property he developed, whereas von Neumann insisted on placing the project's intellectual property developments in the public domain. Instead, Julian Bigelow, who had had some limited relevant experience at MIT, was hired as chief engineer at the recommendation of Norbert Wiener. Bigelow did not get along with Goldstine, and Bigelow's perfectionist tendencies slowed progress so much that von Neumann had to replace him with James Pomerene in the midst of construction.

Funding proved to be adequate, but it was a patchwork from various sources, with several periods of uncertain funding during the course of the project. Space of all kinds was inadequate until a new building was constructed. Even then, there were no well-equipped laboratories. This forced a strategy of using off-the-shelf components for in-house construction and contracting out all work on hardware that required development, most notably the main storage unit to be developed by RCA. In the end, IAS staff did have to do on-site research on both storage and input-output equipment. The Institute's buildings were located in an upscale residential community, and the project management was faced with answering public outcries about the safety of having a machine that consumed so much power in their neighborhood.

7. Technology transfer

While historians today know what was going on at each of the various computer projects that were scattered around North America and Europe in the decade following the war, it does not mean that the computer pioneers themselves were aware of what was going on at other sites. Indeed, communication was weak and technology transfer was fairly difficult at the time. At the end of the war, there were no regular professional computing societies, conferences, journals, or formal training programs. Special conferences and sessions on computing at the meetings of other professional groups, such as the Institute of Radio Engineers, began to be held soon after the war ended; and in the early 1950s the first computer journals were formed.[6]

The IAS computer project was an important source of information for the European and North American communities about computer design and usage. Von Neumann and his colleagues Herman Goldstine and Arthur Burks wrote reports on the logical design and programming of computers, which were distributed to all the major computer projects. Progress reports, circuit diagrams, and blueprints from the IAS project also circulated widely. Von Neumann was a frequent speaker about his computer project at professional conferences, and Goldstine and Bigelow occasionally made presentations.

A number of people also came to visit the project. Scientists came from all over the world to spend a few months or a year as visiting scholars doing scientific computations or research on numerical methods. The military services and the U.S. Weather Bureau seconded scientific personnel to the Institute in order to learn about numerical weather prediction practices. A steady stream of scientists, engineers, military personnel, and others briefly visited the Institute to learn more about the project. As the project neared its end date, the staff scattered to various research organizations around the

---

[6] See, for example, William Aspray's introduction *to Proceedings of a Symposium on Large-Scale Digital Calculating Machinery* (1947), jointly sponsored by The Navy Department Bureau of Ordnance and The Harvard University Computation Laboratory, Charles Babbage Institute Reprint Series in the History of Computing, Volume 7, Cambridge, MA: MIT Press and Tomash Publishers, 1985; Eric A. Weiss, Publications in Computing: An Informal Review", *Communications of the ACM* 15 (July 1972), pp. 492–497; Anon., *A World List of Computer Periodicals*, The National Computing Centre Limited, 1970, Manchester, England; Harry Polachek, "History of the Journal *Mathematical Tables and Other Aids to Computation, 1959–1965*", *IEEE Annals of the History of Computing*, Vol. 17, No.3, 1995, pp. 67–74; Charles Concordia, "In the Beginning There Was the AIEE Committee on Computing Devices," *Computer* 1976, pp. 42, 44; Franz L. Alt, "Fifteen Years ACM," *Communications of the ACM* 30 (October 1987), pp. 850–859; Lee Revens, "The First 25 Years: ACM 1947–1962," *Communications of the ACM* 30 (October 1987), pp. 860–865; Anita Cochran, "ACM: The Past 15 Years, 1972–1987," *Communications of the ACM* 30 (October 1987), pp. 866–872; Eric A. Weiss, "Commentaries on the Past 15 Years," *Communications of the ACM* 30 (October 1987), pp. 880–883; Eric A. Weiss, "Publications in Computing: An Informal Review," *Communications of the ACM* 15 (July 1972), pp. 491–497; Morton M. Astrahan, "In the Beginning There Was the IRE Professional Group on Electronic Computers," *Computer*, December 1976, pp. 43–44; Walter Anderson, "The Middle Years," *Computer* (December 1976), pp. 45–53.

world. For example, Jule Charney joined the MIT faculty and formed its numerical meteorology program. Goldstine left to become director of IBM's new T.J. Watson Research Laboratories, and Pomerene went with him to become one of IBM's leading computer engineers and a future IBM Fellow. Burks formed an academic program in computing at the University of Michigan. Gerald Estrin moved to Israel, where he built that country's first computer. These are only a few examples. Together with Penn, Harvard, and MIT, the Institute became one of the leading sources of early computing personnel.

8. Values and impacts

Any particular computer has little value in and of itself. Its value comes instead from its uses, the proof of principles that it demonstrates, the design ideas that it embodies, the theory that is developed in connection with the machine, the people who are trained upon it, and its role as a cultural icon. For example, the ENIAC was perhaps most important as a proof of principle that a large-scale electronic calculating machine could be built and operated successfully, and the UNIVAC may have had its greatest value as a cultural icon, as indicated by the popular use of the word "Univac" to refer to all computers in the 1950s, much like some people speak today of "Xerox machines" instead of "photocopiers." Of course, the meaning of an artifact varies from individual to individual and from organization to organization. The ENIAC was an important tool to the weapons designers at Los Alamos, not just a proof of principle; and the UNIVAC was an important business commodity for Sperry Rand, not just a cultural icon.

The IAS computer was valuable in several ways. The numerical analysts, computer engineers, and computational scientists (to use modern terminology) who were trained on the IAS computer project and who became early leaders in the computing field have already been discussed. Various scientific computations and theories were developed based upon use of the IAS computer. Martin Schwarzschild, for example, did important work on the internal composition of stars. But numerical meteorology was regarded by von Neumann as the crucial test of the scientific value of the IAS computer, and the results were extraordinary. While the IAS computer was barely powerful enough to calculate a numerical weather forecast, the team of meteorologists and mathematicians working on the IAS computer project developed both a theory of numerical meteorology and a practice of numerical weather forecasting. These rapidly spread around the world – to universities and research centers for further study and to the military weather bureaus and U.S. Weather Service, where numerical forecasting techniques were rapidly put into practice. Some interesting early work was done on numerical analysis, but the IAS was only one of several places redefining this discipline for use with the computer. Von Neumann's own work on automata theory showed considerable promise, but it was incomplete at the time of his early death. The parallel arithmetic and storage architecture used in the IAS machine was copied on many computers of the 1950s, although the contributions to computer engineering were undoubtedly greater at some other places, notably MIT. It is not surprising that the IAS project contributed more to scientific computing than to computer engineering because that is how von Neumann defined the project.

9. Technological and other obstacles

The current way in which many historians understand technology is in terms of systems that include not only technological artifacts and knowledge, but also – seamlessly – the people and social organizations that are associated with them. Thus it is appropriate not only to consider technological obstacles to the achievement of some project mission, but other human and institutional factors.

What were the main obstacles to achieving the objective of the IAS computer project, of demonstrating the value of the computer as a scientific instrument? One was to find an adequate device to serve as the main memory of the computer. This was a problem that all early computer designers faced. The problem was perhaps more difficult at the IAS because of the desired speed characteristics and the parallel architecture that had been chosen to accommodate scientific computation. It was also more difficult because of the lack of adequate laboratory facilities and experienced technical staff at the Institute, and the limited buy-in from RCA. The storage problem was overcome, as indicated above, by adopting the Williams tube technology first developed in Manchester.

A second obstacle was the reliability and the maintainability of the computer. It was a struggle to get a computer working at all. The IAS staff did not have the time, the funding, the facilities, the expertise, or the motivation to build a computer that was highly reliable and easily maintained. Many of the one-of-a-kind computers of the early years had reliability and maintainability problems, and the IAS computer was no exception. Once it was regarded as finished and had been placed in full-time operation, the IAS computer was operable about 70 percent of the time. Maintenance remained a problem. The tool most commonly used by the engineering staff after the computer was commissioned was wire-cutters. The computer was not laid out in a way that made maintenance easy, and generally when a vacuum tube or some unit failed, the engineers had to cut away wires from some units to gain access to the inoperable equipment.

A third obstacle was adequate numerical methods and algorithms. Numerical methods had traditionally focused on issues that were of concern to calculators using pencil and paper or desk calculators. These included methods for solving small systems of linear equations and inverting small matrices, approximating roots of polynomials, and solving a few kinds of ordinary differential equations. Concern was traditionally given to truncation errors in numerical methods because of the inability to carry out large numbers of calculations. The numerical issues the IAS scientists addressed had a different emphasis: finding methods of inverting matrices and solving linear systems that were efficient for large systems, methods for solving partial rather than ordinary differential equations (especially nonlinear PDEs), accumulation of round-off errors instead of truncation errors, and generating an adequate supply of random numbers for Monte Carlo methods.

A fourth obstacle was producing an adequate theoretical understanding of the capabilities and possible organizational structures of computing equipment. Presper Eckert was the first of a long line of engineers who have questioned the value added by von Neumann's use of the McCulloch-Pitts neural net concepts in the description of the stored program concept, as given in the Draft Report on EDVAC. Von Neumann used this terminology to focus on the logical design elements of computers, separated from any engineering implementation. He continued this line of inquiry on the IAS computer project in a series of studies on the theory of automata. These studies compared the computing process and power of the computer with the human nervous system, and explored such questions as how complex an automaton must be in order to learn or self-replicate. Von Neumann achieved promising but not extraordinary results in this area before his death, and for many years these theoretical studies had only limited significance for computer design. In recent years, one line of theoretical inquiry begun by von Neumann (cellular automata) has been relevant to the construction of neural nets and massively parallel computers.

10. First-mover advantages and organizational continuity

Business historians have noted the considerable advantage a firm has in being an early entrant to a market. These "first movers" have the opportunity to define the characteristics of the products that are offered in this market, build a customer base, hire the most talented employees, build a supplier network, and take marketing advantage of the absence of other competitors. In order to build upon first-mover advantages, organizations need organizational continuity, a willingness to commit to building on their early start.

How do the concepts of early entry, first-mover advantages, and organizational continuity apply to the Institute for Advanced Study? It is clear that the Institute was not organizationally committed to computers and that it terminated the computer project soon after von Neumann's departure, never to enter this field again. Although the Institute and Princeton University are two separate institutions, there were many ties between them. The Institute was housed on the university campus for most of its first decade of existence and then moved to premises only a half-mile from campus. Three of the five original Institute faculty members were drawn from the university mathematics department. Colloquia and facilities were open to graduate students and faculty from both institutions. Original plans had called for the university physics department to have a hand in building the computer (although there is little evidence that this happened). A few university faculty members were active users of the Institute computer. The university purchased the computer from the Institute and operated it for its final three years (without moving it from the Institute premises). Thus it is reasonable to ask what advantages Princeton University, which did have a continuing interest in computing and which is today moderately strong in computer science, gain from the early activities at the Institute for Advanced Study.

The answer is that Princeton is much like the other major research universities in the United States that became involved with computing in the 1940s: Harvard, MIT, Columbia, and Pennsylvania.[7] With the exception of MIT, none of these institutions took advantage of its early lead to become and continue to be a leading school in computer science. Even MIT, which has been a leading institution in the computing field since the 1920s, had a sharp discontinuity in its computing programs just after the war ended, abandoning its support for Samuel Caldwell and his operation in favor of work by Jay Forrester on real-time digital computing.

In Princeton, once the IAS computer project was terminated, almost the entire computer staff left to build up computer programs at other institutions. The only principal remaining at the Institute was Julian Bigelow, the original chief engineer, but he did little substantial work on computers after the project was terminated. Princeton University hired none of the people who had worked on the IAS computer project. Instead, the university left the development of its computing program to Forman Acton, a junior faculty member in applied mathematics. Acton had tried to use the IAS computer during its hey-day. While he had not been prevented from doing so, he had been discouraged by the fact that nobody at the Institute would give him the extra assistance they gave to the preferred users in learning the craft knowledge that is essential in operating an experimental machine.

Acton moved into the electrical engineering department at the university to build up a computer science program, but it did not have its first real strength until Edward McCluskey was hired from Bell Laboratories. In the 1960s and 1970s Princeton attracted a number of strong computer scientists, including John Hopcroft, Jeff Ullman, and Peter Denning. But McCluskey and Ullman

---

[7] The issue of early entry of universities in the United States into computer science is explored in depth in a forthcoming book by this author on the history of academic computer science.

left for Stanford, Hopcroft for Cornell, and Denning for Purdue because computer science could not seem to grow strong for a variety of reasons, including the hostile environment of the electrical engineering department. It was not until the 1980s that Princeton was able to build up a strong and stable computer science program.

11. Conclusions

This paper is not intended to be a history of the computer project at the Institute for Advanced Study in Princeton. Instead it uses this computer project as a case study to illustrate concepts employed by historians of technology which this author believes can be profitably employed in writing the history of computing. The list of concepts described in this paper is by no means complete. It barely touches, for example, on many of the concepts that have been employed by the social constructionist school. Some concepts are of course more applicable than others in a given situation, but a judicious application of these concepts would undoubtedly enrich the scholarship on the history of computing. It will also build connections between the study of the history of computing and the study of the history of technology more generally. Computer history has been studied in a somewhat insular manner. This has meant it has not attracted as much historical talent as the subject merits, and also that the history of computing has not been integrated as much into the teaching and research in the history of technology or general history as it might have. Scholars of the history of computing have an opportunity to change this, if they choose to do so.

WILLIAM ASPRAY is Executive Director of Computing Research Association, a non-profit organization that serves the computing research community in North America. He holds a BA in philosophy and an MA in mathematics from Wesleyan University, and a Ph.D. in history of science from the University of Wisconsin. He has previously held teaching positions in mathematics, computer science, history, and history of science at Williams, Harvard, Minnesota, and Rutgers; and he has worked for the Charles Babbage Institute for the History of Information Processing and the IEEE Center for the History of Electrical Engineering. His books include *Computer: A History of the Information Machine*, co-authored with Martin Campbell-Kelly (1996), *and John von Neumann and the Origins of Modern Computing* (1990). He is currently working on a history of academic computer science in the United States.

"Nothing New Since von Neumann": A Historian Looks at Computer Architecture, 1945–1995

Paul Ceruzzi

Abstract. It is sometimes said that computer architecture has hardly changed since the publication of the "First Draft of a Report on the EDVAC" by John von Neumann in 1945. Obviously computing has advanced by great strides since 1945, a fact that might call that statement into question. Or perhaps not. This paper examines to what extent such a statement is true, and to what extent its assertion blinds one to genuine advances in computer architecture since 1945. It concentrates on a few selected events and developments, and therefore is not a comprehensive history of the evolution of computer architecture since 1945. These events include: a) the emergence of a stabile, "mainframe" architecture by 1960, b) the IBM System/360 and its architectural innovations, c) the development of the minicomputer, d) the microprocessor, e) RISC.

1. Introduction

What follows is a history of the evolution of computer architecture, and a "road map" for historians who may at times feel lost in the "trackless jungle," to use Mike Mahoney's phrase, of all that has happened in computing since 1945. This history will be familiar to many of you. I propose it as a middle ground between the two contradictory truths in the history of computing. These are, on the one hand, that computing has progressed by orders of magnitude since 1945 – a progress unmatched by nearly any other technology in human history. On the other hand, that computer architecture continues to feel the influence of the 1945 report on the EDVAC, by von Neumann, and the 1946–47 reports on the Institute for Advanced Studies Computer (IAS), by Burks, Goldstine, and von Neumann.

2. First plateau: 1955–1965

By the end of 1960 there were about 6,000 general-purpose electronic computers installed in the United States. Nearly all of them were descendents of the EDVAC and IAS computer projects of the 1940s, where the concept of the stored program first appeared. But in many ways there were significant modifications to those designs. Some of these innovations became selling points as different vendors sought to establish their products in the marketplace. The most important of them are summarized here.

Word length – core memory

The introduction of reliable core memory made it practical to fetch data in sets of bits, rather than one bit at a time as required by a delay line. For a computer doing scientific work, it seemed natural to have this set correspond to the number of digits required for a typical computation – say, from seven to twelve decimal digits. That meant a block size, or *word length*, of 30 to 50 bits. Longer word lengths were preferred for scientific calculations but increased the complexity and cost of the design.

Computers intended for commercial use did not require handling numbers with many digits. Quantities of money handled in the 1950s seldom exceeded a million dollars, and two digits to the right of the decimal place were sufficient. Business-oriented computers could therefore use a shorter word length, or they could use a variable word length, if there was a way to tell the processor when the end of a word was reached. The IBM 702, IBM 1401, RCA 301 and RCA 501 all had variable word lengths, with the end of a word set by a variety of means. The 1401 used an extra bit appended to each coded character indicating whether it was the last one of a word or not, the 702 used a special character that signified that the end was reached. Although popular in the 1950s, computers with variable word lengths fell from favor in the following decades and are no longer common.

Register structure

Processing units of early computers contained a set of circuits that could hold a numeric value and perform rudimentary arithmetic on it – usually nothing more than simple addition. This became known as an *accumulator*, since sums could be built up or "accumulated" in it. Another set of circuits made up the *program counter*, which stored the location of the program instruction that the processor was to fetch from memory and execute.

The typical cycle of a processor was to fetch an instruction from memory, carry out that instruction on data in the accumulator, and update the program counter with the address of the next instruction. In the simplest case the program counter was automatically incremented by one (hence the term "counter"), but branch instructions could specify that the counter point to a different memory location.

A computer program orders the processor to perform arithmetic or logic (e.g. add, compare, etc.), tells the processor where the relevant data are to be found, and tells it where to store results. As with the sequential fetching of instructions, the processor often requires pieces of data that are stored close to one another in memory. A program that performs the same operation on such a list of data might therefore consist of a long list of the same operation, followed by only a slight change in the address. Or the program could modify itself and change the address each time an operation is executed. Neither process is elegant.

Beginning with an experimental computer at the University of Manchester in 1948, designers added to the processor an extra *index register* to simplify working with arrays of data. In early descriptions of the Manchester computer, the designers called this register a "B-line," the symbol "A" was used for accumulator, and "C" for control. This term persisted into the 1950s. By specifying a value to increment the address field of an instruction, programs no longer had to modify themselves as envisioned by von Neumann and other pioneers. That greatly simplified the already difficult process of programming.

These three types of registers: accumulator, program counter, and B-line or index register, made up the processing units of most large computers of the 1950s. For example, the IBM 704, announced in 1954, had a 36-bit word length, a core memory holding 4,096 words, and a processor with an accumulator, program counter, and three index registers. Another register was coupled to the accumulator and dedicated to multiplication and division (e.g., to store the extra bits that are generated when two 36-bit numbers are multiplied together).

In 1956 the British firm Ferranti, Ltd. announced a machine, called Pegasus, whose processor contained a set of eight registers, seven of which could be used as accumulators or as index registers. That inaugurated the notion of providing general-purpose registers that a program could use for any of the above functions, as needed by a specific program. Other companies were slow to adopt this philosophy, but by the end of the next decade it became the most favored design.

Number of addresses

Instructions for an accumulator-based machine had two parts: the first specified the operation (add, subtract, compare, etc.), the second the address of the data to be operated on. If an operation required two pieces of data, the other operand needed to be present in the accumulator. It could be there as the result of the previous operation, or as a result of an explicit instruction to load it into the accumulator from memory. Because each instruction contained one memory reference, this was called a *single-address* scheme. Its virtue was its simplicity; it also allowed the address field of an instruction to be long enough to specify large portions of memory. Many computers built in the 1950s used it, including the original UNIVAC and IBM's series of large scientific computers, the 701, 704, 709, and 7090.

There were alternatives to the single-address scheme. One was to have an operation followed by two addresses, for both operands. A third address field could be added, to store the results of an operation in memory rather than assume they would remain in the accumulator. The UNIVAC 1103, RCA 601, and IBM 1401 used a two-address scheme, while the UNIVAC File Computer and the Honeywell H-800 used a three-address scheme.

These schemes all had address fields that told where data were located. One could also include the address of the next *instruction*, instead of going to the program counter for it. Drum computers like the IBM 650 and Librascope LGP-30 minimized the time spent searching the drum for the next instruction – each instruction could direct the computer to the place on the drum where the desired data would be, after executing the previous instruction. Programming this way was difficult, but it got around the inherently slow speeds of drum machinery. With the advent of magnetic core, this scheme fell from necessity.

Finally, one could design an instruction that specified no addresses at all: both operands were always kept in a specified set of registers, in the correct order. Results likewise went into that place, in the proper order for further processing. That required organizing a set of registers (or memory locations) in a structure called a *stack*, which presented data to the processor as Last-In, First-Out. Computers with this scheme first appeared in the 1960s, but they never seriously challenged the single-address design. The stack did not prevail as a method of designing large system processors. However, it would have its day in the sun, as it reappeared first of all in minicomputers, and then most dramatically in the design of the first microprocessors, which will be discussed later.

I/O channels

One of the UNIVAC's innovations was its use of a storage area that served as a "buffer" between the slow input and output equipment, such as card readers and electric typewriters, and the much faster central processor. Likewise the UNIVAC 1103A introduced the concept of the "interrupt," which allowed the machine's processor to work on a problem, stopping to handle the transfer of data to or from the outside world only when necessary. These concepts became well established, and further extended, for large commercial machines of the 1950s.

As the requirements for matching the speeds of input/output with the central processor grew more complex, so too did the devices designed to handle this transfer. With the introduction of the IBM 709, IBM engineers designed a separate processor, called a "channel," to handle input and output. This channel was in effect a small computer of its own, dedicated to the specific problem of managing a variety of I/O equipment operating at different data rates. Sometimes, as designers brought out improved models of a computer, they would add to the capabilities of this channel until it was as complex and powerful as the main processor – and one now had a two-processor system. At that point the elegant simplicity of the von Neumann architecture was in danger of being lost, possibly leading to a baroque and cumbersome system.

The complexity of I/O channels drove up the cost of systems that used them. For customers who used a computer for problems that handled large quantities of data, they were necessary. In time, the use of channels became a defining characteristic of what became known as the *mainframe* computer: one that was physically large and expensive, and which contained enough memory, flexibility, I/O facilities, and processing speed to handle the needs of large customers. The mainframe computer became the standard of the 1960s, although other classes would arise both at the higher end, where faster processing but simpler I/O was required, and at the lower end, where low cost was a major design goal.

Floating-point hardware

One final design feature needs to be mentioned. Although it was of concern primarily to scientific applications, it had an impact on commercial customers as well. In the words of one computer designer, it is the "biggest and perhaps only factor that separates a small computer from a large computer," that is, whether or not a machine handles floating-point arithmetic in its hardware.

One can program any computer to operate as a floating-point machine. But the programming is complex, and it slows the machine down. Or one can design the electronic circuits of the machine to handle floating-point in hardware. Such a design will calculate much faster but it makes the processor more complicated and expensive. As electronic engineering advanced to a point where circuits became more reliable, and as memory capacities and access times improved, the balance tilted in favor of hard-wiring floating-point arithmetic into the processor. The IBM 704, delivered in 1956, included it, and that feature played a major role in propelling IBM ahead of UNIVAC in sales of large machines.

Manufacturers felt that commercial customers did not need floating-point and would be unwilling to pay for it. They typically offered two separate lines of machines, one for commercial customers and the other for scientific or engineering applications. The former, which included the UNIVAC, the UNIVAC File, and IBM 702 and 705, had only fixed-point arithmetic, and often a variable word length. The latter, like the 704 and 709, had floating-point and a relatively long, fixed word length. Input/output facilities were often more elaborate for the business-oriented machines.

This parallel development of two, almost similar lines of equipment persisted through the late 1950s into the next decade. For many customers the distinction was not that clear. For example the IBM 650, intended for commercial customers, was often installed in university centers, where professors and students often labored to develop floating-point software for it. Likewise, the IBM 704 had better performance over the 705, because of what many felt was a superior architecture, and customers who ordered a 704 for scientific work soon found themselves using it for commercial applications as well. The preference for the 704 increased even more as the programming language FORTRAN became available on it.

3. IBM 7094: the canonical architecture

Introduction of the 7090

The 7090 had the same architecture as the vacuum tube-based 709 and was introduced only a year after deliveries of the 709 had begun (ca. 1960). That was an admission by IBM that vacuum tube technology was obsolete, and it had to take a financial loss on the 709 product line. Other, smaller companies had already introduced transistorized products that were getting praise from the trade press. According to folklore, IBM submitted a bid to the U.S. Air Force to supply solid state computers for the Ballistic Missile Early Warning System (BMEWS) around the Arctic Circle. At the time IBM had just announced the 709, but the Air Force insisted on transistorized machines.

IBM designers planned to meet the deadline by designing a computer that was architecturally identical to the 709, only using transistors. They were thus able to use a 709 to develop and test the software that the new computer would need. The 709 could be programmed to make it behave as if it were the new computer. Even that technique did not guarantee that IBM would meet the deadline. IBM delivered computers to a site in Greenland in late 1959, but "IBM-watchers" claimed that the machines, as delivered, were not finished. According to them, the company dispatched a cadre of up to 200 engineers to Greenland to finish the machine as it was being installed.

Whether or not that story is true, the company did deliver a transistorized computer, which it marketed commercially as the Model 7090. The 7090 (and a later upgrade called the 7094, which had four additional index registers) is regarded as the "classic" mainframe, from its combination of architecture, performance, and from its success: hundreds of machines were installed at an equivalent price of over $3 million each.

A description of a 7094 installation

The term "mainframe" probably comes from the fact the circuits of a mainframe computer were mounted on large metal frames, housed in the cabinets. The frames were on hinges and could swing out for maintenance. A typical installation consisted of a number of these cabinets standing on a tiled floor. The floor was a "false-floor" – the tiles were raised a few inches above the real floor, leaving room for the numerous, thick connecting cables that snaked from one cabinet to another, and for the circulation of conditioned air. A cabinet near the operator's console housed the main processor circuits. These were made up of discrete transistors, mounted and soldered along with resistors, diodes, jumper wires, inductors, and capacitors, onto printed circuit boards. The boards in turn were plugged into a "backplane," where a complex web of wires carried signals from one circuit board to another. Some mainframes were laboriously wired by hand, but most used a technique called "wire wrap": it required no soldering, and for production machines the wiring could be done by a machine, thus eliminating errors. In practice there would always be occasional pieces of jumper wire soldered by hand to correct a design error or otherwise modify the circuits. The density of these circuits was about 10 components per cubic inch.

The 7094 was delivered with a maximum of 32,768 words of core memory. In modern terms, that corresponds to about 150 Kilobytes: about what came with the IBM Personal Computer when it first appeared in the early 1980s. Although marketed as a machine for science and engineering, many customers found it well suited for a variety of tasks. It could carry out about 50 to 100 thousand floating-point operations per second, making it among the fastest of its day. Comparisons with modern computers are difficult, as the yardsticks have changed, but it was about as fast as a personal computer of the late 1980s vintage. Its 36-bit word length made it well-suited for scientific calculations that require many digits of precision, and it had the further advantage of allowing the processor to address a lot of memory directly.

The console itself was festooned with an impressive array of blinking lights, dials, gauges, and switches. It looked like what people thought a computer should look like. The rows of small lights indicate the status of each bit of the various registers that make up the computer's central processor. In the event of a hardware malfunction or programming error, the operator could read the contents of each register directly in binary numbers. He or she could also execute a program one step at a time, noting the contents of the registers at each step. If desired, he could directly alter the bits of a register by flipping switches. Such "bit twiddling" was exceedingly tedious, but it gave the op-

Figure 1: "Computing" at North American Aviation, Los Angeles, California, ca. 1952. It was this room, and the flow of information that went through it, that was replicated by the architecture of the stored program digital computer. Photo credit: National Air and Space Museum

erator an intimate command over the machine that few since that time have enjoyed.

Most of the time, the operator had no need to do those kinds of things. The real controlling of the computer was done by its programmers, few of whom were ever allowed into the computer room. Programmers developed their work on decks of punched cards. These in turn were read by a small IBM 1401 computer and transferred to a reel of tape. The operator took this tape and mounted it on a tape drive connected to the mainframe (although there was a card reader directly attached for occasional use). Many programmers seldom saw the machine that actually ran the programs. In fact, many programmers did not even use a keypunch, but rather wrote out their programs on special coding sheets, which they gave to keypunch operators. The operator's job consisted of mounting and unmounting tapes, pressing a button to start a job every now and then, occasionally inserting decks of cards into a reader, and reading status information from a printer. It was not that interesting or high-status a job, though to the uninitiated it looked impressive.

A 7094 installation rented for about $30,000 a month, or an equivalent purchase price of from $1.6 to $3 million. With that cost it was imperative that the machine never be left idle. In later decades one might have one's personal computer run a "screen saver" while going to a meeting or to lunch; the number of computer cycles wasted by this practice would have been scandalous in 1963. On the 7094, programs were gathered onto reels of tape and run in batches. The programmer (whose monthly salary might be $400) had to wait until a batch was run to get his results, and if he then found that he had made a mistake or needed to further refine his problem, he had to submit a new deck and wait once more. However tempting, the idea of gaining direct access to the machine – to submit a program to it and wait a few seconds while it ran – was out of the question, given the high costs of letting the processor sit idle for even a few minutes. That method of operations was a defining characteristic of the mainframe era.

Besides the processor circuit cabinets, magnetic tape drives dominated a mainframe installation. These tapes were the medium that connected a mainframe computer to the outside world. Programs and data were fed into the computer through tapes; the results of a job were likewise sent to a tape. If a program ran successfully, an operator took the tape and moved it to the drive connected to a 1401 computer, which handled the slower process of printing out results on a chain printer (unlike a modern printer, there was typically no direct connection). Results were printed, in all capital letters, on 15" wide, fan-folded paper.

A few mainframes had a video console, but there was none on the 7094's main control panel. Such a console would have been useful only for control purposes, since the sequential storage on tapes prevented direct access to data anyway. In general they were not used because of their voracious appetite for core memory.

4. IBM 360: shift to sets of general purpose registers

Origins

In early 1965, IBM delivered the first of a series of mainframes that would propel that company into an even more commanding position in the industry. That was the System/360, announced in April 1964. It was so named because it was aimed at the full circle of customers, from business to science, for customers who did a lot of mathematical calculation and for those who did simpler arithmetic on large sets of data. System/360's primary selling point was that IBM was offering not one but a whole line of computers, with a promise that programs written for one model would work on larger models, thus saving a customer's investment in software as business grew. IBM announced six models on April 7, 1964. Later on it announced others, while dropping some of the original six by the time deliveries began. The idea was not entirely new: computer companies had tried to preserve software compatibility as they introduced newer models, as IBM had done with their 704, 709, and 7090 machines. But the 360 was a *series* of computers, all announced at the same time, offering about a 25:1 performance range. Except for a small run of machines delivered to the Army in the late 1950s, that had never been attempted before.

In an often-repeated phrase, first used in a *Fortune* magazine article, an IBM employee said, "you bet your company" on this line of computers. Besides the six computer models, IBM introduced "over 150 different things – new tapes, new disks, the 029 card punch" on the same day. Had the 360 failed, it would have been a devastating blow, although IBM would still have survived as a major player in the business. The company could have introduced newer versions of its venerable 1401 and 7090-series machines, and it still had a steady stream of revenue from pre-computer punched card installations. But such a failure would have restructured the computer industry.

System/360 did not fail. Within weeks of the product announcement in April 1964 orders began coming in: "Orders for System/360 computers promptly exceeded forecasts: over 1100 were received in the first month. After five months the quantity had doubled, making it equal to a fifth of the number of IBM computers installed in the U.S." The basic architecture served as the anchor for IBM's product line into the 1990s.

Manufacturing and delivering the line of computers required enormous resources. The company expanded its production facilities, but delivery schedules slipped, and shortages of key components arose. The success of the 360 threatened the company's existence almost as much as a failure might have. For those employees driven to the breaking point – and there were many – the jump in revenues for IBM may not have been worth the physical and mental stress. From 1965 to 1970, thanks mostly to System/360, IBM's gross income more than doubled. Net earnings also doubled, surpassing $1 billion by 1971. IBM had led the U.S. computer industry since the mid 1950s. By 1970 it had an installed base of 35,000 computers, and by the mid-1970s it made sense to describe the U.S. computer industry as having two equal parts: IBM on one side and everyone else *combined* on the other.

The problems IBM faced in trying to meet the demand – employee burnout, missed shipping dates, quality control on the production lines – were problems its competitors might have wished for. Obviously many customers found this line of machines to their liking. Most NASA centers, for example, quickly switched over to 360 (Model 65 or higher) from their 7090 installations to meet the demands of putting a man on the Moon. Commercial firms, who used computers for business data processing, likewise replaced their 7030s and other systems with models of the System/360. There was some resistance to replacing the venerable 1401 with the low-end 360 intended to replace it, but in general the marketplace gave overwhelming approval to the notion of a compatible family of machines suitable for scientific as well as business applications.

The decisions that led to System/360 came from an IBM committee known as "SPREAD," which met daily in a motel in Cos Cob, Connecticut for two months in late 1961. Their report, issued internally on December 28, 1961 and published 22 years later, reveals much about the state of computing, as it then existed and as key engineers and executives at IBM thought it would become.

Their deliberations began with a survey of the company's existing products. In 1961 IBM was fielding a confusing tangle of machines, few of which were compatible with one another. The SPREAD Committee, composed of members from both of IBM's product lines, did not agree at first on a unified product line, but eventually they recognized its advantages and incorporated that as a recommendation in their final report. As with many great ideas, the notion of having a unified product line seems obvious in retrospect, but that was not the way it seemed at first to those assembled in the rooms of the motel.

Less obvious was scalability. Though the SPREAD Committee agreed that this was needed, at the early stages both Fred Brooks and Gene Amdahl – later two of the 360's principal architects – argued that "it couldn't be done." Few other technologies, if any, scale simply. Civil engineers use different criteria when designing large dams than they use for small ones. The engine, transmission, power train, and frame of a large sedan are not simply bigger versions of those designed for a subcompact. What the SPREAD Committee was proposing was a range of 25:1 in computing – more like comparing a subcompact to an 18-wheeler. By 1970, after IBM had announced an upgrade to the 360 line, they were offering compatible computers with a 200:1 range.

Microprogramming

What changed Brooks's and Amdahl's mind was the rediscovery of a concept almost as old as the stored program computer itself. In 1951, at a lecture given at a ceremony inaugurating the Manchester University digital computer, Maurice Wilkes argued that "the best way to design an automatic calculating machine" was to build its control section as a little stored program computer of its own, wherein each control operation is broken down into series of "micro-operations" directed by a matrix of components that stored a "micro-programme [sic]." By adding a layer of complexity to the design, one in fact simplified it. The design of the control unit, typically the most difficult, could now be made up of an array of simpler circuits, like those for the computer's memory unit. Wilkes made the bold assertion that this was the "best way" because he felt it would give the design more logical regularity and simplicity; almost as an afterthought he mentioned that "the order code need not be decided on finally until a late stage in the construction of the machine." He did not say anything about a series of machines or computers having a range of power.

The idea was kept alive in later activity at Manchester, where John Fairclough, a member of the SPREAD Committee, studied electrical engineering. Through him came the notion of using microprogramming as a way of implementing a common set of instructions across the line of 360s, while allowing the engineers charged with the detailed design of each specific model to optimize the design for low cost and adequate performance. The microprogram, in the form of a small read-only memory built into the control unit of each model's processor, would be written to ensure compatibility. Microprogramming gave the 360's designers "...the ability to separate the design process...from the control logic that effectively embodied the instruction-set characteristics of the machine we were trying to develop."

IBM's adoption of this concept extended Wilkes's original insight. In essence it is a restatement of the fundamental property of a general-purpose, stored program computer – that by accepting complexity at one level (computers require very large numbers of components), one gains power and simplicity at another level (the components are in the form of regular arrays that can be analyzed by tools of mathematics and logic). Some understanding of it appears inchoate in the earliest of the digital machines. Wilkes himself may have been inspired by the Bell Labs relay computer Model VI, which he probably inspected during a visit to America in 1950. On the Model VI a set of coils of wire stored information that allowed the machine to execute complex subsequences upon receiving one simple instruction from a paper tape.

By adopting microprogramming IBM gained one further advantage, which some regard as the key to the 360's initial success. That was the ability to install a microprogram that would allow the processor to understand instructions written for an earlier IBM computer. In this way IBM salesmen could convince a customer to go with the new technology without fear of suddenly rendering an investment in applications software obsolete. Larry Moss of IBM called this ability *emulation*, implying that it was "as good as" (or even better than) the original, rather than mere "simulation" or worse, "imitation." The 360 Model 65 sold especially well because of its ability to emulate the large business computer 7070, and IBM devoted extra resources to the low-end models 30 and 40 to emulate the 1401.

In theory any stored program computer can be programmed to act as if it were another – a consequence of its being a "Universal Turing Machine," after the mathematician Alan M. Turing, who published on this concept in the 1930s. In practice that usually implies an unacceptable loss of performance, as the extra layers of code slow things down. Trying to emulate one computer with another usually lands the hapless designer in the "Turing Tar-Pit," where anything is possible but nothing is practical. The 360 avoided that pit because its emulation used a combination of software and the microprogram of each machine's control unit. When combined with the faster circuits it also used, the combination permitted the new machines to run the old programs as much as *10 times* faster than the same program would have run on, say, a 1401. By 1967, according to some estimates, over half of all 360 applications were emulations of older hardware.

1401 emulation was especially crucial to IBM's 'bet the company' gamble: in December 1963 Honeywell introduced the H-200 computer, with a program they called "Liberator" that allowed it to run 1401 programs. H-200 sales were immediately brisk, just as IBM announcing the 360 line with its implied incompatibility with the 1401. The IBM division that sold the 1401 went through a Slough of Despond in early 1964, but it climbed out after orders for the lower-end models of the 360 came rolling in. The success of emulation demonstrated a paradox of computer terminology: software, despite its name, is more permanent and harder to modify than hardware. To this day there are 1401 programs running routine payroll and other data processing jobs on modern computers from a variety of suppliers. When programmers coded these jobs in the early 1960s, using keypunch machines, they had no idea how long-lived their work would be. (The longevity of 1401 software is a major cause of the "Year-2000" bug.)

Every System/360 except for the smallest Model 20 contained 16 general-purpose registers in its central processor. In the 360 any of the 16 registers could be used for any operation (with a few exceptions, i.e. extra registers for floating-point numbers).

The 360's word length was 32 bits – 4 bits shorter than word length of the 7090/7094 scientific computers, but because 32 was a power of 2, it simplified the design. Most early computers used sets of 6 bits to encode characters; System/360 IBM used 8 bits, which IBM called a *"byte,"* (the term was coined in 1956 by Werner Buchholz of IBM). Eight bits allowed 256 different combinations for each character: more than adequate for upper and lower case letters, the decimal digits 1 to 10, punctuation, accent marks, etc. with room to spare. And since 4 bits were adequate to encode a single decimal digit, one could "pack" two decimal digits into each byte, compared to only one decimal digit in a 6-bit byte. (The 360's memory was addressed at the byte level; one could not fetch a sequence of bits that began in the middle of a byte.)

ASCII and EBCDIC

To encode the 256 different combinations, IBM chose an extension of a code they developed for punched card equipment. This "EBCDIC" (Extended Binary Coded Decimal Interchange Code) was well designed, complete, and offered room for future expansion. It had one unfortunate characteristic – incompatibility with another standard being developed at the same time. This standard, known as ASCII and supported by the American National Standards Institute in 1963, standardized only seven bits, not eight. One reason was that at the time punched paper tape was still in common use, and the committee felt that punching eight holes across a standard piece of tape would weaken it too much (there were a few other reasons as well). The lack of an 8-bit standard made it inferior to EBCDIC, but with its official status ASCII was adopted everywhere but at IBM. The rapid spread of minicomputers using ASCII and Teletypes further helped spread the code. With the dominance by IBM of mainframe installations, neither standard was able to prevail over the other. IBM had representatives on the committee that developed ASCII. The System/360 had a provision to use either code, but the ASCII mode was later dropped as it was little used. The simultaneous adoption of two incompatible standards within a few years of each other was unfortunate but probably inevitable. In ASCII, the ten decimal digits were encoded with lower numerical values than the letters of the alphabet; with EBCDIC it was the opposite. Therefore a sorting program would sort "3240" before "Charles" if the data were encoded in ASCII, but "Charles" before "3240" if EBCDIC. Because of its beachhead in minicomputers, ASCII would prevail in the personal computer and workstation environment beginning in the 1980s.

The 360's designers allowed for 4 bits of a word to address the 16 general-purpose registers and 24 bits to address the machine's core memory. That allowed direct access to 2^{24}, or 16 million addresses, which seemed adequate at the time. Like nearly every other computer design, the address space was eventually found to be inadequate, and in 1981 IBM extended the number of address bits to 31, allowing for access to 2 billion addresses.

For the cheaper models, even allowing 24 bits was extravagant, as these were intended to do their work with a much smaller memory space. Carrying the extra address bits would impose an overhead penalty that might allow competitors like Honeywell to offer machines that were more cost-effective. IBM's solution was to carry only 12 of the possible 24 address bits in an instruction. This number would then be added to another number stored in a "base" address register to give the full 24-bit address. If a program required fewer than 2^{12} or 4 thousand bytes of memory, going to the base register was not necessary. That was the case for many smaller problems, especially those that the cheaper models of the 360 were installed for. For longer problems there was of course the additional penalty incurred when going to the base register to obtain an address, but in practice this was not a severe problem.

Finally, the System/360 retained the concept of having channels to handle input and output. With a standard interface, IBM could offer a single line of tape, card, and printing equipment that worked across the whole line of machines – a powerful selling point whose advantages easily offset whatever compromises had to be made to provide compatibility. The trade press called I/O devices "peripherals"; but they were central to the System/360 project: a new model keypunch, new disk and tape drives, and even the Selectric typewriter with its famous golf-ball print head and classic keyboard layout.

To sum up, the architectural design of the 360 was one of creative and sometimes brilliant compromises to achieve compatibility across the range of performance. Initially it had a fairly simple design, which over the years grew ever more complex, baroque, and cumbersome. The fact that it could grow as it did, enough to remain viable into the 1990s, is testimony to the strength of the initial effort.

5. The minicomputer: invention, definition, architecture

At the same time as the System/360's announcement, the minicomputer was being developed. With hindsight one can see that the minicomputer revealed a segment of computing not covered by the System/360, in spite of its name. A number of factors define the minicomputer: architecture, packaging, and the role of third parties in developing applications, price, and financing. It is worth discussing the first of those, architecture, in some detail to see how the minicomputer differed from what was prevalent at the time.

Minicomputer: creative ways to get around short word length

First of all, minicomputers used short word lengths, which lowered costs. Above all, they found ways to get around the drawbacks of a short word length. They did that by making the computer's instruction codes more complex. Besides the operation code and memory address specified in an instruction, minicomputers used several bits of the code to specify different "modes" that extend the memory space. One mode of operation might refer not directly to a memory location but to another register, in which the desired memory location is stored. That of course adds complexity; operating in double precision also is complicated, and both might slow the computer down. But with the newly-available transistors coming on the market in the late 1950s, one could design a processor that, even with these added complexities, remained simple, inexpensive, and fast.

The CDC 160 and the origins of the minicomputer

The Whirlwind (a computer prototype built at MIT) had a word length of only 16 bits, but the story of commercial minicomputers really begins with an inventor associated with very large computers: Seymour Cray. While at UNIVAC Cray worked on the Navy Tactical Data System (NTDS), a computer designed for navy ships and one of the first transistorized machines produced in quantity. Around 1960 Control Data, the company founded in 1957 that Cray joined, introduced its model 1604, a large computer intended for scientific customers. Shortly thereafter CDC introduced the 160, designed by Cray ("almost as an afterthought," according to a CDC employee), to handle input and output for the 1604. For the 160, Seymour Cray carried over some key features he pioneered for the Navy system, especially its compact packaging. In fact, the computer was small enough to fit around an ordinary-looking metal desk – someone who chanced upon it would not even know it was a computer unless told beforehand.

The 160 broke new ground by using a short word length (12 bits) combined with ways of accessing memory beyond the limits of a short address field. It was able to directly address a primary memory of 8 thousand words, and it had a reasonably fast clock cycle (6.4 microseconds for a memory access). And the 160 was inexpensive to produce. When CDC offered a stand-alone version, the 160A, at a price of $60,000, the computer found a ready market. Control Data Corporation was concentrating its efforts on very-high performance machines, for which Cray became famous, but it did not mind selling the 160A along the way. What Seymour Cray had invented was in fact a minicomputer.

Other architectural features defined the Mini. These include direct memory access, which handled I/O without resorting to the channels of the mainframe. It was remarked that a single channel for an IBM System/360 cost more than an entire PDP-1, an early mini from Digital Equipment Corporation. A third feature was the refinement of the bus. This was nothing new—it is found in the Harvard Mark I, where it was spelled "buss," and the Whirlwind used it. But minicomputers, especially second generation minis like the Data general NOVA and the PDP-11, refined this concept.

6. The microprocessor

In 1964 Gordon Moore, then of Fairchild and soon a cofounder of Intel, noted that from the time of its invention in 1958, the number of circuits that one could place on a single integrated circuit was doubling every year. By simply plotting this rate on a piece of semilog graph paper, "Moore's Law" predicted that by the mid 1970s one could buy a chip containing the equivalent logic circuits as those used in a 1950s-era mainframe. By the late 1960s Transistor Transistor Logic (TTL) was well established, but a new type of semiconductor called "Metal-Oxide Semiconductor," (MOS), emerged as a way to place even more logic elements on a chip. MOS was used by Intel to produce its pioneering 1103 memory chip, and it was a key to the success of pocket calculators. The chip density permitted by MOS brought the concept of a computer-on-a-chip into focus among engineers at Intel, Texas Instruments and other semiconductor firms. That did not mean that such a device was perceived as useful. If it was generally known that one could place enough transistors on a chip to make a computer, it was also generally believed that the market for such a chip was so low that its sales would never recoup the large development costs required.

Who invented the microprocessor?

The story of the microprocessor's invention at Intel has been told many times. In essence, it is a story encountered before: Intel was asked to design a special-purpose system for a customer. It found that by designing a general-purpose computer and using software to tailor it to the customer's needs, the product would have a larger market. It was a reversal of the approach taken by calculator (and game) makers, and it allowed the Intel design to initiate a new class of personal computers.

Intel's customer was Busicom, a Japanese company that was a top seller of hand-held calculators. Busicom sought to produce a line of products with different capabilities, each aimed at a different market segment. It envisioned a set of custom-designed chips that incorporated the logic for the advanced mathematical functions. Intel's management assigned Marcian E. Hoff, who had joined the company in 1968 (Intel's 12th employee), to work with Busicom.

Intel's focus had always been on semiconductor memory chips. It shied away from logic chips like those suggested by Busicom, since it felt that markets for them were limited. Hoff's insight was to recognize that by designing fewer logic chips with more general capabilities, one could satisfy Busicom's needs elegantly. Hoff was inspired by the PDP-8, which had a very small set of instructions, but which its thousands of users had programmed to do a variety of things. He also recalled using an IBM 1620, a small scientific computer with an extremely limited instruction set that nevertheless could be programmed to do a lot of useful work. The 1620 went by the informal name CADET – "Can't Add, Doesn't Even Try."

Hoff proposed a logic chip that incorporated more of the concepts of a general-purpose computer. A critical feature was the ability to call up a subroutine, execute it, and return to the main program as needed. He proposed to do that with a register that kept track of where a program was in its execution and saved that status when interrupted to perform a subroutine. Subroutines themselves could be interrupted, with return addresses stored on a "stack."

With this ability, the chip could carry out complex operations stored as subroutines in memory, and avoid having those functions permanently wired onto the chip. Doing it Hoff's way would be slower, but in a calculator that did not matter, since a person could not press keys that fast anyway. The complexity of the logic would now reside in software stored in the memory chips, so one was not getting something for nothing. But Intel was a memory company, and it knew that it could provide memory chips with enough capacity. As an added inducement, sales of the logic chips would mean more sales of its bread-and-butter memories.

That flexibility meant that the set of chips could be used for many other applications besides calculators. Busicom was in a highly competitive and volatile market, and Intel recognized that (Busicom eventually went bankrupt). Robert Noyce negotiated a deal with Busicom to provide it with chips at a lower cost, giving Intel in return the right to market the chips to other customers for non-calculator applications. From these "unsophisticated" negotiations with Busicom, in Noyce's words, came a pivotal moment in the history of computing.

The result was a set of four chips, first advertised in a trade journal in late 1971, which included "a microprogrammable computer on a chip." That was the 4004, on which one found all the basic registers and control functions of a tiny, general purpose stored program computer. The other chips contained a Read-Only Memory (ROM), Random-Access Memory (RAM), and a chip to handle output functions. The 4004 became the historical milestone, but the other chips were important as well: especially the ROM chip that supplied the code that turned a general-purpose processor into something that could meet a customer's needs. Also at Intel, a team led by Dov Frohman developed a ROM chip that could be easily reprogrammed and erased by exposure to ultraviolet light. Called an EPROM (Erasable Programmable Read-Only Memory) and introduced in 1971, it made the concept of system design using a microprocessor practical.

Stan Mazor did the detailed design of the 4004. Federico Faggin also was crucial in making the concept practical. Masatoshi Shima, a representative from Busicom also contributed. Many histories of the invention give Hoff sole credit; all players, including Hoff, now agree that is not accurate. Faggin left Intel in 1974 to found a rival company, Zilog. Intel, in competition with Zilog, felt no need to advertise Faggin's talents in its promotional literature, although Intel never showed any outward hostility to its ex-employee. The issue of whom to credit reveals the way people think of "invention": Hoff had the idea of putting a general purpose computer on a chip, Faggin and the others "merely" implemented that idea in silicon. At the time, Intel was not sure what it had invented either: Intel's patent attorney resisted Hoff's desire at the time to patent the work as a "computer." Intel obtained two patents on the 4004, covering its architecture and implementation; Hoff's name appears on only one of them.

The move to 8 bits

The 4004 worked with groups of four bits at a time – enough to code decimal digits but no more. At almost the same time as the work with Busicom, Intel entered into a similar agreement with Computer Terminal Corporation (later called Datapoint) of San Antonio, Texas, to produce a set of chips for a terminal to be attached to mainframe computers. Again, Mazor and Hoff proposed a microprocessor to handle the terminal's logic. Their proposed chip would handle data in 8-bit chunks, enough to process a full byte at a time. By the time Intel had completed its design, Datapoint had decided to go with conventional TTL chips. Intel offered the chip, which they called the 8008, as a commercial product in April 1972.

In late 1972, a 4-bit microprocessor was offered by Rockwell, an automotive company that had merged with North American Aviation, maker of the Minuteman Guidance System. In 1973 a half dozen other companies began offering microprocessors as well. Intel responded to the competition by announcing the 8080 in April 1974: an 8-bit chip that could address much more memory and required fewer support chips than the 8008. The company set the price at $360 – a somewhat arbitrary figure, as Intel had no experience selling chips like these, one at a time. (Folklore has it that the $360 price was set to suggest a comparison with the IBM System/360.) Although a significant advance over the 8008, the 8080 could execute programs written for the other chip: this compatibility would prove crucial to Intel's dominance of the market. The 8080 was the first of the microprocessors whose instruction set and memory addressing capability approached those of the minicomputers of the day.

Significance of this invention

Obviously this invention has been a key to the "revolution" not just in computing but in embedded, smart machinery that has swept across the world in recent decades. For this study, its invention has another significance: The Microprocessor was the embodiment of mini concepts onto silicon. It was a significance that was missed by DEC and Data General, with disastrous consequences for those two companies. In short: architecture became a mass-produced commodity. Companies like DEC or DG, that were founded on and based their revenues on innovative architecture, had either to adapt to this new world, or go under.

7. RISC

Since the invention of the 8080 there has been a steady progression of Intel microprocessors, which have steadily encroached on the market to a point where they are almost synonymous with the term architecture. That is not quite true, though, and the following will discuss one significant parallel development.

The VAX

Through the 1980s the dominant mainframe architecture continues to be a descendent of the IBM Sytem/360, while the dominant mini was the DEC VAX, which evolved as a 32 bit extension of the 16-bit PDP-11. Although a mini, the VAX architecture had a lot in common with the IBM System/360 and its descendants. Like the 360, its instruction set was contained in a microprogram, stored in a read-only memory. Like the 360, the VAX presented its programmers with a rich set of instructions that operated on data in almost every conceivable way. The 370/168 had over 200 instructions, the VAX 11/780 over 250. There were sets of instructions for integers, floating-point numbers, packed decimal numbers, and character strings; operating in a variety of modes. This philosophy had evolved in an environment dominated by magnetic core memory, to which access was slow relative to processor operations. Thus it made sense to specify in great detail what one wanted to do with a piece of data before going off to memory to get it. The instruction sets also reflected the state of compiler technology. If the processor could perform a lot of arithmetic on data with only one instruction, then the compiler would have that much less work to do. A rich instruction set would reduce the "semantic gap" between the English-like commands of a high-level programming language and the primitive and tedious commands of machine code. Cheap read-only memory chips meant that the designer could create these rich instruction sets at low cost if the computer was microprogrammed.

John Cocke at IBM

Those assumptions had been long accepted. But computer science was not stagnant. In the mid-1970s John Cocke of IBM looked at the rapid advances in compilers and concluded that a smaller set of instructions, using more frequent commands to load and store data to and from memory, could operate faster than the System/370. Thomas Watson, Jr. once wrote a memo describing IBM's need to have "wild ducks" among its ranks – people who were not content to accept conventional wisdom about they way things were done. Cultivating such a person in the conservative culture of IBM was not easy, but Watson knew, perhaps better than any other computer executive, that IBM could not survive without them. John Cocke, with his then-radical ideas about computer design, fits that description.

Cocke's ideas led to an experimental machine called the IBM 801, completed in 1979. For many reasons, including the success and profits of the 370 line and its successors, IBM held back introducing a commercial version of the design (the IBM-RT, introduced in 1986, was a commercial failure and did not exploit the idea very well). Still, word of the 801 project got out, along with a rumor that it could execute System/370 programs at much faster speeds although it was a smaller computer. By the late 1970s magnetic core had been replaced by semiconductor memory, whose access times matched the speeds of processors. Frequent load and store instructions no longer exacted a speed penalty. Finally, some researchers looked at the VAX and concluded that one could not extend its design any further; they began looking for alternatives.

RISC: Hennessy & Patterson

In 1980 a group at Berkeley led by David Patterson, after hearing "rumors of the 801," started a similar project called RISC – Reduced Instruction Set Computer. Another project, called MIPS, began in 1981 at Stanford under the leadership of John Hennessy. As they publicized their work they were met with skepticism: RISC looked good in a university laboratory but did not address the real needs of actual customers. One trade journal even worried that RISC, from the start associated with UNIX, was not well suited for data processing jobs written in COBOL. Meanwhile, sales of Intel-based PCs, the VAX, and the System/370 family – all complex instruction set processors – were booming. With a massive build up of the Defense Department under President Ronald Reagan, Wall Street was enjoying another round of Go-Go years. Those watching the trajectory of their stocks in DEC, Data General, IBM, and Wang were not worried about RISC.

SUN Microsystems' products initially used the Motorola 68000 microprocessor, whose design was very much in the spirit of the PDP-11 and VAX. Beginning in 1987 and probably owing to Bill Joy's influence, SUN introduced a workstation with a RISC chip based on Patterson's research at Berkeley. Called SPARC (Scalable Processor Architecture), this design did more than anything else to overcome skepticism about RISC. Hennessy and Patterson became evangelists for RISC, buttressed by some impressive quantitative measurements that showed how a RISC design could squeeze much more processing power out of a piece of silicon than conventional wisdom had thought possible. More telling, their data showed that RISC offered a way of improving microprocessor speeds much more rapidly than mini and mainframe speeds were improving – or could improve. The unmistakable implication was that the puny, cheap microprocessor, born of a pocket calculator, would soon surpass minicomputers, mainframes, and even supercomputers in performance. If true, their conclusions meant that the computer industry as it had been known for decades, and over which the U.S. Justice Department fought IBM throughout the 1970s, was being driven to extinction by its own offspring.

SUN went a step further to promote RISC: they licensed the SPARC design so that other companies might adopt it and make SPARC a standard. The combination of a license to copy the SPARC processor, plus Berkeley UNIX, made it almost as easy to enter the workstation market as it was to make an IBM compatible PC. SUN gambled that it too would benefit by continuing to introduce products with high performance and a low price. They succeeded, although such a philosophy meant slim profit margins, since SUN could not own the architecture.

The Stanford MIPS project also spawned a commercial venture, MIPS Computer Systems, which also helped establish a commercial market for RISC microprocessors. Digital Equipment Corporation bought a chip from MIPS for one of their workstations in 1989 – even DEC now admitted that RISC was not going away (an internal RISC project at DEC, called "Prism," had been canceled in 1988). Silicon Graphics also based its newer workstations on MIPS microprocessors. Hewlett-Packard converted its line of workstations to a RISC design called Precision Architecture. After failing with the RT, IBM introduced a successful RISC workstation in 1990, the R/6000. In the early 1990s Apple and IBM joined forces with Motorola to produce a RISC microprocessor called Power PC, which they hoped would topple the Intel 8086 family. IBM's role in the design of the Power PC was a fitting vindication of the ideas of John Cocke, the "wild duck" who started the whole phenomenon.

8. Conclusion

The above brief survey reveals that much has indeed happened in computing since von Neumann. But it also leads one to a paradoxical conclusion that as more and more systems are built around mass-produced microprocessors supplied by Intel, computer architecture no longer is so important. The focus has now shifted to software. Architectural questions now arise in the discussion of internetworking of individual computer systems. Unfortunately, there is no equivalent to the EDVAC Report that serves to guide one through networking, as that report guided one through processor design in the past 50 years. The conclusion seems therefore to be that if there is ever to be something new since von Neumann, it will be in an arena that von Neumann did not foresee – networking.

PAUL E. CERUZZI is Curator of Aerospace Electronics and Computing at the Smithsonian's National Air and Space Museum in Washington, DC. Dr. Ceruzzi received a B.A. from Yale University in 1970 and his Ph.D. in American studies from the University of Kansas in 1981. His graduate studies included a year as a Fulbright Scholar at the Institute for the History of Science in Hamburg, West Germany. He received a Charles Babbage Institute Research Fellowship in 1979. Before joining the staff of the National Air and Space Museum, he taught history of technology at Clemson University in Clemson, South Carolina. Dr. Ceruzzi is the author or co-author of several books on the history of computing and related issues: *Reckoners, The Prehistory of The Digital Computer* (1983), *Computing Before Computers* (1990), *Smithsonian Landmarks in the History of Digital Computing* (1994), *Beyond the Limits: Flight Enters the Computer Age* (1989), and *A History of Modern Computing, 1945–1995* (1998). He has also served as a consultant on several projects devoted to this subject, including the Time-Life Books series on *Understanding Computers*, and the BBC/PBS television series "The Machine That Changed the World," broadcast in the U.S. in the spring of 1992.

Part III: The German Scene

Hoelzer's analog
machine (Pennemunde)

Z1, Z2, Z3, Z4
(Berlin)

G1, G1a
(Gottingen)

Z4
(Berlin, Harz,
Hinterstein, Zurich)

The DEHOMAG D11 Tabulator – A Milestone in the History of Data Processing

Friedrich W. Kistermann

Abstract. The DEHOMAG D11 tabulator has been overlooked for too long in the history of data processing. This is due to the scarcity of literature about this machine and its usual classification as a tabulator, even though its internal structure corresponds to an automatic calculator. In general, very little attention has been paid to the time period preceding the electronic age, that is, there has been a neglect of the pre-computer era. The D11 tabulator, however, had a decisive influence on the diffusion of punched card data processing in Germany.

1. Setting the stage

The development of the DEHOMAG D11 cannot be understood without looking at the development of the Hollerith tabulating machines in general, and at the differences in the application environment on both sides of the Atlantic in particular.

It is helpful to understand the purpose of a tabulator, which becomes apparent if we look at its position in the data processing work flow. Fig. 1 is an abstract representation of the processing stages. Whenever someone performs any kind of data processing work, for instance when a database management system is used for some task, the same basic steps which are outlined in Fig. 1 must be followed.

True enough, the steps have different names these days, because punched cards are no longer used, but the work flow is generic and independent of the equipment used. In this example, the punched card contains important data concerning a person, event, or product, that is to be processed by the system. The punched card is, therefore, the unit record required for the specific application, let's say for the computation of gross and net wages.[1] For this process to function properly, the individual steps have to be carefully planned.

---

[1] Friedrich W. Kistermann, "The Invention and Development of the Hollerith Punched Card: In Commemoration of the 130th Anniversary of the Birth of Herman

Right from the beginning the data to be processed or just stored on a card must be verified, and corrected if necessary. After sorting, which is the process of grouping data in the predefined order necessary to obtain a specific result, the tabulator can do its work. Normally, it was used to print a report and it often produced summary cards as an important by-product. The summary cards made further processing of the data possible, especially in the case of cumulative processing. Because of this logical sequence, the tabulator stands at the end of the processing chain.

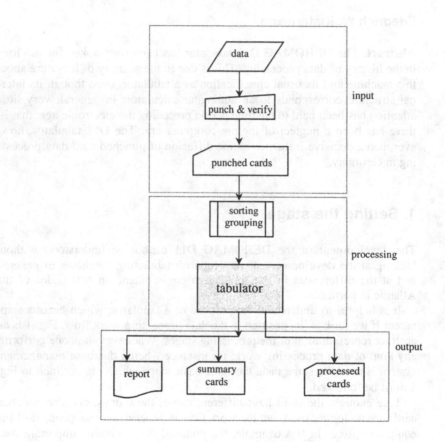

Figure 1: Role of the tabulator in the DP work flow

Hollerith and for the 100th Anniversary of Large Scale Data Processing," *Annals of the History of Computing* 13 (1991) No. 3, 245–259.

The work flow diagram shows that every punched card installation needed at least one card punch, one verifier, one sorter and one tabulator, assuming that the customer was not contracting a service bureau.

In 1895, Herman Hollerith used his punched card system for cost accounting in the railroad industry. This happened in parallel to his work for the United States census bureau and census bureaus in some other countries.[2] In 1908, after approximately ten years of development and a major change in construction, the Hollerith punched card system was ready to be widely used in industry and trade.

The Hollerith punched card system

The Hollerith punched card system consisted of a key punch, a gang punch, a vertical sorter and a tabulator.[3] The Hollerith tabulator of 1908 formed the basis for the successful introduction of the punched card into many application areas.[4] The card feeding and sensing device were on the left of this non-printing tabulator. The plugboard, which allowed the machine operation to be defined by wiring, was on the front. The electrical impulses went from the card sensing brushes to the hubs of the plugboard, and from there by wire connection to a specific counter, where the amount punched into the card was added. The counters, with nine digits each, were located above the plugboard and were visible from the front.

The tabulator underwent some further development after its initial introduction in 1908. In 1911, a manually operated clutch was introduced, which allowed the counters to disengage from the counter clearing shaft and thus retain a result until it was recorded. This created the possibility of two-level group control. The groups were separated by so-called stop cards, which were not punched. If the read brushes detected a stop card, the tabulator would stop feeding cards. The results were read from the counter wheels and were transcribed onto prepared forms. Thereafter, the operator had to clear the counters and restart the machine.

---

[2] Friedrich W. Kistermann, "Ein Kapitel Industriegeschichte: Herman Hollerith (1860–1929) - Ein Pfälzer Emigrantensohn gründet einen Weltkonzern," in *Nachbar Amerika: Verwandte – Feinde – Freunde in drei Jahrhunderten*, ed. Gudrun Schäfer (Landau, 1996) 233–267.; and Friedrich W. Kistermann, "Was the Father of Herman Hollerith a Revolutionary?," *Annals of the History of Computing* 19:4 (1997) 69–70.

[3] Friedrich W. Kistermann, "The Way to the First Automatic Sequence-Controlled Printing Calculator: The 1935 DEHOMAG D11 Tabulator," *Annals of the History of Computing* 17:2 (1995) 33–49; and Friedrich W. Kistermann, "Locating the Victims: The Nonrole of Punched Card Technology and Census Work," *Annals of the History of Computing* 19:2 (1997) 31–45.

[4] Kistermann, *The Way*, Fig. 3.

Hollerith's printing tabulator

The next step in the development of the Hollerith system was to add a printing device. This was also invented by Herman Hollerith. He was granted the patents in 1912, although he had formulated his ideas in 1899. The engineering model was ready in 1917.[5] The printing device was attached to the nonprinting tabulator on a pedestal extending to the right of the counters. In this model, the stop cards were replaced by an automatic group control. Two sets of card sensing brushes allowed two successive cards to be compared. If the group identifiers were not equal, the tabulator stopped feeding cards, printed the counters' contents, cleared the counters if they were coupled with the clearing shaft, whereafter the machine resumed card feeding. This provided an uninterrupted work flow at the printing tabulator.

The ability to print out the contents of every card was especially advantageous. It was necessary when statements of accounts, inventory lists, detailed sales analysis reports, etc., had to be processed.

With every new function on the tabulator, the plugboard increased "in size," that is, more hubs had to be wired.

The Hollerith printing tabulator, announced at the end of 1920, marked the transition of the tabulator from a "statistical" machine into a machine that could support business management in many more applications. However, there was still one shortcoming in the 1920 model. Negative results were shown at the counters as complementary numbers and were printed as such. A digit 9 in the highest counter position indicated a negative result. To do subtraction, the operators also had to punch complementary numbers into the cards.

Table 1: Application environment: Germany (DEHOMAG) vs. USA (IBM)

Field of application	Germany	USA
Banks	*** Current account business, account statement *Required*: printing of true numbers, compound interest calculations, balancing	Check transactions
Wage accounting (gross pay to net pay) 5 or more wage fields; 10 or more deductions & additional charges	*** 50 deductions per 5 workers *Required*: cross-footing	* 1 deduction per 5 workers
Cost accounting (since 1895 resp. 1902)	**	**
Business/factory statistics (since 1902)	**	**
Market analysis	*	**

---

[5] Kistermann, *The Way*, Fig. 4.

We now come to the second half of the 1920s, which saw an increase in demand on the part of existing and prospective customers for more functions in the tabulator and more capacity on the punched card. There was a particular need for a balancing feature, which would allow account balances to be obtained, not only in the banking business, but also in factory accounting and other applications.

The DEHOMAG engineers in Germany became aware of the fact that the Hollerith/IBM tabulators had some deficiencies with regard to application environments in that country, which were quite different from those in the United States. Table 1 contains some examples of the differences between applications in Germany and the USA. The importance of the application is given by the number of stars, from one to three.

Banks wanted to be able to print out complete bank account statements which would show all transactions. In other words, they disliked having one form for credits, another for debits, and a third, which was handwritten, for the balance. Accounting in industry and trade had the same requirements. Cross-footing would be the answer.

If an account has a variable number of entries, which is usually the case, and the balance is to be printed in the last line of the form, then some type of forms control is needed. Also, banks wanted to print true numbers, marked by an appropriate sign, when a negative balance was present. They also wanted to avoid punching complementary numbers instead of negative values. And last, but not least, they wanted to avoid the burden of manual interest calculation by having the tabulator calculate compound interest.

The second line in the application environment table shows the demand on the part of large companies for help in wage accounting. This was due to the more complicated tax reduction procedure that existed in Germany at the time compared to the system in the United States. Cross-footing would also be the answer to this problem.

In wage accounting, cross-footing is cross-addition of all deductions and all additional charges to obtain two sums, which are added with the corresponding sign to the gross wage, printing all details and partial sums on one pay slip.

Some of these requirements were met by introducing additional features in the Hollerith/IBM tabulators then installed. This work was done by DEHOMAG engineers and technicians under the leadership of Ulrich Kölm, who was the most prolific inventor of the group.[6] Details of his work cannot be given here, due to space limitations.[7]

Since the plugboard of the 1920 tabulator could not be extended, the engineers had to add a number of smaller plugboards, placing them just beneath the printing device.[8]

---

[6] Kistermann, *The Way*, Fig. 6.

[7] Kistermann, *The Way*, 43.

[8] Kistermann, *The Way*, Fig. 5.

The period between 1926 and 1931 is very interesting as far as the inventions at DEHOMAG in Berlin are concerned. When the summary punch was added to this tabulator in 1931, another small plugboard had to be added. However, the wiring of so many plugboards became something of a problem. The important improvements in these upgraded machines were:

- the balancing device
- introduction of 9's-complement instead of the 10's complement arithmetic which had been used since Hollerith's time
- introduction of 9's-complement arithmetic for punching negative values

This is remarkable because it made the punching of negative numbers easier and less error-prone. From then on, negative numbers were flagged using an X-punch in the highest column of the card field.

To accomplish this distinction, selectors were added to the tabulator. Selectors are multicontact relays which allow the transfer of negative data to another counter, for example, thereby separating them from positive data when the responsible selector is engaged by the X-punch. That is, selectors are used especially for control purposes. This "field upgrade" could be installed in the tabulators at the customer's site.

Although it was not possible to install the cross-footing feature in these machines because that would have demanded a fundamental change in the construction of the tabulator, DEHOMAG's customers knew that cross-footing was possible in principle. The book-keeping machines of that time, such as the Elliott-Fisher, the Moon-Hopkins or the National, all had this feature.

These developments, coupled with customer demands, caused the DEHOMAG engineers to think about designing a new tabulator, independent of the IBM Corporation. The reasons for this were varied, but the driving forces were, firstly, trying to overcome the wiring intricacies of the enhanced IBM IIIA tabulator and, secondly, the addition of the cross-footing feature. At that time, the IBM technicians were in the process of developing a universal tabulator, a multiplier, and were struggling with the alphabetic printing feature, not to mention the introduction of the 80-column punched card in 1928. Therefore, the IBM colleagues had no time to deal with the requirements confronting the DEHOMAG.

The result of the construction efforts in Germany was the DEHOMAG BK tabulator, which was announced in March 1933.[9] This machine was a breakthrough for punched card data processing in Germany. The long requested cross-footing feature finally made this tabulator capable of going automatically through the complex wage accounting procedure, to give just one example.

---

[9] Kistermann, *The Way*, Fig. 7.

Cross-footing

Let me explain cross-footing. If numbers in different fields of a punched card are to be added (totaled), each of them must be transferred into its own counter (accumulator) during one machine cycle. The next machine cycle then reads the next card and the new data in each field is added only to its assigned counter. This is the same procedure as footing a column of numbers. Conventional tabulators were not able to transfer and add numbers contained in different punch fields on the same card, because reading these fields was a parallel process, as was the addition in the assigned counters. Once the card had been read, there was no way to alter the results.

Controlling the way the card was processed after it had been read, actually required additional cycles, called intermediate cycles (later named program steps), between the reading of two successive cards. This could only be achieved by temporarily stopping the card feed. Since this was caused anyway by automatic group control when a new group identifier was detected, this card-stop was used in the newly constructed machine as the ideal occasion to start intermediate cycles during which data would be transferred from one counter to another. The data was added when transferred. A number of these transfers, that is, cross-footing operations, could be done in parallel by the new machine.

More flexibility in printing was achieved by breaking up the permanent connection between the printing devices and the counters (which had been in existence since 1908). The printing devices could now be wired separately on the plugboard. It was also no longer necessary to punch negative numbers as complements. Cross-subtraction allowed the automatic conversion of the counter's content into its complement during the transfer to another counter.

The new wiring possibilities resulted in more complex wiring on the built-in plugboard. This would not have been efficient in the long run. Fortunately, a short time after the DEHOMAG BK tabulator was announced, the exchangeable plugboard was introduced. This allowed new applications to be wired apart from the tabulator.

The DEHOMAG BK tabulator was so great a success that DEHOMAG decided to enhance it once more. One year later the BK was finally able to compute the compound interest on bank accounts, previously a very tedious type of work for bank clerks. This machine, called DEHOMAG BKZ, also encouraged the company to continue with this development line, which eventually led to the DEHOMAG D11.[10]

---

[10] Friedrich W. Kistermann, "Multiplication, Division, and Printing with Punched Card Machines," *Annals of the History of Computing* 19:4 (1997) 67–69.

2. The DEHOMAG D11 tabulator

The D11 is the culmination of forty years of evolution of the Hollerith punched card system.[11]

It began in 1895 with the freight account application in the New York Central and Hudson River Railroad, when Herman Hollerith transformed his 1890 electric tabulating system (pure counting and tabulation) into a business-oriented one, the Hollerith punched card system. This can also be called the Hollerith data processing system to emphasize that there is, in principle, no difference between using electromechanical and electronic devices.

The D11 is an operator-oriented machine with the printing device in front of the operator, the card feed on the left and some counters visible on the right side of the machine. The exchangeable plugboard, which holds the "program," is also on the right side of the machine. The summary punch stands on its left.

Let me turn to a man, Hans Gross, who played an important part in the construction of the D11 after he had worked with Ulrich Kölm on the BK. Although he was not an inventor like Kölm, he assisted him in his work, and complemented him during the development of the tabulator. Unfortunately, we know very little about Gross short life. He died at the age of 34. Because he was the manager of the circuit design department in 1932, it can be assumed that Gross was instrumental, not only in the construction of the BK, but also in the construction of the D11. At that time, Kölm was deputy manager of production and in charge of manufacturing the BK and later the D11. He set up a factory, which was opened in January 1934. Unfortunately, very few traces could be found to support our conjectures concerning Hans Gross. So they may be seen as "informed speculation."

In principle, the D11 was designed with commercial applications in mind.[12] Only one motor drives the machine through the main shaft, which in turn drives the aggregates, such as the card reader, the counters, the print unit, and other parts. The transfer of power occurs with a gear transmission which drives the cam shaft, among other items. This shaft provides the timing that controls the electrical actions that can occur in the various cycles (card feed cycle, intermediate cycles, etc.), provided these actions have been "ordered" by the program on the plugboard.

The card feed is connected to the main shaft via a clutch, which provides the means to disconnect the card reader, if the work of the machine is to be stopped by the group control unit. Then, intermediate cycles, instead of card cycles, are started to execute a program, which is wired on the exchangeable plugboard. These intermediate cycles are mechanical as well as electrical, because relays are being switched and counters turned, as directed by the plugged program. During machine operation, the exchangeable plugboard is a

---

[11] Kistermann, *The Way*, Fig. 10.

[12] Kistermann, *The Way*, Fig. 8.

permanent part of the machine and controls its entire operation. Consequently, this plugboard is comparable to a *stored program.*

All machine cycles (the card cycle, the intermediate cycles etc.) are mechanical motions, that is, the motor makes exactly one rotation during one cycle. Intricate timing mechanisms, such as cams, clutches, gears and other devices, ensure that all operations follow in orderly sequence, that is, a counter will not accept data outside of its allotted time, the printer will move only when results are available, and so on.

With the help of a sorter, the cards are grouped into an order suitable for the report to be generated. The feeding of the card deck is controlled by the group control. A change in the group identifier starts up to nine intermediate cycles, during which cross-addition and cross-subtraction can be done, even several in parallel. It is also possible to print intermediate results.

The tabulator can start a multiplication with up to eight digits in the multiplicand and six digits in the multiplier, and the result can be used in subsequent operations. Certain operations can be made dependent on a condition, for instance, the punching of a zero-balance can be prevented or a negative total can be printed in red. And a summary punch, connected via multi-wire cable, can create a record of a specific result.

The flexibility of the DEHOMAG D11 was so great that it is not possible to go into any greater detail. An indication of this flexibility can be seen in the number of hubs in the plugboard. In 1929, the DEHOMAG III B (the extended tabulator IBM III A) with its enhancements, had 286 hubs and 14 switches. By 1933, the DEHOMAG BK had 1258 hubs and 36 switches. This was increased to 2040 hubs and 20 switches in the DEHOMAG D11, in September 1935.

This wiring/programming flexibility required a radical change in the skills required of the employees at customers' Hollerith departments. DEHOMAG educated the customers' personnel as well as its own technicians. Programming manuals with complete programming examples were provided.

Right from the beginning, the DEHOMAG staff developed a means to help program the machine. Fig. 2 shows a punched card with a three column punch field for the account number and a seven column punch field for credit (Soll) and debit (Haben) amounts, which are punched in the same field but differentiated by an X-punch for the debits.[13] This is the basis for the report shown just beneath the card and which is the result of the tabulator's processing operation.

One customer has two credit amounts and three debit amounts on his account number 511. The credit and debit sums, marked with an asterisk, are printed just beneath the single entries. These credit and debit sums are deducted and the result, the negative balance, marked with a special sign (*), is

---

[13] "Haben" (debit) is negative because this is the view of the bank clerk. From the customer's point of view it is positive, because he has money on his account. It simply depends on the beholder's position: behind or in front of the counter.

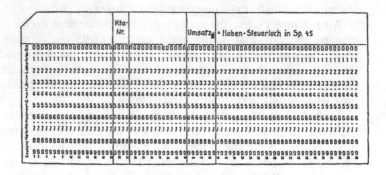

Figure 2: DEHOMAG D11 tabulator – programming (I)

Konto-Nr.	Soll-Umsatz	Haben-Umsatz	Soll-Saldo	Haben-Saldo
5 1 0	2	1 2 5		
	7 2 0			
		1 2		
		6 0		
	7 2 2 •	4 9 7 •	2 2 5 •	
5 1 1	3 0			
		3 4 0		
		1 9		
		1		
	3 0 0			
	3 3 0 •	3 6 0 •		3 0 ²
5 1 2	1 4 0			
	5 6 0			
		5		
		2 5 0		
	5			
	7 0 5 •	2 5 5 •	4 5 0 •	
5 1 3		5 0		
		1 5		
		3		
	2 3 0	7 5		
	2 3 0 •	1 4 3 •		8 7 •

printed on the same line in a third column. This application is known as "bal-ancing in two counters with balance printing."

Fig. 3 shows the printed form that was used as a planning guide to prepare the wiring (or programming) of the plugboard (or control panel).

Because of the great flexibility of the D11, the wiring preparation was standardized right after initial experience with the machine. The wiring (or programming) sheet consisted of three parts:

a)   the cycle overview
b)   the control area
c)   the transfer field

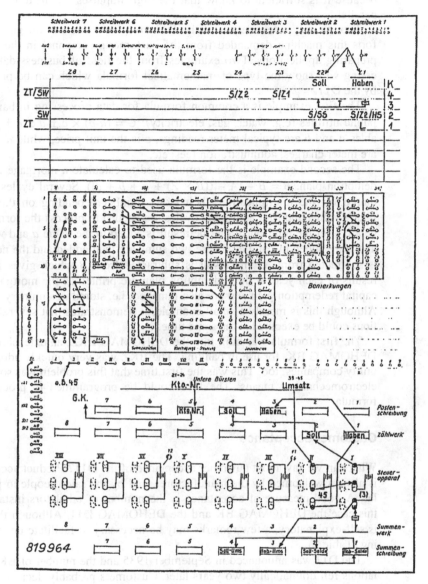

Figure 3: DEHOMAG D11 tabulator – programming (II)

Unfortunately, this is not the place to go any deeper into the programming details beyond just showing the programming form. Let me just add that even wired plugboards of great complexity could be documented on such a sheet. The wires which transferred the electrical impulses were omitted for clarity, because it is sufficient to know that the digit impulses are all transferred in parallel.

Though only a few details of the D11 are given, one can imagine the efforts this machine demanded from the customer's personnel in the way of preparing applications. Two examples from the field of business data processing will show the type of mathematical formula which can be processed by the D11.

The first example is a common formula for data processing in banks and public utilities (e.g. water, gas, electricity): $C = (a\text{-}b) \times c \pm d \pm e \pm f$. Variables a and b are typical meter readings, c is the unit price, while d, e and f are meter charges, administration fees, etc.

The second example is the formula used to calculate a mortgage loan redemption plan: $B' = B - (A - ((B \times Z) + a + b + c))$. Several cycles are required for the multiplication, their exact number depending on the multiplier's value, and four additional cycles are needed to process the formula. In detail: after the multiplication, the result is added to the value a and values b and c are added in parallel. These two sums are then added and the new sum is subtracted from A. The result is subtracted from B which gives B'. The results of each year of the redemption plan are printed as the mortgage loan capital redemption plan and are also punched, i.e. stored in summary cards.[14] Although this is merely a small example, it demonstrates that several operations could be executed at the same time.

The first formula was built into the DEHOMAG BKZ, an extension of the DEHOMAG BK, for use in banking. The BKZ was very quickly adopted by utility companies, too. This was the first time that this problem was solved by electromechanical means. The D11 could be programmed to process this formula.

Customer acceptance

What can we say about customer acceptance of the D11? Product acceptance is always welcomed by the manufacturer, and by the sales people in particular. A good measure of acceptance is the number of tabulators installed, in this case the DEHOMAG BK and the DEHOMAG D11. Although the table is self-explanatory, a few remarks may be necessary to facilitate understanding (Table 2).

The D11 was announced in September 1935 and the number of BK installations fell dramatically two years later. Customers probably decided to wait for the availability of the D11. The first D11 was installed in July 1936. By 1937, the BK model was obsolete, while installation of the D11 soared. In 1941, a total of 834 D11s had been installed. At the beginning of 1943 the

---

[14] Erwin Alte, "Die Anwendung des Hollerith-Verfahrens im Aktivgeschäft einer Hypothekenbank," *Hollerith-Nachrichten*, Berlin, 1937, 1033–1044.

Table 2: Number of installations per year of DEHOMAG tabulators

	DEHOMAG BK	DEHOMAG D11
1934	32	-
1935	90	-
1936	93	11
1937	18	101
1938	5	138
1939	0	179
1940	1	232
1941	1	173
1942	-	140
1943	-	71
1944	-	25
1945	-	4

number of D11s installed rose to 1075. After WW II the yearly installation
increased again. This means that the D11 saw 25 years of production, and a
much longer time in use. The last D11s were shut down at the beginning of
the 1970s. The remarkable long life of this machine is further evidence of its
usefulness and efficiency, which could not be provided by IBM tabulators.
Some D11s have survived in German museums, e.g. in Munich, Paderborn,
and Dresden. The D11 which is installed in the House for the History of IBM
Data Processing in Sindelfingen was shut down by the customer and returned
to IBM in 1972.[15]

Characteristics of the DEHOMAG D11

Let me summarize the functions and characteristics of the DEHOMAG D11.
The machine featured:

- Electromechanical technology
- Decimal base
- Parallel data transfer during input, processing, and output
- Parallel data processing (data-flow structure)
- 11-digit counters with algebraic sign (also used for storage)
- Zero-reset of the counters (adding the complement of the counter to
  itself)

---

[15] Friedrich W. Kistermann, Karl-Otto Reimers, "The House for the History of IBM
Data Processing, Sindelfingen," *Annals of the History of Computing* 19:4 (1997)
73–76.

- Stored-program via exchangeable control panel (plugboard)
- Three-level group control
- Nine program steps, which can be repeated indefinitely
- Conditional branching by punch position and/or by processing results, e.g. zero balance, positive or negative sign, red/black printing, etc.
- Multiplication via special hardware or plugboard wiring (multiple multiplications in parallel operation)
- Division with special hardware or plugboard wiring or reciprocal multiplication (multiple divisions in parallel)
- Formula calculations
- Numerical printing of single values, intermediate results, and final sums
- Printing of forms with forms control
- Summary card punching (including group identification)
- Counting capability
- Printing possible at every program step
- Use of nine's complement system with automatic end-around carry
- Negative data input as regular numbers (with X-punch indication)
- Lasting tabulator: announced in September 1935, retired in 1960.

An appropriate, if cumbersome designation of the D11 would be the following: automatic sequence-controlled, stored-program, forms-controlled printing and summary punching calculator. However, one important feature must be added to this description: the D11 is, definitely, an early type of data-flow machine.

So why was it called a tabulator? The DEHOMAG engineers were well aware of the fact that they had actually constructed a calculator. The DEHOMAG papers include a letter which shows the outcome of a competition for naming the new machine. The first prize was awarded for the name "Hollerith Allrechner," which can be translated as "Hollerith General Purpose Calculator." Another proposal was "Master Calculator." The final decision was to call it "Hollerith Tabelliermaschine, Type D11." This choice was natural for the marketing experts, because what would customers have said, had the salesmen offered them the new machine merely as a calculator instead of a true tabulating machine?

Whatever the definition of a "computer" may be, and at the moment there is no consensus about this, the DEHOMAG D11 *is a computer*.

3. Summary

The development of the Hollerith tabulator has revealed a rather slow pace of improvement. The addition of new functions came if and when customers made demands based on their experience with the Hollerith data processing system and on the expected improvement to their business. The system helped primarily to streamline the work flow. It made bureaucratic organizations more efficient and made it easy to obtain summarized data, which are the basis for getting information quickly.

The DEHOMAG D11 tabulator is not comparable with any other machines of this kind. The intention of this contribution is to present a machine which has been hitherto almost unknown, although it can be considered a milestone in the development of data processing.

Don't look too closely at the speed and the capacity of the D11, capacity meaning the number of counters, the number of program steps and the number of selectors. The hardware of the machine reflects 1930s technology: clumsy counters and slow relays. But the ideas realized in the D11 were new and brilliant, and outstanding for that time.

Acknowledgments

I would like to thank Rolf Ziegler, my colleague in the House for the History of IBM Data Processing, for carefully reading this paper, for his suggestions, and for polishing my English.

FRIEDRICH W. KISTERMANN graduated in chemistry at Johann Wolfgang Goethe University in Frankfurt/Main obtaining a Ph.D. in 1960. He worked with Henkel & Cie, Düsseldorf, in the field of patent documentation and information retrieval from 1960 until 1964, leading to some journal publications and basic work for the World Patent Index. In 1964, he transferred to IBM Germany, where he worked in DP application development, special libraries, information retrieval, DP planning and company communications. During this time he wrote some DP book translations and journal publications, and lectured in systems analysis, decision tables and information retrieval. After taking early retirement in 1987 he did consulting work for a few years on data base management systems and history of technology. History of data processing has been one of his fields of interest since 1976, and culminated in three exhibitions (*Development of Data Processing*, Hannover Fair, 1979; *Calculating machines of Philipp Matthäus Hahn*, 1989/90, and *Wilhelm Schickard*, 1993) and journal publications. In 1988 he helped to establish the House for the History of IBM Data Processing in Sindelfingen, where he is still doing voluntary work.

The Architecture of Konrad Zuse's Early Computing Machines

Raúl Rojas

Abstract. This paper provides a detailed description of the architecture of the computing machines Z1 and Z3 which Konrad Zuse designed in Berlin between 1936 and 1941. The necessary basic information was obtained from a careful evaluation of the patent application Zuse filed in 1941. Additional insight was gained from a software simulation of the machine's logic. The Z1 was built using purely mechanical components; the Z3 used electromechanical relays. However, both machines shared a common logical structure, and their programming model was the same. I argue that both the Z1 and the Z3 possessed features akin to those of modern computers: the memory and processor were separate units; the processor could handle floating-point numbers and compute the four basic arithmetical operations as well as the square root of a number. The program was stored on punched tape and was read sequentially. In the last section of this paper, I show that, surprisingly, the Z3 can emulate any modern computer.

1. The Z1 and Z3

Konrad Zuse is popularly recognized in Germany as the father of the computer, and his Z1, a programmable automaton built from 1936 to 1938, has been called the world's "first programmable calculating machine." Zuse was born in Berlin in 1910 and died in December of 1995.

While still a student at the Berlin Polytechnic, Zuse started thinking about computing machines in the 1930s. He realized that he could build an automaton capable of executing a sequence of arithmetical operations like those needed to compute mathematical tables. Coming from a civil engineering background, he had no formal training in electronics and was not acquainted with the technology used in conventional mechanical calculators. This nominal deficit worked to his advantage, however, because he had to rethink the whole problem of arithmetic computation and thus hit on new and original solutions.[1]

Zuse decided to build his first experimental calculating machine exploiting two main ideas, namely that the machine would work with binary numbers and that the computing and control unit would be separate from the storage. Years before John von Neumann explained the advantages of a computer architecture in which the processor is separated from the memory, Zuse had already arrived at the same conclusions. In 1936, Zuse completed the memory of the machine he had planned. It was a mechanical device, but not of the usual type. Instead of using gears (as Babbage had done in the previous century), Zuse implemented logical and arithmetical operations using sliding metallic bars. The bars could move only in one of two directions (forward or backward), making them appropriate for a binary machine. The processor of the Z1 was completed a few months after the storage unit, using the same kind of technology. It worked in concert with the memory but was never very reliable. The main problem was the precise synchronization that was needed in order to avoid applying excessive mechanical stress on the moving parts. It is interesting to note that in the same year in which Zuse completed the

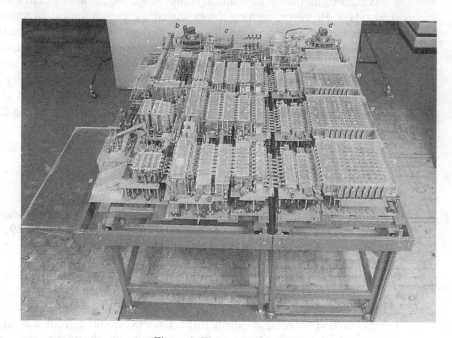

Figure 1: The reconstructed Z1

[1] K. Zuse, *Der Computer – mein Lebenswerk*, Springer-Verlag, Berlin, 1970. K.-H. Czauderna, *Konrad Zuse, der Weg zu seinem Computer Z3*, (Oldenbourg Verlag, Munich, 1979).

memory of the Z1, Alan Turing wrote his ground-breaking paper on computable numbers in which he formalized the intuitive concept of computability.
Fig. 1 is a photograph of the reconstruction of the Z1 that can be seen today
in Berlin's German Technology Museum. In the 1980s, Zuse directed the
reconstruction of the machine (which was destroyed in a bombing raid during
World War II).

The Z1, although unreliable, showed that the architectural design was
sound, which compelled Zuse to start investigating other kinds of technology.
Following the advice of his friend Helmut Schreyer, he considered using
vacuum tubes, but gave up the idea in favor of electromechanical relays
which were easier to obtain before and during the war. Zuse built an "intermediate" simpler model (the Z2) using a hybrid approach (a processor built
out of relays and a mechanical memory). In 1938, Zuse started building the
Z3, a machine consisting purely of relays but with the same logical structure
as the Z1. It was ready and operational in 1941, four years before the ENIAC.
Fig. 2 shows a photograph of the reconstructed Z3 in Munich's *Deutsches
Museum*. Like the Z1, the Z3 was destroyed during the war; it was reconstructed by Zuse in the 1960s.

Figure 2: The reconstructed Z3

This paper offers a detailed discussion of the architecture of the Z1 and Z3. Although the Z1 was reconstructed for a museum, the information available describes only the design of the mechanical memory.[2] Zuse documented the Z3 in his patent application Z-391 of 1941, which is rather difficult to decrypt due to the non-standard notation and terminology.[3] Since Z1 and Z3 were practically equivalent from the logical and functional points of view, I will refer only to the Z3 from now on. The main architectural difference between the Z1 and Z3 was the fact that the square root operation was left out of the Z1. There were also minor differences in the number of bits used for arithmetical operations in the processor (the Z1 used one bit less for the mantissa of floating-point numbers) and the number of cycles needed for each instruction. With this minor caveat, and taking only the architectural features into account, one can speak of the Z1 and Z3 as nearly equivalent machines.

2. Architectural overview of the Z3

This section summarizes the most relevant architectural features of the Z3. The paper goes from the simple to the complex: first I provide an overview of the architecture, then I go into more detail. In order to avoid awkward sentences, I will refer to the Z3 in the present tense.

Block structure

The Z3 is a floating-point machine. Whereas other early computing automata like the Mark I, the ABC, and the ENIAC worked with fixed-point numbers, Zuse decided very early on to adopt what he called "semi-logarithmic" notation, which corresponds to the modern floating-point representation.

Fig. 3 is an overview of the main building blocks of the Z3. The first relevant feature is the separation between processor and memory. The Z3 consists of a binary memory unit (capable of storing 64 floating-point numbers), a binary floating-point processor, a control unit, and I/O devices. The memory and the arithmetical unit are connected by a data bus, which transmits the exponent and mantissa of the floating-point representation. The control unit contains the microsequencers needed for each instruction. Control lines going from the control unit to the processor, the memory, and the I/O devices enforce the correct synchronization of all units. The tape reader provides the

---

[2] U. Schweier and D. Saupe, "Funktions-und Konstruktionsprinzipien der programm-gesteuerten mechanischen Rechenmaschine Z1," *Arbeitspapiere der GMD* 321, (Bonn, 1988).

[3] K. Zuse, "Patentanmeldung Z-391", in R. Rojas (ed.), *Die Rechenmaschinen von Konrad Zuse*, (Springer-Verlag, Berlin 1998).

punched tape

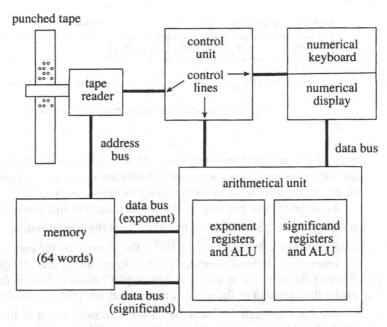

Figure 3: Building blocks of the Z3

opcode of each instruction as well as the address for memory access. The I/O devices are connected by a data bus to the computing unit.

Floating-point representation

Fig. 4 shows the representation used in the memory of the Z3. The first bit is used to store the sign of the number, the following seven bits are for the exponent, and the last 14 bits for the mantissa (only the 14 places to the right of the binary point). The bits of the exponent are called Part A of the number and are denoted by a_6 to a_0. The bits of the mantissa are called Part B of the number and are denoted by b_0 to b_{-14}. The exponent is coded as a two's complement number. The range of possible values therefore runs from −64 to 63. The mantissa is stored in *normalized* form, that is, the first digit before the decimal point (b_0) must always be a 1. This digit does not need to be stored (and therefore does not appear in Fig. 4), so that the effective range of the numbers in the memory unit is equivalent to a mantissa of 15 bits. However, there is a problem with the number zero, which cannot be expressed using a normalized mantissa. The Z3 uses the convention that any mantissa with exponent −64 is to be considered equal to zero. Any number with expo-

sign   exponent   significand

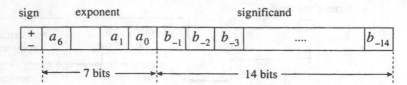

Figure 4: Floating-point representation in memory

nent 63 is considered infinitely large. Operations involving zero and infinity are treated as exceptions, and special hardware monitors the numbers loaded in the processor in order to set the exception flags (see Section 4).

According to this convention, the smallest number that can be stored in the memory of the Z3 is $2^{-63} = 1.08 \times 10^{-19}$, and the largest number that can be represented is $1.999 \times 2^{62} = 9.2 \times 10^{18}$. The arguments for computations can be entered as decimal numbers on the keyboard of the Z3 (four digits). Pushing the appropriate button in a row of 17 buttons labeled from −8 to 8 enters the exponent of the decimal representation. The original Z3 could only accept input between 10^{-8} and 10^8. Zuse's reconstruction of the Z3 for the Deutsches Museum in Munich provides enough buttons for larger exponents – this arrangement allows the whole numerical capacity of the machine to be reflected on the acceptable input and output. However, the Z3 does not print the numerical results the program produces. A single number is displayed on an array of lamps representing the digits from 0 to 9. The largest number that can be displayed is 19,999. The smallest is 00001. The largest exponent that can be displayed is +8, the smallest −8.

Instruction set

The program for the Z3 is stored on punched tape. One instruction is coded using 8 bits for each row of the tape. The instruction set of the Z3 consists of the nine instructions shown in Table 1. There are three types of instructions: I/O, memory, and arithmetical instructions. The opcode has a variable length of two or five bits. Memory operations encode the address of a word in the lower six bits, that is, the addressing space has a maximum size of 64 words, as mentioned above.

Table 1: Instruction set and opcodes of the Z1 and Z3

	Instruction	Description	Opcode
I/O	Lu	read keyboard	01 110000
	Ld	display result	01 111000
memory	Pr z	load address z	11 $z_6z_5z_4z_3z_2z_1$
	Ps z	store address z	10 $z_6z_5z_4z_3z_2z_1$
arithmetic	Lm	multiplication	01 001000
	Li	division	01 010000
	Lw	square root	01 011000
	Ls1	addition	01 100000
	Ls2	subtraction	01 101000

The instructions on the punched tape can be arranged in any order. The instructions Lu and Ld (read from keyboard and display result, respectively) halt the machine, so that the operator has enough time to input a number or write down a result. The machine is then restarted and continues processing the program.

The instruction most conspicuously absent from the instruction set of the Z3 is conditional branching. Loops can be implemented by the simple expedient of bringing together the two ends of the punched tape, but there is no straightforward way to implement conditional sequences of instructions. However, we will show later that conditional branching can be simulated on this machine.

Number of cycles

The Z3 is a clocked machine. Each cycle is divided into five "stages" called I, II, III, IV, and V. The instruction read from the punched tape is decoded in stage I of a cycle. The two basic arithmetical operations carried out by the machine are the addition and subtraction of exponents and mantissas. These operations can be executed in the first three stages of each cycle. Stages IV and V are used to prepare arguments for the next operation or to write back results.

The instructions of the Z3 require the following number of cycles:

- Multiplication: 16 cycles
- Division: 18 cycles
- Square root: 20 cycles
- Addition: 3 cycles
- Subtraction: 4 or 5 cycles, depending on the result

- Read keyboard: 9 to 41 cycles, depending on the exponent
- Display output: 9 to 41 cycles, depending on the exponent
- Load from memory: 1 cycle
- Store to memory: 0 or 1 cycle

According to Zuse, the time required for a multiplication was three seconds. Considering that a multiplication operation needs 16 cycles, one can estimate that the operating frequency of the Z3 was 16/3 = 5.33 Hz.

The number of cycles needed for the *read* and *display* instructions is variable, because it depends on the exponent of the arguments. Since the input has to be converted from decimal to binary representation, the number of multiplications with the factor 10 or 0.1 is dictated by the decimal exponent (see Section 4).

Addition and subtraction require more than one cycle because, in the case of floating-point numbers, care has to be taken to set the size of the exponent of both arguments to the same value. This requires some extra comparisons and shifting.

A number can be stored in memory in zero cycles when the result of the last arithmetical operation can be redirected to the desired memory address. In this case, the cycle needed for the store instruction overlaps the last cycle of the arithmetical operation.

Programming model

It is very important to describe the Z3 programming model, that is, the part of the machine visible to the programmer. From the point of view of the software, the Z3 consists of 64 memory words that can be loaded into two floating-point registers, which I simply call R1 and R2. These two registers contain the arguments of the arithmetical operations. The programmer can write any sequence of instructions, but has to keep in mind the state of the machine's registers.

The important point to remember is the following: the first load operation in a program (Pr z) transfers the contents of address z to R1. Any other subsequent load operation transfers a word from memory to R2. The first read keyboard instruction loads the numerical input into register R1, a second read instruction loads register R2.

Arithmetical operations do not specify their arguments in the opcode. Their implicit semantics is the following:

Multiplication:	R1:= R1×R2
Division:	R1:= R1/R2
Addition:	R1:= R1+R2
Subtraction:	R1:= R1−R2
Square root:	R1:= SQRT(R1)

R2 is cleared after an arithmetical instruction, whereas the result is stored in R1. Load operations following an arithmetical instruction refer to R2. The store and display instructions always refer to R1, which also contains the result of the previous arithmetical operation. After a store or display operation, R1 is cleared. The next load operation refers to R1.

The programming model of the Z3 is best clarified by an example. Assume that we want to compute the following expression:

$$x^2 + bx$$

Assume further that we have stored the constant b in address 2 of the memory unit. The value x is stored in address 1. The program that performs the desired computation is the following:

Pr 1   load x in R1
Pr 1   load x in R2
Lm   multiply R1 and R2, result in R1
Ps 3   store R1 in address 3, clear R1
Pr 1   load x in R1
Pr 2   load b in R2
Lm   multiply R1 and R2, result in R1
Pr 3   load x^2 in R2
Ls1   add R1 and R2, result in R1
Ld   display result

After the last instruction has been executed, the processor is reset to its initial state. A new program sequence can then be started.

3. The Z3 block diagram

In this section, I take a closer look at the structure of the Z3 and describe its main building blocks in more detail. The main issue is how to enforce the correct synchronization of the available components.

The processor

Fig. 5 shows a simplified representation of the arithmetical unit of the Z3. The unit has two parts: the left side is used for operations with the exponents of the floating-point numbers, the right side for operations with the mantissas. Af and Bf are registers used to store the exponent and mantissa of register R1, seen from the programmer's point of view. I will refer to R1 as the register pair [Af:Bf]. The register pair [Ab:Bb] stores the exponent and mantissa of R2. The pair [Aa:Ba] contains the exponent and the mantissa of a third temporal floating-point register invisible to the programmer. The two arithmetic logical units (ALUs), A and B, are used to add or subtract exponents and mantissas, respectively. The result of the operation in the exponent part is put into Ae. In the mantissa part, the result of the operation is put into Be. The pair [Ae:Be] can be considered an internal register invisible to the programmer. In Part B, a multiplexer allows selection of Ba or the output of the ALU as the result of the operation. The multiplexer is controlled by a relay Bt (if Bt is equal to zero, then Be is set equal to Ba).

The small boxes labeled Ea, Eb, Ec, Ed, Ef, Fa, Fb, Fc, Fd, and Ff, are switches that open or close the data bus. If the content of register Af is to be transferred to Aa, for example, the box of relays Ea is set to 1 and the result is Aa:=Af. As can be seen in Fig. 5, the content of Af can only be transferred to Aa or Ab, whereas the content of Ae can be transferred to Aa, Ab, or Af, according to the states of the switches. The structure of Part B of the arithmetical unit is very similar, but, in addition to the multiplexer controlled by the relay Bt, there is also a shifter between Bf and Ba and a shifter between Bf and Bb. The first shifter can displace the mantissa up to two positions to the right and one position to the left. This amounts to a division of Bf by 4 or a multiplication by 2. The second shifter can displace the mantissa in Af between 1 and 16 positions to the right, and between 1 and 15 positions to the

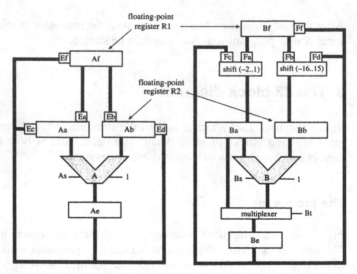

Part A: operations with the exponents   Part B: operations with the significands

Figure 5: The registers and datapath

left. These shifts are needed for addition and subtraction of floating-point numbers. Multiplication and division with powers of two can therefore be performed when the operands for the next arithmetical operation are fetched and, in this sense, do not consume time.

The number of bits used in the registers is as follows:

Af	7 bits	Bf	17 bits
Aa	7	Ba	19
Ab	7	Bb	18
Ae	8	Be	18

As can be seen from this list, Ae uses one extra bit to handle the addition of the exponents of the arguments. Part B of the processor uses two extra bits for the mantissas (b_{-15}, b_{-16}), and makes explicit b_0, which is not stored in memory. The extra bits at positions b_{-15} and b_{-16} are included to increase the precision of the computations. Therefore, the total number of bits needed to store the result of an arithmetical operation in Bf is 17. Registers Ba and Bb require more extra bits (ba_2, ba_1, and bb_1) to handle the intermediate results of some of the numerical algorithms. In particular, the square root algorithm can lead to partial computations in Ba requiring three bits to the left of the binary point.

The basic primitive operation of the datapath is the addition or subtraction of exponents or mantissas. When the relay As (Bs) is set, the negation of the second argument Ab (Bb) is fed into the ALU. Therefore, if the relay As is set to 1, the ALU in Part A subtracts its arguments, otherwise they are added. The same is true for Part B and the relay Bs. The constant 1 is needed to build the two's complement of a number.

Assume that two numbers with the same exponent are to be added. The first exponent is stored in Af, the second in Ab. Since they are equal, no operation has to be performed on this side of the machine. In Part B, the mantissa of the first number is stored in Bf and the mantissa of the second in Bb. The first step consists of loading Ba with Bf by setting the relay box Fa to 1. The addition is performed next, the relay Bt is set to one so that the result Ba+Bb is assigned to Be. The relay box Ff is now set to 1, and the result is stored in Bf. As can be seen, the information can move between registers and so flow through the datapath. The computer architect has to provide the correct sequence of activation of the relay boxes in order to get the desired operation. This is done in the Z3 by using a technique very similar to microprogramming.

The control unit

Fig. 6 shows a more detailed diagram of the control unit and the I/O panels. The circuit Pa decodes the opcode of the instruction read from the punched

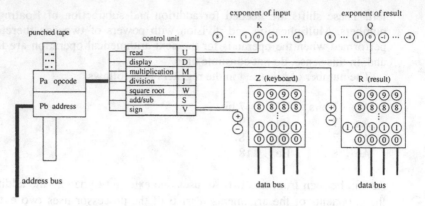

Figure 6: The control unit and I/O panels

tape. If it is a memory instruction, circuit Pb sets the address bus to the value of the lower six bits of the opcode. The control unit determines the correct microsequencing of the instructions. There are special circuits for each of the operations in the instruction set.

Circuit Z represents the panel of buttons used to enter a decimal number in the machine. Only one button in each of the four columns can be activated. The exponent is set by pressing one of the buttons labeled –8 to 8 in circuit K. The output display is very similar to the input panel, but here lamps illuminate the appropriate decimal digits, the exponent of the number (circuit Q), and its sign. Note that there is a fifth digit for the output (which can only be one or zero).

Once a decimal number has been set, a data bus transmits the digits to register Ba and a complex series of operations is started. The decimal input must be transformed into a binary number. This requires a chain of multiplications, shorter or longer according to the absolute magnitude of the exponent. If the exponent is zero, the whole transformation requires nine cycles, but if it is –8, the operation requires $9+4\times8 = 41$ cycles.

Microcontrol of the Z3

The heart of the control unit is made up of its microsequencers. Before I describe the way they work, it is necessary to take a closer look at the chaining of arithmetical instructions in the Z3. Fig. 7 shows the main idea. Each cycle of the Z3 is divided into five stages. Stages IV and V are used to move information around in the machine. During stages I, II, and III, an addition/subtraction is computed in Part A and another in Part B of the Z3. I call this the "execute" phase of an instruction. A typical instruction fetches its

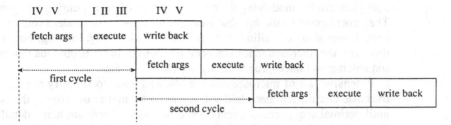

Figure 7: The execution pipeline of the Z3

arguments, executes, and writes back the result. Zuse took great care to save execution time by overlapping the fetch stage of the next instruction with the write-back stage of the current one. One can think of an execution cycle as consisting of just two stages, as shown in Fig. 7, where the first two cycles of a series of instructions have been labeled. I have adopted this convention in the tabular diagrams of the numerical algorithms discussed later on.

Special control wheels provide the microsequencing. There is one wheel for the multiplication algorithm, another to control division, and yet another for the square root instruction. The moving arm shown in Fig. 8 starts moving clockwise as soon as the control unit decodes the corresponding instruction. In each cycle, the arm moves from one position to the next. The arm conducts electricity and activates the circuits with which it comes into contact. In the example shown in Fig. 8, the moving arm sets the relay box Ea to 1 in the first cycle. This leads to the transfer of the contents of register Af into Aa. In the next cycle, the relay boxes Ec and Fc are activated. In this way, the results of the operations in Parts A and B are written back into the registers Aa and Ba, respectively. As can be seen, such control wheels provide a comfort-

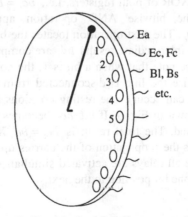

Figure 8: Control wheel for microsequencing

able platform for modifying the exact sequence of events during an operation. They correspond to the microsequencers used today in modern microprocessors. I stop short of calling them a form of microprogramming, because, in this case, the microsequence has been hardwired, but it is obvious that microsequencing and microprogramming are closely related.

Extensive use of microsequencing allowed Zuse to simplify the Z3. Once the basic circuits had been laid out, it was just a matter of refining the control until optimal sequences of events could be found. There are many details that the engineer designing the "microprogram" must keep in mind, otherwise short circuits can destroy the hardware. The Z1, with its mechanical design, was still more sensitive in this respect than the Z3. Even after it was completed, there were sequences of instructions that the programmer had to avoid in order not to damage the hardware. One of those sequences was inadvertently tried at the Berlin Technology Museum in 1994 and damaged the reconstructed Z1.

The adders

An important feature of the Z3 is the design of the adders, which compute additions and subtractions using a method called *carry look-ahead*. If binary addition is implemented in a straightforward way, carries have to be passed from one bit position to the next. In the case of the mantissa, one would need 16 cycles just for transmission of the carry bits. The adders Zuse designed are much faster than that – they perform an addition or subtraction in stages I, II, and III of a single cycle. Subtraction is computed by complementing the second argument and adding an extra 1 at the lowest bit position.

Consider addition of the registers Ba and Bb. I will refer to the i-th bit of register Bb by bb_i or Bb[i], as is convenient. I will use the same notation in the case of other registers. First of all, a partial result is computed which is the bitwise XOR of both registers, i.e., $bc_i = ba_i$ XOR bb_i. A second partial result is the bitwise AND operation applied to both registers, i.e ba_i AND bb_i. The next operation locates the bit positions at which a carry is needed. The intermediate results bd_i are computed using the circuit shown in Fig. 9. Please note that when a bit is 1, the corresponding line carries a current, otherwise the line is disconnected from the power source (so that no short circuit can occur). The resting positions of the relays $bc_1,\ldots,bc_{-16}$ are the ones shown in Fig. 9. If bit bc_i becomes equal to 1, the corresponding relay is closed. The final result is $be_i = bd_i$ XOR bc_i. Note that the use of relays makes the propagation of the carries up to the last bit position needed easier. Since all relays are activated simultaneously, the carry is not delayed going from one bit position to the next.

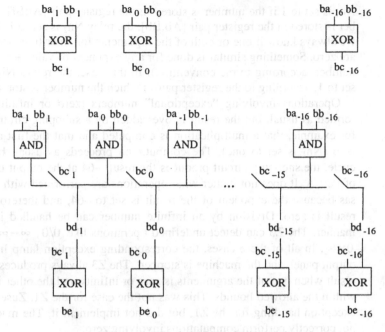

Figure 9: Circuit for carry-look-ahead

4. Numerical algorithms

In this section, I describe the floating-point algorithms used by the Z3. They are, without exception, the same as those normally used in small sequential floating-point processors.[4]

Floating-point exceptions

The problem with normalized floating-point notation is that special conventions have to be used to deal with the number zero. The Z3 solves this problem and deals with other exceptions (overflow and underflow) by monitoring the value of the exponent after any arithmetical operation or a load from memory. A special circuit looks at the state of the bus Ae and captures exceptions. Any number with exponent –64 is flagged as zero: a relay denoted

---

[4] I. Koren, *Computer Arithmetic Algorithms*, (Prentice Hall, Englewood Cliffs, NJ, 1993).

Nn_1 is set to 1 if the number is stored in the register pair [Af:Bf]. If the number is stored in the register pair [Ab:Bb], the relay Nn_2 is set to 1. In this way, we always know if one or both of the arguments for an arithmetical operation are zero. Something similar is done for any exponent of value +63 (an infinite number, according to the convention). In this case, the relays Ni_1 or Ni_2 are set to 1, according to the register pair in which the number is stored.

Operations involving "exceptional" numbers (zero or infinity) are performed as usual, but the result is overridden by a snooping circuit. Assume, for example, that a multiplication is computed and that the first argument is zero (Nn_1 is set to one). The computation proceeds as usual, but, in each cycle, the snooping circuit produces the result −64 at the output of the adder of Part A. It does not matter what operations are performed with the mantissas because the exponent of the result is set to −64, and therefore the final result is zero. Division by an infinite number can be handled in a similar manner. The Z3 can detect undefined operations like $0/0$, $\infty - \infty$, ∞/∞ and $0 \times \infty$. In all of these cases, the corresponding exception lamp lights on the output panel, and the machine is stopped. The Z3 always produces the correct result when one of the arguments is zero or infinity and the other is a number within the allowed bounds. This was not the case for the Z1. Zuse considered exception handling for the Z1, but did not implement it. The machine could not correctly perform computations involving zero.

An additional circuit looks at the exponent of the result at the output of the exponent's adder. If the exponent is greater than or equal to 63, overflow has occurred and the result must be set to infinity. If the exponent is lower than − 63, underflow has occurred and the result must be set to 0. To achieve this, the appropriate relay (Nn_1 or Ni_1) is set to one.

Zuse managed to implement exception handling using just a few relays. This feature of the Z3 is one of the most elegant in the whole design. Many of the early microprocessors of the 1970s did not include exception handling and left it to the software. Zuse's approach is sounder, since it frees programmers from the tedium of checking the bounds of their numbers before each operation.

Addition and subtraction

In order to add or subtract two floating-point numbers x and y, their representation must be reduced to the same exponent. After this has been done, only the mantissas have to be added or subtracted. If the exponents are different, the mantissa of the smaller number is shifted to the right as many places as necessary (and its exponent is incremented correspondingly to keep the number unchanged) until both exponents are equal. It can, of course, happen that the smaller number becomes zero after 17 shifts to the right.

The signs of the two numbers are compared before deciding on the type of operation to be executed. If an addition has been requested and the signs are the same, the addition is executed. If the signs are different, a subtraction is executed. If a subtraction has been requested and the signs are different, an addition is executed. If the signs are the same, the subtraction is executed. A special circuit sets the sign of the result according to the signs of the arguments and the sign of the partial result.

A chain of relays (not a control wheel) controls addition and subtraction, since the maximum number of cycles needed is low. Initially, the arguments for the addition are stored in the register pairs [Af:Bf] and [Ab:Bb]. In Cycle 1, the exponents are subtracted. In Cycle 2, the mantissa with the larger exponent is loaded into register Ba and the mantissa with the smaller exponent into register Bb. The mantissa in register Bb is shifted as many places to the right as the absolute value of the difference of the exponents (exception handling takes care of the case when the smaller number becomes zero after the shift). In stages I, II and III of Cycle 2 the mantissas are added, and finally the processor tests if the result is greater than 2. If this is the case, the mantissa is shifted one position to the right and the exponent is incremented by 1.

In the case of a subtraction, four or five cycles are needed. The first two cycles are almost identical to the first two cycles of the addition algorithm, but now the mantissas are subtracted. Cycle 3 is executed only when the difference of the mantissas is negative. The effect of Cycle 3 is just to make the mantissa of the result positive. Cycle 4 is very important: the difference of two normalized mantissas can have many zeros in the first bit positions to the left. Shifting Be to the left as many places as necessary (this is done with the shifter between the relay box Fd and register Bb) normalizes the result. The number of one-bit shifts is subtracted from the exponent in Part A of the processor. In Cycle 5, the result is stored in the register pair [Af:Bf].

Multiplication

The multiplication algorithm of the Z3 is like the one used for decimal multiplication by hand, that is, it is based on repeated additions of the multiplicator according to the individual digits of the multiplicand. At the beginning of the algorithm, the first argument is stored in the register pair [Af:Bf]. The second argument is stored in the register pair [Ab:Bb]. The temporary register pair, [Aa:Ba], is set to zero. The algorithm takes 16 cycles to run. Note that only the bits of the multiplicand from position -14 to position 0 are used. The exponents are added in the first cycle and the result is kept in a loop in subsequent cycles in Part A of the arithmetical unit. The mantissas are handled in Part B of the unit. Register Ba contains the partial result of the computation. The basic multiplication loop has the following form:

```
Be:=0
FOR i = –14 to 0 DO
      Ba:= Be/2
      Be:= Ba + Bb×(i-th bit of Bf)
```

Note that in each iteration the partial result Be is shifted one position to the right to produce Ba:=Be/2. This is done with the shifter connected to Fc.

The result of the multiplication is a number r, such that $1 \leq r < 4$ (for arguments within bounds). In the last cycle, there is a check to see if $r \geq 2$. If this is the case, the result is shifted one position to the left and a 1 is added to the exponent of the result.

Division

The division algorithm is similar to the multiplication algorithm, but using repetitive subtraction instead of addition. At the beginning of the algorithm, the dividend is stored in the register pair [Af:Bf]. The divisor is stored in the register pair [Ab:Bb]. The algorithm takes 18 cycles to run.

The main idea behind the algorithm is very simple. The exponent of the result is obtained by subtracting the exponents of dividend and divisor. Now for the mantissa: assume that we want to compute x/y for the mantissas x and y. Since we are dealing with normalized numbers, the first digit of the result is 1 if $x \geq y$ and zero if $x < y$. In the first case, we set the first digit of the result to 1 and compute the remainder, which is $x - y$. The remainder is divided recursively by y. If the result bit is zero, the remainder is just x and the recursive division is continued as in the first case. The basic division loop has the following form:

```
Ba:=Bf
FOR i=0 to –14 DO
      IF (Ba–Bb ≥ 0)   THEN    Be:=Ba–Bb
                                Bf[i]:=1
                       ELSE     Be:=Ba
                                Bf[i]:=0
      Ba:=2×Be
```

In each iteration, the partial result Be is shifted one position to the left to produce Ba:=2×Be. This is done with the shifter connected to the relay box Fc.

The result of the division of mantissas is a number r greater or equal to ½ and smaller than 2. If r is smaller than 1, a 1 is subtracted from the exponent, and the result is shifted one position to the left in order to get a normalized number.

The square root algorithm works in a very similar way to division. The details of the algorithm have been published.[5]

Read and display instructions

The two most complex instructions of the Z3 are those related to the input and output of decimal numbers. A decimal number of four digits entered on the keyboard is converted into a binary integer. This is done by reading each digit sequentially, transforming it into a binary number, and storing it in the four bits Ba[−10], Ba[−11], Ba[−12], and Ba[−13] of register Ba. The number in register Ba is multiplied by 10 and the procedure is repeated for the other digits. After 4 iterations, the decimal input has been transformed to a binary number (the exponent is adjusted to the correct value). The difficult part is handling the exponent. If the exponent e is positive, the mantissa has to be multiplied e times by 10. If it is negative, it must be multiplied $-e$ times by 0.1. Multiplying by 10 is relatively easy: The mantissa in Be can be shifted one bit to the left and then stored in Ba (i.e., Ba:=2×Be). At the same time Be can be shifted three places to the left and can be stored in Bb (i.e., Bb:=8×Be). The addition of Ba and Bb then provides the desired result: the multiplication of the original number in Be by the constant 10. The process takes 4 cycles for each multiplication, that is, 32 cycles for the decimal exponent +8. Since a read operation needs a minimum of 9 cycles, this means that a decimal number with exponent +8 is read in 41 cycles.

In the case of negative exponents, multiplication by the constant 0.1 is performed using the shifters and the adders as well. This multiplication is somewhat more complex, because 0.1 is a periodic number in the binary system. A description of the microsequencing used would take us too far away from the main topics, so I will omit it here.[6]

The display instruction works by multiplying or dividing iteratively by 10. If the binary exponent of the number in register R1 is positive, the number is multiplied by 0.1 as many times as needed to make the binary exponent equal to 2 and until the first four bits on the left of register Bf contain a number between 0 and 9 (0000 and 1001). This is the decimal digit that can be displayed in the next column of the output panel. The number is subtracted from the mantissa in Bf, and the process continues for the following digits. If the binary exponent of the number in register R1 is negative, the process is similar, but multiplication by the constant 10 is used.

[5] R. Rojas, "Konrad Zuse's Legacy: The Architecture of the Z1 and Z3," *IEEE Annals of the History of Computing*, 19:2, (1997), 5–16.

[6] R. Rojas (ed.), *Die Rechenmaschinen von Konrad Zuse*, (Springer-Verlag, Berlin, 1998).

5. Complete architecture of the Z3

We are now in a position to make sense of the detailed diagram of the Z3 shown in Fig. 10. I discussed the control unit and the I/O earlier. Notice that the four decimal digits of the input keyboard are transferred to register Ba using the relay boxes Za, Zb, Zc, and Zd, which are activated one after the other.

The relay boxes Eg and Ei are used to set some useful constants directly into the exponent registers (+13 and –4). The shifter Ee between register Af and register Aa is used for the square root algorithm. The exponent of the result (Aa) becomes half the exponent (Af) of the original number.

Ah_1 is a relay acting as a flip-flop. When it is set to 0, the register pair [Af:Bf] is accessed by load operations. When it is set to 1, the register pair [Ab:Bb] is accessed. This relay is reset to 0 by the control line a_i. The control lines a_l, a_j, b_l, and b_j are used to clear the registers Af, Ab, Bf, and Bb when needed.

The box labeled "zero, infinite" below Ae represents the circuits for exception handling. They snoop permanently on the data bus (results of operations and data from memory) and raise the corresponding exception flags when needed. The shifter below Be is used to displace the mantissa one bit to the right. This provides the normalization needed for the mantissa whenever Be exceeds or equals 2.0.

Fp and Fq are the relays that control the number and direction of one-bit shifts in the shifter below the relay boxes Fc and Fa. Fh, Fi, Fk, Fl, and Fm have the same function in relation to the other shifter. Using these five bits, the numbers between –16 and 15 can be represented, which is also the range of the second shifter. When such a shift is performed, the number represented by the relays Fh to Fm is transferred through the relay box Bn to register Ab, in order to modify the exponent of the result. If the number is shifted 10 positions to the left, +10 is subtracted from the exponent of the result. Such drastic shifts are mostly needed after subtractions.

Look again at the diagram of the Z3. Everything makes sense now and looks as conventional as any modern small floating-point processor. It is indeed amazing how Konrad Zuse was able to find the adequate architecture right from the beginning. The Z3 processor employs just 600 relays; the memory needs three times as much. By having to optimize the design, by having to save hardware everywhere, Zuse was *forced* to think and rethink the logical structure of his machine. He was not allowed the luxury of the almost unlimited funding allocated by the U.S. military for the development of the ENIAC or by IBM for the Mark I. He was all alone. While this may have worked to his advantage on the conceptual side, it may also have worked to his disadvantage, considering the negligible impact that the Z1 and Z3 had on the emerging U.S. computer industry after World War II.

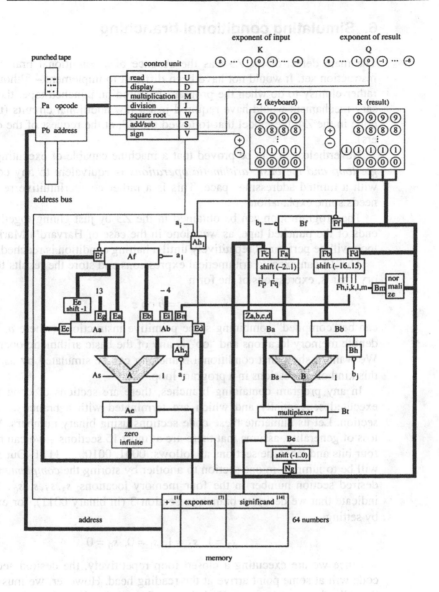

Figure 10: Complete architecture of the Z3

6. Simulating conditional branching

The main defect of the Z3 was the absence of a conditional branch in the instruction set. It would not have been difficult to implement – although it is rather clumsy to do when the program is stored on punched tape, the necessary mechanism would have required just a few additional circuits (this was done in the Z4, the model that followed the Z3, at the request of the customers).

Nevertheless, it can be proved that a machine capable of executing *a single loop and the basic arithmetic operations* is equivalent to any computer with a limited addressing space. This is a rather counterintuitive result that needs some explanation.

The program loop can be obtained in the Z3 by just gluing together both ends of the punched tape, as was done in the case of Harvard's Mark I. The loop will be performed repetitively until a halting condition is reached.

The Z3 can execute arithmetical expressions and store the results to memory, that is, expressions of the form

$$a := b \text{ op } c$$

can be "compiled" combining some primitive instructions (where a, b and c denote memory locations and "op" is any of the basic arithmetic operations). We want to show that conditional branching can be simulated by using only this kind of expressions in a program loop.

In any program containing branches, there are sections of code that are executed sequentially and which are terminated with a branch to another section. Let us numerate these code sections using binary numbers. Without loss of generality, assume that there are at most 15 sections – we can then use four bits and label the sections as follows: 0001, 0010, ..., 1111. Our strategy will be to jump from one section to another by storing the *complement* of the desired section number in the four memory locations s_3, s_2, s_1, s_0. We can indicate that we desire to branch to section 3 (in binary 0011), for example, by setting

$$s_3 = 1, \ s_2 = 1, \ s_1 = 0, \ s_0 = 0$$

Since we are executing a closed loop repetitively, the desired section of code will at some point arrive at the reading head. However, we must ensure that all other sections of code being read until the desired section appears, and which are always being executed, do not store the results of their operations in memory. In this way, it does not matter how many operations are performed until the desired section is reached, since the state of the memory is not changed.

Implementing this idea requires putting a *guard* at the beginning of each code section. This is done by using the auxiliary memory location t and computing the expression

$$t = ((s_3 - a)(s_2 - b)(s_1 - c)(s_0 - d))^2$$

at the beginning of each code section with the 4-bit binary label *abcd*. Since this computation involves only basic arithmetical operations and fixed memory addresses, it can be performed by the Z3. Now, the variable t is zero if we are in the desired code section, and one if we are not. We can therefore rewrite all expressions of the form "$a = b$ op c" as

$$a = at + (1 - t)(b \text{ op } c)$$

If we are in the desired code section, memory location a is set to the new value "b op c" (since $1 - t = 1$). If we are not, memory location a remains unchanged (since $1 - t = 0$).

Of course, we must take care to put the code necessary for computing t at the beginning of each section and to write only programs which use expressions of the form given above. Each section of code is closed by a branch to another code section. Since the results of arithmetical operations can be used to set the values of the variables s_3, s_2, s_1, s_0, all kinds of conditional branches can be executed. It can be proved that a Turing machine with a tape of limited size can be simulated by the Z3 using this approach.[7] Thus, the Z3 can, in fact, simulate any other computer.

Only one problem remains: since the program loop is executed repetitively, how do we stop the machine? This can done easily in the Z3 by causing an arithmetical exception. We can reserve a section of code as the "stop" section. When this section of code is called (by setting the locations s_3, s_2, s_1, s_0 to the appropriate section number), the auxiliary memory location t will be zero in this code section. The only operation that we include in this section is $0/t$. Whenever t is zero, the machine stops and signals the arithmetical exception 0/0. If t is not zero, the machine just goes through this computation and proceeds to the next section. Had Zuse not included arithmetical exceptions in his Z3, we would not be able to stop the loop and this whole approach would not work!

The invention of the computer

From the theoretical point of view, it is interesting to see that limited precision arithmetic embedded in a WHILE loop can compute anything that computers can. The result seems counterintuitive, until we realize that operations like multiplication and division are iterative computations in which branching decisions are taken by the hardware. The conditional branchings we need are

---

[7] For details see: R. Rojas, "How to Make Zuse's Z3 a Universal Computer," *IEEE Annals of the History of Computing*, Vol. 20:3 (1998) 51–54.

embedded in these arithmetical operations, and the whole purpose of the transformations used is to lift the branches up from the hardware in which they are buried to the software level, so that we can control the program flow. The whole magic of the transformation consists in making the hardware branchings visible to the programmer.

The approach discussed in this paper could, perhaps, be criticized because it greatly slows down the computations. From a purely theoretical point of view, this is irrelevant. From a practical point of view, nobody would program the Z3 as I have just described, in the same way that nobody solves industrial problems using Turing machines. Also, the large loop of punched tape needed for the program would pose extraordinary and most likely unsolvable mechanical difficulties.

We can therefore say that, from an *abstract theoretical perspective*, the computing model of the Z3 is equivalent to the computing model of today's computers. From a practical perspective, and in the way the Z3 was really programmed, it is not. However, it is clear to me from the study of Zuse's unpublished manuscripts (held in the archives of the Heinz-Nixdorf Museum in Paderborn, Germany) that after completing the Z3, Zuse realized (between 1943 and 1945) that he could "lift" the decisions taken in hardware to the software level, so as to give the programmer full control of the computation.

Sometimes the dividing line between calculating machines and universal computers is drawn by differentiating between machines with externally or internally stored programs. I have argued elsewhere[8] that this is not a valid criterion. An external program can work as an interpreter of numerical data. The external program becomes a fixed part of the processor, and the data becomes the program, much in the same way as a universal Turing machine works as an interpreter. I have argued that what is needed for universal computation is a minimal instruction set and indirect addressing. Self-modifying programs can simulate indirect addressing, so that the instruction set becomes the defining criterion. A machine with enough addressable memory and an accumulator, that is capable of executing the instructions CLR (clear), INC (increment), LOAD, STORE, and BZ (branch if zero) is a universal computer. We have seen in this paper that a single WHILE loop and the four basic arithmetical operations suffice to implement a universal computer.

To summarize: the Z1 and Z3 were not fully-fledged computers in a practical sense, but neither were any of the other early machines. Atanasoff's ABC was a special purpose machine for Gauss elimination, the Harvard Mark I lacked conditional branching, although it featured loops; the ENIAC was not even programmable through software – the building blocks had to be hardwired in dataflow fashion. Conditional branching was available in the ENIAC only in a limited way, and self-modifying programs were, of course,

[8] R. Rojas, "Who invented the computer? The debate from the viewpoint of computer architecture," in W. Gautschi (ed.), *Fifty Years Mathematics of Computation, Proceedings of Symposia in Applied Mathematics*, AMS (1993) 361–366.

out of the question. Zuse's machines, however, embody many of the concepts of today's computers and seem more modern than their American counterparts – an astonishing achievement for someone working in relative isolation, and who was inventing and reinventing everything he needed on the way to what he called his life's achievement: the computer.

Acknowledgments

Deciphering the sketchy documentation available was possible only with the collaboration of several of my students at the universities in Halle and Berlin. I thank specially Alexander Thurm who implemented a gate-level simulation of the Z3 processor and a Java simulation for the Internet. Konrad Zuse provided invaluable help for understanding his patent applications.

RAÚL ROJAS was appointed professor of computer science at the Free University of Berlin in 1997. He received his B.Sc. and M.Sc. in mathematics from the Technical University of Mexico, and also a M.Sc. in economics from the National University of Mexico. He then went to Germany and received a Ph.D. and Habilitation (an additional German degree) from the Free University of Berlin. His main field of research is the theory and application of artificial neural networks. He is the author of many papers on this subject and of the well-known book *Neural Networks*, Springer-Verlag, 1996. Dr. Rojas recovered the architecture of Konrad Zuse's computing machines from the available documentation and published the details for the first time in his book *Die Rechenmaschinen von Konrad Zuse*, Springer-Verlag, 1998. He was a visiting professor at the Technical University of Vienna in 1993, and professor of computer science at the University of Halle from 1994 to 1997.

Konrad Zuse's Z4: Architecture, Programming, and Modifications at the ETH Zurich

Ambros P. Speiser

Abstract. Konrad Zuse built the Z4, a relay computer with a mechanical memory of unique design, during the war years in Berlin. Eduard Stiefel, a professor at the Swiss Federal Institute of Technology (ETH), who was looking for a computer suitable for numerical analysis, discovered the machine in Bavaria in 1949. Despite considerable doubts regarding the machine's operability, he decided to acquire the Z4 for his Institute in Zurich. The machine had a number of unique features which were convincing evidence of Zuse's admirable creativity. The Z4 went into operation in September 1950. It functioned satisfactorily, and in the following years several significant results in numerical analysis were obtained with its help.

1. Discovery and Acquisition of the Z4 by the ETH

In 1948, the *Eidgenössische Technische Hochschule* (ETH, or Swiss Federal Institute of Technology) in Zurich established the Institute (Department) of Applied Mathematics, on the recommendation of Professor Eduard Stiefel. The declared goal of the new institute was the advancement of numerical analysis. From the beginning, Stiefel started looking for ways of gaining access to computing power beyond what simple desktop calculators could offer. He soon realized that commercial punched card machines were not adequate for mathematical work, and that the electronic computer projects already under way, mainly in the US, but also in Britain and in Germany, would not fill this gap for several years to come. He thus decided that the ETH should build its own electronic computer. He sent two of his assistants, Heinz Rutishauser and myself, to the United States with the assignment of studying the new technology in order to start a similar project at the ETH. We spent most of the year 1949 with Howard Aiken at Harvard and John von Neumann at Princeton, but we also looked at other installations, among them the ENIAC at Aberdeen and the Mark II at Dahlgren. We gratefully acknowledge the hospitality with which we were received and the openness with which we were given information.

Before we returned, that is, in the middle of 1949, Stiefel was informed about the existence of Konrad Zuse's Z4. At that time Zuse was living in Hopferau, a German village near the Swiss border. Stiefel was told that the machine might be for sale. He visited Zuse, inspected the device, and reviewed the specifications. Despite the fact that the Z4 was only barely operational, he decided that the idea of transferring it to Zurich should by all means be considered. Stiefel wrote a letter to Rutishauser and me (we were at Harvard at the time), describing the situation and asking us to get Aiken's opinion. Aiken's reply was very critical – the future belonged to electronics and, rather than spending time on a relay calculator, we should now concentrate our efforts on building a computer of our own. We reported Aiken's opinion to Stiefel stating, however, that we did not fully agree with him and that, in our opinion, the proposition should certainly not be flatly rejected. Stiefel acted swiftly. He persuaded the ETH President, Hans Pallmann, to provide the necessary funds. A five-year rental contract was then signed. It was agreed that the rent of 50,000 Swiss Francs (about $12,000) would be paid in advance. Further, when the contract expired, the ETH would have the option of purchasing the machine for the additional amount of 20,000 Swiss Francs (about $5,000). Zuse used this sum to move to Neukirchen, a village about 70 Km from Göttingen. He opened a small workshop where he refurbished the machine and made the modifications that we had requested. Since the Z4 was originally built in Berlin, the question of whether Zuse was really the owner of the machine arises. And, if he was not the owner, who was? As far as I know, this question has never been answered.[1]

Stiefel must receive credit for a wise decision. There were enough uncertainties that could have made him shy away from acquiring the Z4. Among them were the following:

1. Relays were a technology of the past, electronics could do the job a hundred, or even thousand times faster, as Aiken had stated.
2. The state of the machine and its demonstration for Stiefel were not fully satisfactory. It was uncertain whether the small group around Zuse would be able to make the machine fully operational again.
3. The mechanical memory was most unusual, nobody had ever seen anything similar before. An informed assessment of its operating performance seemed impossible, although failure of the memory would have rendered the entire machine useless.

---

[1] The early history of the Z4, along with the political background, is described in: Hartmut Petzold, *Rechnende Maschinen* (Düsseldorf, 1985); Konrad Zuse, *Der Computer – mein Lebenswerk* (Berlin, 1984). Cf. also Konrad Zuse, "Installation of the German Computer Z4 in Zurich in 1950," *Annals of the History of Computing* 2 (1980): 239–241; Paul E. Ceruzzi, "The early computers of Konrad Zuse, 1935 to 1945," *Annals of the History of Computing* 3 (1981): 241–262.

4. The installation and operation of the Z4 would absorb the small group around Stiefel, which really ought to concentrate on the design of the electronic computer. Building this computer would, of course, remain the institute's goal, the Z4 being only a temporary solution.

Despite these negative aspects, Stiefel decided to rent the Z4. His decision was clearly dictated by his priorities: His foremost desire was to have an instrument for numerical calculations at his disposal as soon as possible, so that he could start his research in numerical mathematics. He preferred to start working with a relay calculator that was available immediately than to wait for an electronic machine that would take at least 3 years to complete. In contrast to many other projects which were in the hands of electrical engineers, Stiefel's ambition was to advance mathematics, not computer technology.[2]

2. Description of the Z4

In the following sections, expressions such as "hardware", "software", "machine language", "compiler", "architecture" and the like are used freely, although they were unknown in 1950. They only arrived a decade later, but the underlying concepts were quite familiar to us.

The architecture of the Z4, which had many similarities with the architecture of the computing machine Z3, can be briefly described as follows:

- It was a relay calculator built using 2200 telephone relays of pre-war design.
- It used a mechanical memory with 64 words.
- The Z4 was a binary, floating-point machine.
- The mantissa was coded using 22 bits for the memory, and 23 for the arithmetic unit. Seven bits were used for the exponent. The sign was coded with 1 bit, and "Sonderzeichen" (special signs like zero or infinite) had a special code.
- The Z4 used efficient algorithms for decimal-binary and binary-decimal conversion.
- Programming could be done using a very convenient machine language, which was better than those used in most projects we had seen elsewhere. The list of instructions was extensive.

---

[2] Ambros P. Speiser, "The Relay calculator Z4," *Annals of the History of Computing* 2 (1980), 242–245; Ambros P. Speiser, "Die Z4 an der ETH Zürich. Ein Stück Technik- und Mathematikgeschichte," *Elemente der Mathematik* 36 (1981), no. 6: 145–153.

- The program control unit had two punched tapes that could be used alternately. The ends of the tapes could be glued together to form a loop.
- There was an intermediate memory for numbers based on a punched tape.
- The results were printed with a typewriter. There was an extensive choice of formats for the fixed-point or floating-point numbers. Output was also possible by means of indicator lamps.

The following timing data give an idea of the typical execution times of some instructions of the Z4:

- Addition   0.5 seconds
- Multiplication   3 seconds
- Division   6 seconds
- Square root   6 seconds

A memory access, which could be overlapped with another instruction, took around 0.5 seconds. The overall performance of the machine was about 1000 instructions per hour.

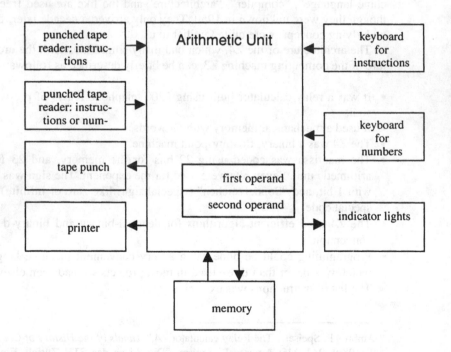

Figure 1: The architecture of the Z4

A partial list of instructions

In this section, we provide a list of the instruction set of the Z4, but some preliminary remarks are necessary.

There are two operand registers, Op_1 and Op_2. When numbers are read from memory, the first number goes into Op_1, the second into Op_2, and the result of an operation goes into Op_1. Store operations take the number from Op_1 which is subsequently cleared, except if the store operation is preceded by the instruction Rh ("Resultat halten", retain result). A store operation is not allowed when Op_1 and Op_2 are both loaded; such an instruction indicates a programming error and it causes the machine to stop. Multiplication with constants was extremely useful, because it saved memory space, and because it was very fast. Together with the instruction that loaded a 1 in Op_1, simple constants could be built up without using the memory.

The instruction set of the Z4 contained the following operations:

- Memory instructions (n is the memory address, between 0 and 63):
- A n (read), S n (store), Rh ("Resultat halten")
- Operations with two operands: + (addition), − (subtraction), ∗ (multiplication), : (division), y/x (division with inverted operands)
- Operations with one operand: √ (square root), x^2 (squaring)
- Multiplication with the constants: 2, 1/2, 1/3, 1/5, 1/7, π
- Load the number 1 into the first operand register
- Output instruction: D (print, "Drucken")
- Programming instructions: Start, Sp (jump, "Sprung"), Stop
- Conditional instructions: Sp' (conditional jump), Stop' (conditional stop)

These specifications, as seen in 1949, were very convincing. It should be borne in mind that, at that time, there were hardly a dozen program-controlled computers in operation, almost all of which were in the US. Less than a handful were being used for research in numerical mathematics, and the others performed routine calculations. There was no doubt that the Z4 could be used for serious mathematical research.

3. Konrad Zuse's Original Contribution

Konrad Zuse must be credited with seven fundamental inventions:

1. The use of the binary number system for mechanical calculation.
2. The use of floating-point arithmetic, along with the formulation of algorithms for conversion from decimal to binary and vice versa, which are

quite complicated in the case of floating-point numbers. These algorithms were, of course, embedded in the hardware.

3. An algorithm for the non-restoring calculation of the square root, with which the square root can be calculated in n steps, where n is the number of digits. This elegant method was still unknown in the U.S. in 1949.

4. Look-ahead execution: The program's instruction stream is read two instructions in advance, testing if memory instructions can be executed ahead of time.

5. Pseudo-memory: If the look-ahead mechanism determines that a number to be stored will be needed again within the next two instructions, the number is transferred to a register of mechanical contacts, where it is available without access delay. For this purpose, the memory has two registers of reading contacts.

6. Special values ("Sonderwerte"): If a result exceeds the capacity of the arithmetic unit, it is coded as a "Sonderwert". There are the following "Sonderwerte": ∞, $<<$ *(very small)*, $>>$ *(very large)*, each of which can occur with or without a sign, and also 0 and ? (? means undetermined). These values can be stored and are used correctly in calculations; for example: 1 : 0 = ∞, 1 : ∞ = 0, ∞ - ∞ = ?, 1 + ? = ?

7. The most unusual feature was undoubtedly the mechanical memory. It had 64 words with 32 bits, making up a total of 2048 bits. This mechanism was completely different from others used in cash registers or desktop calculators. The mechanical elements not only could be used for storage, but also for calculation, for example for address decoding. A relay memory would have required about 2500 relays, which would have more than doubled the size and weight of the machine.

All of these were Zuse's personal achievements. In one or two cases he was perhaps not the first inventor, but he certainly had no knowledge of what was being done elsewhere. Up until 1950 he lived in complete isolation from the world outside Germany. Creative power of this dimension, under such circumstances, demands the highest respect! The non-restoring square-root algorithm had not even been discovered by von Neumann. Anyone who competes with von Neumann deserves to be admired!

4. Modifications to the Z4

Early in 1950, Zuse moved the machine to Neukirchen. He used the payment from the ETH to build up his company and to put the machine into operating condition. He also had to make a number of modifications which Stiefel, in consultation with his assistants, had requested. They were:

1. The use of punched tapes, not only for instructions, but also for numbers, in order to allow the intermediate storage of data beyond the capacity of the memory.
2. The addition of a second tape reader (for subroutines and for numbers).
3. The addition of a typewriter (in the original machine, the output was only from a lamp field).
4. By far the most important change, (and, at the same time, the one that was easiest to implement) was the inclusion of two conditional instructions: A conditional jump and a conditional stop, both to be executed only if the number in the first operand register was negative.

The important fact was that Zuse immediately understood the significance of these changes. He proceeded to implement them, although some implied considerable modifications, mainly the use of punched tapes for numbers.

The conditional instructions can be explained as follows: The list of instructions of the Z4, as originally provided by Zuse, included the instructions Start, Stop, and Jump. When a tape was inserted in the tape reader, the instructions were read, but not executed, until a Start instruction was found. The program then started executing. The Jump instruction caused the tape reader to skip all ensuing instructions until a new Start instruction was found. Finally, the Stop instruction caused the machine to halt, indicating the end of the task. At this point the operator had to intervene, for example, by inserting a new program. The newly introduced conditional instructions worked as follows: Conditional jump as well as conditional stop were executed only when the number in the 1st operand was negative.

It was obvious that the existence of conditional instructions greatly expanded the range of tasks to be handled, or, conversely, without these two instructions the Z4 was a computer of very limited use in mathematics. In this context, there was one question that we asked ourselves repeatedly: Why did Zuse not include conditional instructions in the first place? Is it possible that, despite his proven creative power, the idea had not occurred to him? Zuse himself never gave a clear answer to this question. Thus, it remains open, despite its importance for the history of computing.

An interesting feature was the relay circuit for the addition of binary numbers. A relay adder, like any mechanical adder, must be able to deal with carries, and, specifically, with the case when a carry to the n-th digit causes an additional carry to the $n+1$st digit, and so on, through many digits. Zuse's circuit required two stages for this process: An intermediate set of relays prepared the circuit for the carries, and the next step transferred the result to the final set of relays. This determined the speed of the entire machine. While we were at Harvard, we studied the description of Aiken's relay calculator Mark II (completed in 1947), and we found that he used a circuit which required only one stage. We showed the circuit to Zuse during one of our visits in 1950. He immediately recognized its significance, and he admitted that the

Figure 2: The Z4 at the ETH Zürich (1951)

thought had not occurred to him. The circuit would have made the Z4 50 percent faster. However, it was agreed that a change was out of the question, since it would have required the rewiring of at least half of the machine.[3]

5. Operation of the Z4

The machine was moved to the ETH in September 1950 and, after a relatively short period, it was operating. The Z4 proved to be quite reliable, and the frequency of breakdowns was well within the limits of what was compatible with satisfactory operation. Quite soon, the machine could be left running unattended overnight, which was quite unusual at the time. Zuse himself was understandably proud of this achievement. He was a man with a good sense of humor, and once stated that the rattling of the relays of the Z4 was the only interesting thing to be experienced in Zurich's night life!

To illustrate the conditions under which we were working, let me repeat that the machine's computing power was 1000 operations per hour. For operational reasons, problems that lasted more than 100 hours could not be considered. Thus, 100,000 operations and 64 places of memory were the

---

[3] These circuits are described in Ambros P. Speiser, *Digitale Rechenanlagen* (Berlin, 1965).

computational limits that were set. In the light of today's technology, where the term Gigaflops is in everyday use, and memory is measured in Gigabytes, it is hard to believe that useful work of any kind could be done with the Z4. And yet, at that time, at least on the European continent, there was no mathematical institute which had access to computing power comparable to ours.

Work with the Z4 was interactive in the true sense of the word. Of course the term "interactive computing" did not exist at the time. The mathematician was programmer and operator at the same time, and he could monitor the running of his program. Intermediate results were printed out, inspected, and the program could be modified if necessary. But the signals that the operator received from the computer were not only optical, they were also acoustical. The clicking of the tape reader was an indication of how fast the program was proceeding, or of whether it had got stuck, and the rattling of the relays signaled the type of operation in progress. This was a great help in spotting both hardware and programming errors.

Soon after the Z4 was put into operation, numerous users appeared on the scene. Its user-friendliness enabled them to learn programming within a day.

Maintenance of the machine was largely my responsibility. Although, as stated, reliability was quite good, I nevertheless remember many hours of searching for errors, which often originated in malfunctioning dusty relay contacts. We also discovered several cold soldering joints that gradually failed to conduct. Finding them was particularly bothersome, because they caused non-deterministic mistakes. On two occasions, I had to disassemble parts of the memory. This meant removing about 1000 pins and replacing them again. There were two kinds of pins: one was 2.5 and the other 2.6 millimeters long. If, due to a mistake, I mixed up one single pin, the entire memory was blocked – a very frustrating experience!

We also made some hardware changes. Rutishauser, who was exceptionally creative, devised a way of letting the Z4 run as a compiler, a mode of operation which Zuse had never intended. For this purpose, the necessary instructions were interpreted as numbers and stored in the memory. Then, a compiler program calculated the program and punched it out on a tape. All this required certain hardware changes. Rutishauser compiled a program with as many as 4000 instructions. Zuse was quite impressed when we showed him this achievement.

It was my job to make the necessary wiring changes. I vividly remember the hours it took me to find out which of the perhaps 30,000 soldering joints had to be changed to implement Rutishauser's ideas!

By the time the Z4 was moved out of Zurich in 1954, programs with about 100,000 instructions had been written. About 6,500 hours of computing time had been registered, during which 6 million operations had been performed. Academic clients could use the machine free of charge, external users were charged 1 Swiss cent per operation, or 10 Francs (about $2.50) per hour.

Acquisition and operation of the Z4 is largely the consequence of an exceptionally fruitful partnership between Zuse and Stiefel. They were both 39 years old when they met. But their backgrounds, both cultural and academic, could hardly have been more different. Each had a powerfully creative mind, as well as an ability to intuitively grasp the essential elements of a complex situation. A lasting friendship developed between the two men.

6. Scientific Work

When the machine was installed, significant scientific work began almost immediately, and within a few years Zurich rose to become one of the foremost centers of numerical analysis. I cannot give a detailed description of the results, a representative selection of some keywords must suffice:[4]

- Method of conjugate gradients
- QD (quotient-difference) algorithm and its relation to continued fractions
- The concept of numerical stability in the solution of differential equations
- The concept of programming languages and the contributions to ALGOL (Heinz Rutishauser, Friedrich L. Bauer, Klaus Samelson)

During these years, Zurich also became one of the leading centers in applied mathematics. Even Stiefel would not have dared to hope for such a degree of success!

The Z4 was also used extensively in education. As early as 1951, we offered students a course in computer programming with practical exercises on the machine. We believe we were the first on the European continent to do so. This should be taken into consideration by those who often claim that Swiss universities were late in recognizing the importance of computer science ("sie hätten die Informatik verschlafen")!

7. Reactions in Germany

The situation in Germany in the early 1950s is worth commenting on. While the Z4 was operating in Zurich, three electronic computer projects were under way in West Germany: at Darmstadt, Munich, and Göttingen. We maintained close contact with our colleagues there, specially during the planning and design phase of the ERMETH (*Elektronische Rechenmaschine an der*

---

[4]   Cf. also Eduard Stiefel, "Rechenautomaten im Dienste der Technik. Erfahrungen mit dem Zuse-Rechenautomaten Z4," *Arbeitsgemeinschaft für Forschung des Landes Nordrhein-Westfalen,* Heft 45 (1955): 29–45.

Figure 3: The Ermeth design team at the ETH Zurich (1953). Standing from left to right: Eduard Stiefel, Peter Läuchli, Alfred Schai, Appenzeller, Messerli, Walther, Sieberling, Engel. Sitting from left to right: Robert Stock, Ambros P. Speiser, Annemarie Hürlimann, Heinz Rutishauser, Hans Schlaeppi.

ETH, our electronic computer).[5] But, to put it mildly, they only had a moderate interest in the Z4. To understand this attitude one must take the fundamental difference in priorities into account: Stiefel wanted computing power at his disposal as fast as possible, even if it was only modest. The three German groups, on the other hand, wanted to build an electronic computer with advanced technology; they were not under pressure from a computer-hungry mathematician.

But when the scientific results started to flow out of Zurich in 1951 and 1952, some criticism was voiced in Germany to the effect that the Z4 should have been kept at home rather than letting it go abroad. In retrospect, the explanation for what happened is quite clear to me: in 1950, universities were still suffering from the consequences of the war. Buildings were badly damaged, equipment was almost non-existent and, accordingly, the limited funds had to be spent on the urgent needs of the day in order to keep university life at an acceptable level. The sum of 50,000 Swiss Francs that we paid for Z4 was clearly beyond the reach of any of the universities. Funds of this amount could only have come from the Federal Ministry (*Bundesministerium*), a state

---

[5] Hans-Rudolf Schwarz, "The early years of computing in Switzerland," *Annals of the History of Computing* 3 (1981): 121–132.

ministry (*Länderministerium*), or from the Marshall Plan. But in these circles the opinion was crystal clear: The future belonged to electronics, and it would be a big mistake to divert limited funds to a relay machine whose technology was already considered to belong to the past. (I would be interested to know how the guild of younger computer historians who are now at work views this question!)

But German interest in what was being done in Zurich had been awakened and our ERMETH plans, in particular, were closely followed. When the first industrial machines appeared in Germany, we found that a number of ERMETH ideas had been adopted. To us, it was a source of grief that our work was followed with much more interest in Germany than in our own country. To be sure, Swiss industry was polite and cooperative when we needed help, but they showed no interest in our results – nobody is a prophet in his own country!

8. Biographical Notes on Eduard Stiefel

To conclude, I wish to include some remarks on Eduard Stiefel and his biography. The discovery, acquisition and operation of the Z4, as well as the remarkable success of its operation, can only be understood in the light of Stiefel's personality, which was, in many respects, unusual. Stiefel was born in 1909. He studied in Zurich, Hamburg and Göttingen. He was a topologist and group theorist, and by 1948, at the age of 38, he had acquired an international reputation. At that time he decided to make a complete change in his scientific career and to move into numerical analysis. I remember one day when he came into the assistants' room and told us he had just thrown away the stock of reprints of his topology publications – he had kept only one or two copies of each article. What prompted him to make this move, unexpected by most of his colleagues? Numerical analysis, or applied mathematics, was regarded with skepticism by the scientific establishment. It enjoyed less social regard than "pure mathematics". What prompted him to make the change? Of course, we will never know – but what happened was proof of Stiefel's remarkable intuitive insight into what was to become important in the near future, an insight to which he remained faithful for the rest of his life.[6]

---

[6] Joerg Waldvogel, Urs Kirchgräber, Hans-Rudolf Schwarz and Peter Henrici, "Eduard Stiefel (1909–1978)," *Zeitschrift für Angewandte Mathematik und Physik* 30 (1979): 133–142.

Appendix

Table 1: List of operations of the Z4[7]

Floating-Point Arithmetic	Sequencing	Supervision	Input/Output
Load	No Operation	Start Execution	Read Switches
Load Immediate 1	Stop On Condition	Start Number	Display
Store	Space		Display and Keep
Keep	Skip on Condition		Print
Negate	Call		Print 0
Make Absolute	Call on Condition		Print 2
Add	Return		Print 4
Subtract			Print 6
Reverse Subtract			Carriage Return
Maximum			Tabulate
Maximum Positive			Read Tape 0
Minimum			Read Tape 1
Multiply			Punch
Multiply By Two			Punch Programm
Multiply By Three			Protocol Right
Multiply By Ten			Protocol Down
Multiply By Pi			Protocol Left
Divide			Protocol Up
Divide by Two			
Divide by Three			
Divide by Five			
Divide by Seven			
Divide by Pi			
Reciprocal			
Square			
Square Root			
Signum			
If Indefinite Then			
One			
Zero Test			
Positive Test			
Infinite Test			
Indefinite Test			
Non-Fraction Test			

---

[7] Gerrit A. Blaauw and Frederick P. Brooks, *Computer Architecture: Concepts and Evolution* (Reading, Mass., 1997), 551.

AMBROS P. SPEISER graduated as an electrical engineer from the Swiss Federal Institute of Technology (ETH), Zurich, in 1948. Subsequently, he was involved in the transfer of Konrad Zuse's Z4 to Zurich and its operation. He was a member of the design team of ERMETH *(Elektronische Rechenmaschine an der ETH)*, a magnetic drum computer somewhat similar to Aiken's Mark III. During this period, he earned a D.Sc. at the ETH. From 1955 to 1966 he was Director of the IBM Research Laboratory, located in Ruschlikon near Zurich, and from 1966 to 1988 he was Director of Corporate Research of BBC Brown Boveri (now ABB), with responsibility for laboratories in Baden (Switzerland), Heidelberg (Germany), and Paris (France). From 1959 to 1965 Dr. Speiser was Secretary-Treasurer, and from 1965 to 1968 President of IFIP. He was President of the Swiss Academy of Engineering from 1987 to 1993. Dr. Speiser is the author of books on computers and electronic circuits, and of numerous articles in computer science and on research and science policy. Distinctions: Fellow of the IEEE, holder of the IFIP Silver Core, honorary member of the Zurich Physical Society, honorary member of the Swiss Academy of Engineering, honorary doctor of the ETH.

The Plankalkül of Konrad Zuse – Revisited

Friedrich L. Bauer

Abstract. The ideas that finally led to creation of the Plankalkül first came to Konrad Zuse (1910-1995) around 1938, while working on the Z3. He wanted to build a *Planfertigungsgerät*, and made some progress in this direction in 1943. Around 1944, he prepared a draft of the Plankalkül, which was meant to become a doctoral dissertation some day. Zuse went back to this work when he had to stay in Hinterstein (Allgäu) after the end of the war. The first document from this period is dated June 14, 1945. The final manuscript was finished early in 1946, but was not fully published until 1972. As a result, only rudimentary information, scattered in a few short publications and contributions to conferences, was available up to that point.

The Plankalkül is the first fully-fledged algorithmic programming language. It was far ahead of its time, although not in the line of development so heavily influenced later by machines of the von Neumann type. It was, in fact, more a precursor of functional and object oriented programming. However, it fell into oblivion, and had no influence on most people studying these topics years later.

1. Introduction

In 1972, together with Hans Wössner, I wrote an article for Communications of the ACM[1], in which Konrad Zuse's Plankalkül was called a "Forerunner of Today's Programming Languages." This was 26 years after Zuse completed his manuscript of 1946. 26 years later, I have been asked to contribute an evaluation of the Plankalkül for this volume. I was first inclined to reject the offer, since I thought I had said all there was to say. On closer inspection, I found my views had changed slightly in the meantime, partly due to the general development of programming languages, partly due to the fact that in

---

[1] Bauer, F. L., Wössner, H.: "The ‚Plankalkül' of Konrad Zuse: A Forerunner of Today's Programming Languages." *Communications of the ACM* 15 (1972), 678–685.

early 1972 I only had at my disposal sketchy publications by Zuse[2] (Fig. 1, Fig. 2) and a short passage in his autobiography,[3] since I was not given access to the main manuscript. In late 1972, however, the GMD, a German research institute, was allowed to publish the manuscript, which contained a very elaborate description of the Plankalkül.[4] There was also a more detailed GMD publication by Zuse in 1977[5] and a short introduction in 1980.[6] In 1976, at the now famous International Research Conference on the History of Computers, held at Los Alamos Scientific Laboratory, Donald E. Knuth and Luis Trabb Pardo presented the paper "The Early Development of Programming Languages," prominently listing Zuse's Plankalkül as the first item in a list of twenty early 'high-level' programming languages.[7] More recently, Hartmut Petzold[8] and Wolfgang K. Giloi[9] took an interest in the history of the Plankalkül. But in general, there has been little resonance.

Therefore, the present paper can supplement the 1972 paper to some degree; the Plankalkül can now be judged with more detachment since another 26 years have passed. Zuse's expectations can be discussed and, considering the present state of the art in programming, the Plankalkül may be placed in historical context.

Incidentally, *Plan* is German for 'program'. Thus, when Zuse speaks of a *Plan*, he means what was later called a program. Zuse used *Kalkül* as it was used in *Aussagenkalkül* and *Prädikatenkalkül*. Thus *Plankalkül* is an instrument for reasoning about programs – quite a modern point of view.

---

[2] Zuse, K.: "Über den Allgemeinen Plankalkül als Mittel zur Formulierung schematisch-kombinativer Aufgaben." *Archiv der Mathematik* 1 (1948/49), 441–449 (submitted December 6, 1948). Zuse, K.: "Über den Plankalkül." *Elektronische Rechenanlagen* 1 (1959), 68–71.

[3] Zuse, K.: *Der Computer, mein Lebenswerk.* Verlag Moderne Industrie Munich 1970.

[4] Zuse, K.: *Der Plankalkül.* BMBW-GMD-63, 1972.

[5] Zuse, K.: *Beschreibung des Plankalküls.* BMBW-GMD-112, Oldenbourg, Munich 1977.

[6] Zuse, K.: "Some Remarks on the History of Computing in Germany." In: N. Metropolis et al., *A History of Computing in the Twentieth Century.* Academic Press, New York 1980, 611–627.

[7] Knuth, D.E., Pardo, Luis Trabb: "The Early Development of Programming Languages." In: N. Metropolis et al. *A History of Computing in the twentieth Century.* Academic Press, New York 1980, 202–208.

[8] Petzold, Hartmut: "Eine Sprache, die jede Maschine versteht. Der Plankalkül."

[9] Giloi, Wolfgang K.: "Konrad Zuse's Plankalkül: The First High-Level, 'non von Neumann' Programming Language." *IEEE Annals of the History of Computing*, Vol. 19, No. 2, 1997.

2. Objects of the Plankalkül and notation

First I shall give an introduction to the philosophy of the Plankalkül, together with an explanation of the somewhat peculiar notation Zuse used. The objects of the Plankalkül are not necessarily numbers, they are lists like the ones John McCarthy introduced independently 13 years later in LISP. In contrast to LISP, Zuse's lists are to be formed from a finite ordered sequence (and not pair) of lists and/or from basic elements; the basic elements, however, are restricted to binary values. The basic binary elements comply with Zuse's philosophy that binary arithmetic is the right one for mechanization, in accordance with Gottfried Wilhelm Leibniz.[10] In his draft, Zuse gives examples of applications in which objects are represented using binary coding.

Thus, Zuse's are lists of lists of binary coded elements, i.e., finite trees of binary coded elements. Correspondingly, the operations Zuse studies on these objects are necessarily isomorphic to the operations of classical binary propositional calculus. This orientation could well be the result of Zuse's close contact with his friend Hans Lohmeyer, a mathematician at the Henkel Aircraft Plant, who was a student of the logician Heinrich Scholz in Munster.

Surprisingly for an engineer, Zuse does not provide many graphical illustrations in his Plankalkül manuscript, and also neglects graphical output.

Figure 1: Zuse's 1948 paper   Figure 2: Zuse's 1959 paper

---

[10] "Wunderbarer Ursprung aller Zahlen aus 1 und 0" and "Unus ex nihilo omnia," Leibniz 1696 in a letter to the Duke Rudolf August of Hanover.

Zuse, who frequently tried to build his machines from the most elementary components, did not arrive at the trick of working with ordered pairs of lists and/or basic elements used by McCarthy in LISP. Consequently, Zuse explicitly needs selectors isomorphic to finite subsets of the set of natural numbers – indeed he is forced to use the natural numbers from 0 to 7 to denote the components (*Komponenten*) K0, K1, K2, K3, K4, K5, K6, K7, where he could have used three bits for selection. This, however, is not a severe theoretical drawback and rather a concession to widespread usage.

Needless to say, Zuse does not require the components to be of a single type; thus he uses, for example, a list of several four-bit numbers and a binary sign. He also frequently introduces data constraints (*Beschränkungen*), for example, using only 10 of the 16 possible four-bit code elements to represent the 10 decimal digits.

Zuse works primarily with fixed structures, which he calls inflexible (*starr*). But he also considers dynamic (*quasistarr*) structures, e.g., a list of n four-bit numbers and a binary sign, where the *Struktur-Variable* n is to be calculated before the program itself, the *Rechenplan*. Zuse speaks of a *quasistarrer Rechenplan* if the structure can be pre-computed. This reflects his idea of building a *Planfertigungsgerät*, which he practically abandoned later on. Potentially infinite structures – infinite sequences, infinite trees – are excluded from the language.

Definition of structures

The introduction of object classes is effected by "structure equations" (*Strukturgleichungen*). Structures are denoted by S followed by other characters. S0 always denotes the fundamental binary set, the set formed by two basic elements denoted O and L (the binary digits 0 and 1).
The structure equation

$$S\,1.4 = 4 \times S0$$

defines a structure S1.4 of four binary components. Another way of introducing it is by forming a tuple

$$S\,1.4 = (S0, S0, S0, S0)$$

A new structure S2 could be defined by

$$S2 = n \times S1.4 \qquad \text{or equivalently} \qquad S2 = n \times 4 \times S0$$

which would be used to represent a number with n decimal digits. For example, for $n = 2$, (LOOO, OOLL) would represent the natural number 83. Another structure S3 could be defined by

$$S3 = (S0, n \times S1.4),$$

useful to represent a number with n decimal digits and a sign. For example, in the case $n = 2$, (L, (LOOO, OOLL)) would represent the integer -83. Clearly, $n \times 4 \times S0$ and $4 \times n \times S0$ are different structures: the first is an n-tuple of quadruples, the second is a quadruple of n-tuples.

Variables and components

Variables are denoted by letters like

V   for parameters
Z   for intermediate results
R   for final results

followed by a numerical index. Component selectors are denoted by K followed by a natural number: K0 selects the first component, K1 the second, K2 the third, and so on.

A notation peculiar to the Plankalkül

Assume Z3 to be a variable of structure

$$m \times S1.n = m \times n \times S0.$$

For K1(Z3), the second component of Z3, Zuse uses the following notation:

Z   (name of the variable)
3   (index of the variable)
1   (component index)
1.n (structure of the component)

For K1.2(Z3), the third component of the second component of Z3, the notation is:

Z   (name of the variable)
3   (index of the variable)
1.2   (component index)
0   (structure of the component)

This is part of Zuse's peculiar "row representation": every (operative) construction is broken into four lines, and has in its left margin (strangely called *Randauszug*) the symbols V, K, S:

	Z	∨	Z	⩾	Z
V	3		4		4
K	1		2.3		2.3
S	0		0		0

where V refers to a "variable index," K to a "component index" and S to a "structure." ALGOL uses the reverse order: Z4[2.3]:=Z3[1] ∨ Z4[2.3].

This example also shows the typical Zuse notation with the *Ergibtzeichen* (for typographical reasons later ⩵ , ⇒), which John Backus introduced independently into FORTRAN as the symbol :=, working in the opposite direction. Zuse clearly explains that a construction like the one above contains a hidden sub-index i which is counted upwards:

	Z	∨	Z	⩾	Z
V	3		4		4
K	1		2.3.i		2.3.i + 1
S	0		0		0

The K line may be left empty if no component is to be singled out. The S line gives only additional information; strictly speaking it could be omitted.

Examples of the Randauszug

The following are some commented examples of the notation used by Zuse:

	V		The variable V3 is a list of m pairs
V	3		of the structure $2 \times S1.n$
K			and enters into the computation
S	$m \times 2 \times 1.n$		as a whole

	V		The i-th pair of the list of pairs V3
V	3		(structure $2 \times S1.n$)
K	i		enters into the computation
S	$2 \times 1.n$		i may be a running index

	V		The first element of the i-th pair
V	3		of the list of pairs V3
K	i.0		(structure $S1.n$)
S	1.n		enters into the computation

	V		The binary digit no. 7 (structure S0)
V	3		of the first element of the i-th pair
K	i.0.7		of the list of pairs V3
S	0		enters into the computation

Zuse also gives examples in which the selector for a structured object is computed separately: the k-th component of Z3, where the number k is the value of Z1, an object of the structure S1.n is denoted by

$$
\begin{array}{c|cc}
\text{V} & \text{Z} & \text{Z} \\
& 3 & 1 \\
\text{K} & & \\
\text{S} & m \times 1.n & 1.n
\end{array}
$$

which we would express in ALGOL as Z3[Z1]. This construction and the use of conditional branching (*bedingte Planteile*), denoted with the help of the conditional symbol $\dot{\rightarrow}$ as in

$$ Kla(Z) \dot{\rightarrow} (\varepsilon + 1 \geqslant \varepsilon) \quad | \quad Klz(Z) \dot{\rightarrow} (\varepsilon - 1 \geqslant \varepsilon) $$

influences the course of the computation (*Kla* and *Klz* represent logical predicates defined by the programmer, see Fig. 6). Note that ALGOL uses the "if ... then" construct for the same purpose. Zuse speaks here of "free programs."

3. Zuse's orientation towards symbolic ('non-numerical') computations

Zuse devotes a large part of his 1972 paper to a detailed discussion of three examples of symbolic computations:

- logical-arithmetical operations, where he gives a partial description of his Z3 calculator in terms of the Plankalkül (Chapter 3)
- the problem of well-formedness and parsing of a parenthesis-free formula (Chapter 4, already sketchily treated in his 1948 paper, see Fig. 3)
- the description of a chess machine (Chapter 5, already mentioned in his 1959 paper, see Figs. 5, 6).

The last two were topics that interested Zuse (the parsing algorithm originated in the 1930s from the Berlin logician Karl Schröter) and seemed to be outside the mainstream of programming language development in the 1950s, which was heavily influenced by numerical calculations. This may explain why the American participants at the ALGOL58 conference were not interested in the Plankalkül. In fact, Zuse's way of building structures was a clumsy way of handling matrices, which are so important in numerical analysis. Efficiency, both with respect to time and storage, was considered most important when using the computers of the late 1950s, when a great deal of work was being invested in numerical problems.

An example of syntactical nature

The example Zuse used not only illustrates the Plankalkül notation, but also points in the direction of the systematic construction of a program out of its specification – quite a modern tendency. Zuse starts his 1948 paper by saying: "Expressions formed in this way contain the following symbols: symbols for variables, for negation, for [binary] operations, parentheses and blanks, needed to separate individual expression for automatic handling. The individual symbols are encoded by sequences of True-False values."

In the Plankalkül program (Fig. 3), σ denotes the structure of 8-bit words used to encode the symbols (*Zeichen*), and $m\sigma$ with arbitrary $m \geq 1$, denotes the structure V0 of the symbol sequences that are to be investigated. A call to this program with an (encoded) symbol sequence x as its actual parameter tests the predicate:

$Sa(x)$: x is a "meaningful," i.e., a syntactically well formed Boolean expression.

This predicate is defined inductively in the following way:

1.   A variable is a meaningful expression.
2.   A meaningful expression, prefixed by a negation symbol, yields a meaningful expression.
3.   Two meaningful expressions, connected by an operation symbol, yield a meaningful expression.
4.   A meaningful expression put in parentheses yields a meaningful expression.

To transform this definition into an algorithm, Zuse defines the auxiliary predicates for the symbols x:

$\qquad Va(x)$:   "x is a variable symbol"
$\qquad Op(x)$:   "x is an operation symbol"
$\qquad Neg(x)$:   "x is a negation symbol"
$\qquad Kla(x)$:   "x is an opening parenthesis"
$\qquad Klz(x)$:   "x is a closing parenthesis"

And the composite predicates

$\qquad Az(x) \quad = \quad Va(x) \vee Neg(x) \vee Kla(x)$
$\qquad Sz(x) \quad = \quad Va(x) \vee Klz(x)$
$\qquad Sq(x,y) \quad = \quad (Sz(x) \wedge \neg Az(y)) \vee (\neg Sz(x) \wedge Az(x))$

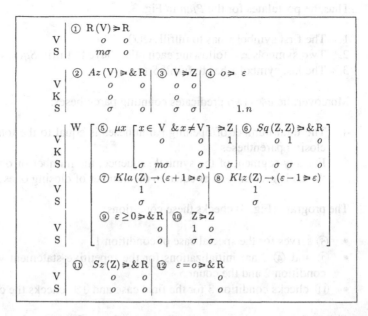

Figure 3: Well-formedness of a propositional formula (from Zuse's 1948 paper)

```
①        proc Sa = ([0 : either] bits V0) bool : begin
②, ③          bits Z0 := V0[0]; bool R := Az(Z0);
④             int eps := 0; if Kla(Z0) then eps := 1 fi;
⑤             for i to upb V0 while R do begin
                  bits Z1 := V0[i];
⑥                 R := R ∧ Sq(Z0, Z1);
⑦                 if Kla(Z1) then eps + := 1 fi;
⑧                 if Klz(Z1) then eps − := 1 fi;
⑨                 R := R ∧ eps ≥ 0;
⑩                 Z0 := Z1                    end
⑪, ⑫          R ∧ Sz(Z0) ∧ eps = 0           end
```

Figure 4: ALGOL 68 equivalent of Zuse's example in Fig. 3, with numbers marking correspondences (from the paper by Bauer and Wössner, 1972)

Thus, he postulates for the *Plan* in Fig. 3:

1. The first symbol x has to fulfill $Az(x)$
2. Two symbols x, y following each other have to fulfill $Sq(x, y)$
3. The last symbol x has to fulfill $Sz(x)$

Moreover, he uses two predicates counting parentheses:

4. The total number of opening parentheses is equal to the total number of closing parentheses
5. For any segment of the symbol sequence, the number of opening parentheses is greater than or equal to the number of closing ones

The program (Fig. 3) checks these conditions:

*   ⑤ serves for the special case of condition 1.
*   ③ and ④   are initializations for the repetitive statement which checks condition 2 and the count 5.
*   ⑪ checks condition 3 for the final case and ⑫   checks the count 4.

The program, by the way, contains some bugs: for example, a count corresponding to ⑦ is missing for the first symbol. More seriously, the condition $x \neq V0[0]$ in ⑤ should be read as $x = V0[i] \wedge i \neq 0$ – a weakness in Zuse's semantics of repetition (see below).

For a direct translation of Zuse's (corrected) program into ALGOL 68, we may assume first that suitable Boolean procedures $Va(x)$, $Op(x)$, etc. have been declared. Using these predicates, we obtain the procedure shown in Fig. 4 in ALGOL 68 (the encircled numbers refer to Fig. 3).

An example from chess

Fig. 5 shows P148, the *Plan* for the predicate "Is the white king in stalemate?," one of the auxiliary procedures for a chess program formulated by Zuse in Plankalkül notation (in his 1959 paper, p. 71) and his related comments. He uses two further auxiliary procedures with the results R17 and R128, which he first defines verbally. The next step would be to state them algorithmically in Plankalkül notation. For R17, a one-line *Plan* can actually be found on page 241 in Zuse's 1972 paper. Likewise, a four-line *Plan* for R128 is there on page 260.

«Der weiße König kann einen Zug machen, ohne dabei in Schach zu kommen.»

$$
\begin{array}{c|cc}
\text{P } 148 & R\,(V) \Rightarrow R\,148 & \\
V & 0 & 0 \\
A & 5 & 0
\end{array}
\tag{1}
$$

$$
\begin{array}{c|c}
\acute{x} & \left[(x \in V) \wedge (x = LO)\right] \Rightarrow Z \\
V & \quad\;\; 0 \quad\qquad\quad\; 0 \\
K & \qquad\qquad 1 \\
A & 4 \qquad\;\; 5 \qquad 3 \qquad\quad 4
\end{array}
\tag{2}
$$

$$
\begin{array}{c|c}
& (Ex) \quad \left[\; (x \in V) \wedge R17\,(Z, x) \wedge (x = O) \vee\; x \right. \\
V & \qquad\qquad\; 0 \qquad\; 0 \\
K & \qquad\qquad\qquad\qquad\; 0\;\; 0 \quad 1 \qquad\quad 1.3 \\
A & 4 \qquad\quad 4\;\; 5 \qquad 2\;\; 2 \quad 3 \qquad\quad 0
\end{array}
\tag{3}
$$

$$
\begin{array}{c|c}
& \wedge \overline{Ey} \;\left[(y \in V \wedge\; y\; \wedge R128\,(V, y, x)\right] \left.\vphantom{]}\right] \\
V & \qquad\quad 0 \qquad\qquad\; 0 \\
K & \qquad\qquad\; 1.3 \qquad\qquad 0\;\; 0 \\
A & 4 \qquad\;\; 5 \quad 0 \qquad\quad 5\;\; 2\;\; 2
\end{array}
\tag{4}
$$

Die hierbei benutzten Unterprogramme sind:

$$
\begin{array}{c|c}
& R17 \;\;(V, V) \\
V & \quad\; 0\;\; 1 \qquad \text{«Die Punkte } V_0 \text{ und } V_1 \text{ sind benachbart.»} \\
A & \quad\; 2\;\; 2
\end{array}
$$

$$
\begin{array}{c|l}
& R128\,(V, V, V) \\
V & \quad\; 0\;\; 1\;\; 2 \qquad \text{«Bei der gegebenen Feldbesetzung } V_0 \text{ ist der} \\
A & \quad\; 5\;\; 2\;\; 2 \qquad \text{Zug von Punkt } V_1 \text{ nach Punkt } V_2 \text{ erlaubt.»}
\end{array}
$$

Das Programm R128 ist verhältnismäßig kompliziert, da untersucht werden muß, welcher Stein auf Punkt V_1, steht, ferner ob der Punkt V_2 zu V_1 in einer solchen geometrischen Relation steht, daß der auf V_1 stehende Stein dorthin setzen kann, und schließlich muß untersucht werden, ob dazwischenliegende Punkte vorhanden sind und ob diese frei sind.

Erklärung der Formel P148 in Worten:

(1) ist ein Randauszug, der besagt, daß über eine Feldbesetzung $(A5)$ eine Aussage gemacht werden soll.

(2) Diejenige Punkt-Besetzt-Angabe $(\acute{x})$, welche in der Liste der Spielbesetzung (V_0) enthalten ist, deren Komponente Nr. 1 = LO ist (Zeichen für König in der Numerierung der Steintypen), ergibt den Zwischenwert Z_0.

(3) Es gibt in der Liste der Spielbesetzung (V_0) einen Punkt (x), der zu Z_0 (Punkt, auf dem der König steht) benachbart ist und der unbesetzt (= O) oder mit einem schwarzen Stein besetzt ist $(x_{1.3})$ (das bedeutet Ja-Nein-Wert Nr. 3 der Besetzt-Angabe x_1; dieser charakterisiert schwarze Steine).

(4) Es gibt keinen weiteren Punkt, der mit einem schwarzen Stein besetzt ist, welcher nach Punkt x gesetzt werden kann.

Figure 5: The chess example: is the white king free to move? (from Zuse's 1959 paper)

Ein Beispiel aus der Schachtheorie

Als Beispiel sei kurz auf die Schachtheorie eingegangen. Zunächst ist der Aufbau der auftretenden Angabenarten interessant.

S 0	Ja-Nein-Wert
S 1. n	n-stellige Folge von Ja-Nein-Werten
A 1 S 1.3	= Koordinate
A 2 $2 \times$ A 1	= Punkt
	(z. B.: LOO, OOL entspricht Punkt e2
	in üblicher Darstellung)
A 3 $\binom{S\,1.4}{B\,3}$	= Besetzt-Angabe
	(z. B.: OOLO, Weißer König)
A 4 (A 2, A 3)	= Punkt-besetzt-Angabe
	(z.B.: LOO, OOL; OOLO
	«Punkt e2 mit weißem König besetzt»)
A 5 $64 \times$ A 3	= Feldbesetzung:
	C 5 Anfangslage
	(Aufzählung der Besetzung der 64 Punkte
	in fester Reihenfolge)
A 6 $64 \times$ A 4	= Feldbesetzung mit Punktangabe,
	C 6 Anfangslage
A 7 $12 \times$ S 1.4	= Anzahlliste der Steine;
	C 7 Anfangslage
	(Angabe, wieviel Steine von jeder Sorte
	auf dem Feld sind,
	z. B. für Bewertungsrechnungen wichtig.)
A 9 (A 5, S 0, S 1.4, A 2)	= Spielsituation;
	C 9 Anfangssituation
	(Feldbesetzung [A 5];
	Angabe, ob Weiß oder Schwarz am Zuge [S 0];
	Angaben über Rochade-Möglichkeiten
	[4 Ja-Nein-Werte] ;
	Angabe der Punkte mit den Möglichkeiten,
	'en passant' zu schlagen).
A 10 (A 6, S 0, S 1.4, A 2)	= Spielsituation mit Punktangabe;
	C 10 Anfangslage
A 11 (A 2, A 2, S 0)	= Zugangabe
	(zwei Punktangaben, gesetzt von ... nach ... .
	Ein Ja-Nein-Wert «Es wird geschlagen»).

Figure 6: Example: chess structures (from Zuse's 1959 paper)

The structures on which the program is based are defined in Fig. 6. A1 = S1.3 serves to encode the rows and columns, a field (*Punkt*) on the chessboard is determined by A2 = 2 × A1. Also a 4-bit structure S1.4 is used to encode the 13-element set {white king, white queen, white rook, white bishop, white knight, white pawn, black king, black queen, black rook, black bishop, black knight, black pawn, empty}, the three unused combinations being given by a predicate B3 that is violated ("data restriction"). Thus, Zuse writes

$$A3 = \begin{pmatrix} S1.4 \\ B3 \end{pmatrix}$$

The remaining structures are defined constructively in a similar way. Zuse's program, literally translated into ALGOL 68, is shown in Fig. 7 (the marks (1) to (4) correspond to those in Fig. 5).

4. Unorthodox elements in the Plankalkül

The non-numerical examples illustrate Zuse's boldness in introducing, in a programming language, not only

- the existence operator $\exists$ in the form $(\exists x)\, F(x)$ and
- the for-all-operator $\forall$ in the form $(x)\, F(x)$, but even
- the determination operator in the form $(\dot{x})\, F(x)$ ("this one, which ...") and the list-forming operator $\hat{x}\, F(x)$ ("the set of all, which ...").

Zuse gives a definition of $\hat{x}\, F(x)$ by an algorithmic implementation and uses

$$V0 = \hat{x}: x \in V0$$

to denote that the list V0 does not contain repetitions. Moreover, he introduces the sieve operator $\tilde{x}\, F(x)$ ("the ordered sublist of all, which ..."). To give an example: for V0=(0,3,5,4,3,6,12,6,4)

$\hat{x}: x \in V0$	is (0,3,5,4,6,12)
$\tilde{x}: x \in V0$	is V0
$\tilde{x}: x \in V0 \wedge \text{even}(x)$	is (0,4,6,12)

He also proposes the μ-operator for lists of the form $\mu x\, F(x)$ ("the first one which"). An example of this would be:

$$\mu x: x \in V0 \wedge \text{odd}(x) \text{ is } 3$$

In this case, he slightly generalizes the idea of Bernays' μ-operator: if there is no suitable element, a special sign ("end symbol") is to be returned (and not zero, as Bernays does – in fact, Zuse's lists may contain elements other than numbers).

Zuse was introduced to the symbolism of mathematical logic, by his own admission, by his friend Hans Lohmeyer. Together with Lohmeyer, he studied, to some extent, the relevant material in the classical books by Schröder, Frege and Hilbert-Ackermann. However, Zuse lacked a thorough mathematical education, so he may have misinterpreted something here and there; in particular, he was innocent and naive enough to bring the concepts of mathematical logic, which were invented for quite another purpose, into the emerging Plankalkül. Zuse's unfamiliarity with mathematical tradition also led him to use non-standard terminology. This, coupled with inconsistent typography, makes reading the Plankalkül difficult and has certainly scared away impatient mathematicians. On the other hand, no mathematician – not even Turing – knew how to build a practical computer around the concepts of mathematical logic in 1942. This makes Zuse unique.

```
mode   A1 = [1 : 3] bool co coordinates 1,...,8 instead of  0,...,7
                                    corresponding to [0: 2] bool co,
mode   A2 = [1 : 2] A1   co point co,
mode   A3 = int      co occupation by 1,...,6 (9,...,14) for white
                    (black) Q, K, R, B, S, P;   8 (instead of 0) for unoccupied co,
mode   A4 = struct (A2 point, A3 occ)  co occupation of the point co
mode   A5 = [1 : 64] A4   co occupation of the board  co;
proc   R17 co adjacent co  = (A2 V0, V1) bool :
           abs(V0[1] − V1[1]) ≤ 1 ∧ abs(V0[2] − V1[2]) ≤ 1;
proc   R128 co move permissible co  = (A6 V0, A2 V1, V2) bool :
           « Corresponding to the occupation occ of V0[i] that belongs to V1,
           where point of V0[i] = V1, the move from V1 to V2 is geometrically
           permissible» ∧ «Intermediate fields, if any, are free»;
proc   R148 co move 2 (wK) permissible co = (A6 V0, ref A2 px) bool :
                     co additional result parameter px for reference to target co   (1)
           begin bool c co if already checked, px refers to permissible target  co
                                                                              := false
           int i := 1; while occ of V0[i] ≠ 2 do  i + := 1;                       (2)
                                                A4 Z0 = V0i];
               for j to 64 while ¬c do                                            (3)
               begin A4 x = V0[j]; px := point of x;
                   c := R17(point of Z0, px) ∧ occ of x ≥ 8;
                   for k to 64 while c do                                         (4)
                   begin A4 y = V0[k];
                       if occ of y > 8 then c := ¬R128(V0, point of y, px) fi
                   end
               end;
               c
           end
```

Figure 7: ALGOL 68 equivalent of Zuse's chess example (from the paper by Bauer and Wössner, 1972)

5. Limitations

In the comments he made in 1972, Konrad Zuse suggests that the Plankalkül represents a machinery equivalent to a stored program computer, i.e., a universal computer[11]. This is not so: a data structure of the Plankalkül cannot be recursive any more than a program can. Certainly, the Plankalkül knows simple and conditional, possibly even nested repetitions (*Wiederholungspläne*), characterized by a prefix W (see for example⑤ , Fig. 3). But that is all. The Plankalkül totally reflects the philosophy of a Babbage-Zuse machine with its finite system of loops and does not cover the possibilities of a von Neumann machine which, if necessary, can perform unrestricted procedures in the way M. S. Mahoney has used this expression in this volume. And the block structure, so important for storage allocation, is not yet carried through methodically – a deficiency which turns up again in APL. The engineer Zuse failed to get the gist of the λ-notation – twelve years later, van Wijngaarden failed to as well. Meanwhile, thanks to Dana Scott, lambda calculus is based on an important mathematical structure in the form of continuous lattices.

6. Conclusion

Konrad Zuse's Plankalkül manuscript, whose origins go back to the early 1940s, was drafted in the second half of 1945. It was an attempt to devise a notational and conceptual system for writing programs, and it was far better than anything that was known up to the middle of the 1950s. However, this very early approach to a high-level programming language did not find practical use. The obstacles were its lack of orientation to numerical computations (or to put it more positively, its adaptation to the field of non-numerical calculation, which was not yet mature), the inefficiency that would have been caused by the unorthodox elements of the Plankalkül and its two-dimensional notation[12]. Nevertheless, it was a remarkable effort, which fully parallels Zuse's achievements with his early machines Z1-Z4. This can be seen most clearly if the Plankalkül is compared to the flow diagram symbolism that originated at about the same time in the US under the influence of John von Neumann.

In his self-righteous way, Zuse blamed the failure of the Plankalkül on others. In the 1993 English version of his autobiography, he wrote (p. 102) "... only occasionally [I] had the opportunity to have discussions with some of the creators of ALGOL, for example Rutishauser and Bauer. Most of the

---

[11] "Wie bereits erwähnt, stellen die freien Rechenpläne im Sinne des PK die heute allgemein übliche Form von Programmen dar."

[12] "The Plankalkül just needs to be made compiler-compatible": K. Zuse, *The Computer, My Life*, Springer, Berlin und Heidelberg 1993, 103.

time we talked past one another. The basic idea of the Plankalkül to systematically construct a programming language from its logical roots, appeared outdated to my partners, or was considered unnecessary ballast." For his efforts towards the practical introduction of the Plankalkül Zuse even saw "a united front of defense on the part of industry and research" (p. 164) and accused the experts in the German Research Council of ignorance and prejudice (p. 158). This made him no friends.

When, in 1972, ALGOL 68 was used to rewrite Zuse's examples of the early 1940s, it was done to facilitate comparison by using a programming language that came closer to the Plankalkül than many others. Should PASCAL or FORTRAN have been used for this purpose? ALGOL 68 had structured elements, too.

In 1997 Giloi wrote "I deem it questionable that the creators of ALGOL were intentionally denying Zuse the credit he deserved as the inventor of important programming concepts, rather I give them the benefit of the doubt that this came from a lack of understanding."[13] I have to answer this criticism (Klaus Samelson and Heinz Rutishauser not being alive any longer) by saying that, while I have on many occasions given Zuse all the credit he deserved ("Plankalkül a forerunner of ALGOL"), I understood only too well the practical limitations of implementing Zuse's ideas in the late 1950s. In fact, I probably understood them better than Zuse, who never wrote a compiler, ever did.

However, on p. 103 in his autobiography, Zuse wrote "Nevertheless, I have not given up all hope that the Plankalkül will once again attain practical importance." But in his 1972 paper, Zuse lived in a dream-world as far as the Plankalkül goes; he had not kept up with the pace of developments in the construction of compilers or in the syntax and semantics of programming languages.

By all means, it can be safely said that Konrad Zuse had, in 1945, more than an inkling of the most important aspects of the emerging field of programming languages, and the time was not ripe until 1956, in a few cases much later. Zuse was ahead of his time, not only with his machines, but also in the Plankalkül, which was his greatest scientific achievement. With good reason, Zuse is the most admired computer pioneer today.

---

[13] Giloi, Wolfgang K.: "Konrad Zuse's Plankalkül: The First High-Level, 'non von Neumann' Programming Language." *IEEE Annals of the History of Computing*, Vol. 19, No. 2, 1997.

FRIEDRICH L. BAUER studied mathematics, theoretical physics, astronomy and logic from 1946 to 1949 at the Ludwig Maximilian University in Munich. He took the state teachers' examination in 1949 and obtained his Ph.D. in 1952. In 1949, he became acquainted with Konrad Zuse and his Plankalkül. His academic career led him to the Munich Institute of Technology in 1952, where he worked under Robert Sauer, taking part in designing the PERM computer. He became a lecturer in 1955. He was appointed Associate Professor of Applied Mathematics at the University of Mainz in 1958. In 1962 he was made a full professor there. He returned to the Technical University of Munich in 1963, where he continued as Professor of Mathematics and Computer Science till his retirement in 1989. He took part in the development of ALGOL and in the creation of the field of software engineering. He has written numerous articles on numerical analysis, the syntax and semantics of programming languages, program development from specification, theoretical computer science and the history of computing, and more than half a dozen books, some translated into foreign languages, the most recent ones being on cryptography, one of his scientific hobbies.

F.L. Bauer has received a number of academic and public honors, among them three honorary doctorates, the IEEE Computer Pioneer Award and membership in the Maximilian Order of Science and Arts of the State of Bavaria. He is member of the Bavarian Academy of Sciences and the *Akademie der Naturforscher Leopoldina* at Halle. He designed the exhibition of computer history at the *Deutsches Museum* and was honored with the *Goldener Ehrenring*. For establishing the discipline *Informatik* at the Technical University of Munich, together with Klaus Samelson, he was recently awarded the Heinz Maier-Leibnitz medal.

The G1 and the Göttingen Family of Digital Computers

Wilhelm Hopmann

Abstract. A small group under the direction of Dr. Heinz Billing constructed four different computers, the G1 (1952), the G2 (1955), the G1a (1958) and the G3 (1961), at the Max Planck Institute in Göttingen. The G1, G2 and G1a were bit-serial machines, which used a magnetic drum for both the main memory and the bit-serial registers. Whereas the G1 and G1a used punched tape programming, the G2 was controlled by stored programs. The G3 was a bit-parallel computer with a ferrite core memory of 4096 50-bit words for numbers and programs, and which operated at a speed of 5000 Flops. The G1 and G2 operated using fixed-point arithmetic, whereas the G1a and G3 had a floating-point arithmetic unit. Each of the G-computers was in operation on a regular basis at the Institute for several years.

1. Introduction

In 1947, a small group of computer scientists from the National Physical Laboratory (NPL) at Teddington visited the British Occupation Zone in Germany to explore what had been done in their field of research over the previous decade. Members of the party were, among others, John R. Womersley, the superintendent of the Mathematics Division of the NPL, Arthur Porter and Alan M. Turing. They met with Alwin Walther from Darmstadt, Konrad Zuse, Heinz Billing and others who were taking part in a colloquium discussing the state of mathematics and computing in Germany.[1] It was after this meeting that Womersley told Billing of the NPL's plans to build a computer much smaller than the famous ENIAC, based on bit-serial numerical processing. Billing wrote about this meeting in his notebook and sketched the main working principle (Fig. 1).[2] This short note marks the beginning of computer development in Göttingen.

---

[1] Whether Turing attended this colloquium is somewhat doubtful and remains an unsolvable "Entscheidungsproblem."

[2] Heinz Billing, private communication to the author.

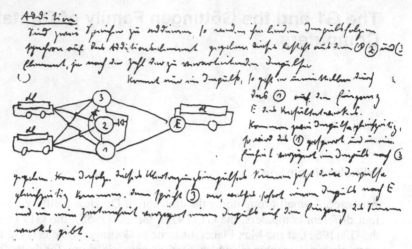

Figure 1: Page in Billing's notebook (Autumn 1947)

In retrospect, the fact that Womersley didn't reveal anything about the realization of the delay lines might have been a stroke of luck. At first, Billing thought of using electrical delay lines, e.g., cables or chains of quadrupoles, but eventually decided to use an endless loop of magnetic tape, a system he had utilized for periodizing analog signals. Then he stuck the tape on the surface of a drum, facing a writing head (W), a reading head (R), an erasing head (E) and a pulse-shaping amplifier, which closed the loop between the reading and the writing head.[3] The drum could rotate at 100 rps,[4] and the circulating *magnetophone* register was thus invented.

These *dynamic* registers were certainly much slower than the ultrasonic mercury delay lines used for the same purpose in the ACE. But all the problems involved in synchronizing them with each other and with the peripherals, caused by changes in the mercury temperature, had been avoided!

In January 1948, Billing successfully tested a drum with a storage capacity of 192 20-bit binary numbers plus the necessary circulating registers, as well as the tracks for the synchronization of the master clock. A first paper on a *Numerical Computer with a Magnetophone-Memory*, was presented at the GAMM (German Society for Applied Mathematics and Mechanics) conference of 1948.[5]

---

[3] See Fig. 4.

[4] Heinz Billing, "Numerische Rechenmaschine mit Magnetophonspeicher," *Zeitschrift für angewandte Mathematik und Mechanik* 29 (1949), 38-42.

[5] It is remarkable that at the same conference Zuse presented a paper on his *Plankalkül* under the title: *The Mathematical Requirements for the Development of*

At the end of 1948, Billing was able to demonstrate a bit-serial binary adder to Professor Werner Heisenberg and others. However, the realization of a complete computer seemed unattainable at that time, as the financial difficulties brought about by the currency reform of June 1948 led to the cancellation of all projects. Thus, a year later, Billing accepted an invitation from the University of Sydney to work there to construct a computer with a magnetic drum memory. Before departing for Australia, however, he left in Göttingen a revised edition of his 1948/49 draft, which had meanwhile been expanded to 34 pages.

The Big Machine

Shortly after his arrival in Sydney, Billing received a letter from Heisenberg offering him much better working conditions than before. Thanks to the Marshall Plan, sufficient money had been made available to finance the hardware of a large computer, which was urgently needed by the Astrophysics Division, headed by Ludwig Biermann. Billing returned to Germany in June 1950 and revised the draft of 1949 for the design of a fully automatic computer (*Voll-Automatische Rechenmaschine*).[6] The proposed system featured a magnetic drum main memory for the storage of 2048 51-bit words capable of storing either a single fixed-point number or two instruction words.

It was estimated that approximately 1200 vacuum tubes would be required. The input/output was to be performed with standard teletype equipment. That system later became the G2. But the estimated three-year construction period needed for such a computer came as a shock to Biermann and his crew. They asked if an additional, much smaller computer could be built in less time.

The Small Machine

The speed of desktop calculations is determined by the time the operator needs to enter numbers into the computer and to subsequently write the results (reading from the display). It seemed, therefore, that a keyboard-operated computer, featuring only a few memory cells to store parameters, constants and temporary results, would be sufficient. Yet the risk of error due to the manual key-in and write-out of parameters and results should also be avoided. The speed of a single operation ought not be much faster than a mechanical desktop calculator.

Based on the experience gained from experiments with the mechanical design of the small drum in 1948 and 1949 and the electronic circuitry working

---

Logistic-Combinatoric Computers. Zuse himself reports that his presentation didn't evoke any response at all!

[6] Heinz Billing, *Vollautomatische Rechenmaschine*, Internal Communication (1950)

at a clock rate of approximately 10 kHz, Billing designed a semi-automatic computer (*Halb-Automatische Rechenmaschine*, HAR),[7] i.e. the G1 computer, during the autumn of 1950. His paper presented the layout of the memory drum, including the 4 circulating registers. It shows how left- and right-shifting of numbers can be performed, and how the transmission of a number from one register to another can be accomplished using logic gates and delays. It also deals with the arrangements for decimal to binary conversion during input and output. Moreover, in the approximately 35 pages of the HAR paper the sets of waveforms controlling the gates for the execution of specific commands, including their timing, are analyzed. Figs. 2, 3, 4 have been copied from that paper[8].

2. The G1 computer

The G1 drum

Fig. 2 shows the G1 memory drum, 8.8 cm in diameter and 17 cm in length. The rotating speed was 50 rps. The track at the far left consisted of 144 mill-cut dents and generated the 7.2 kHz clock pulses which, after amplification and shaping, were used to synchronize the whole system. The other 13 tracks were coated with magnetite.

The first group of 4 tracks was used for the circulating registers, or *dynamic memories*. The angular distance between the writing and reading heads, which was almost 180°, resulted in a double word length of 72 bits,

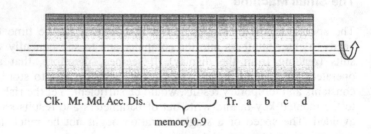

Clk. Mr. Md. Acc. Dis. ⎰⎱ Tr. a   b   c   d

memory 0-9

Figure 2: G1 memory drum

---

[7] Heinz Billing, *Plan für eine halbautomatische numerische Rechenmaschine*, Internal Communication (1950).
[8] All diagrams have been scanned from the referenced German publications and have only been translated into English for this paper.

taking into account the delay in the electronic circuitry. Due to the use of *return-to-zero* writing, an erasing head was needed between the reading- and writing-heads.

The other 9 tracks formed the *static memory*. They are subdivided into 4 sectors of 36 bits each. The *odd* sectors of the next 5 tracks correspond to high-words in the dynamic memories and are addressed as memory locations 0 to 9. The even sectors of these 5 tracks were needed for the decimal to binary conversion. The last 4 tracks provided the 16 static memory locations a0 to a3; b0 to b3, c0 to c3, and d0 to d3.

Each of the 9 write/read heads of the static memory can be connected through an associated electromechanical relay to the *input/output* of the common *read/write* amplifier. Only one relay at a time is actuated according to the appropriate address number or letter used in the command referencing the memory. In contrast to the dynamic memories, a modified Manchester phase encoding technique was used, thus allowing a simple overwriting of stored numbers.

Block diagram of the G1

Fig. 3 shows a simplified block diagram of the G1: in addition to the 3 standard registers of an arithmetic unit, a fourth register, *distributor* (*DIS*), serves for the data flow between that unit, the memory drum and the numerical input/output from/to the user interface.

When I joined the group headed by Billing on January 2nd, 1951, it consisted of himself, H. Öhlmann, and myself. Öhlmann was the physicist responsible for the design and construction of the G2, while I was responsible of the implementation of the G1 circuits. The group was complemented by a designer for the mechanical parts, (the magnetic drums), two mechanics in the workshop and one electronics technician for the assembly and wiring of

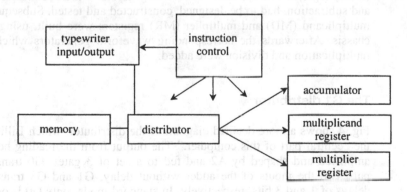

Figure 3: Block diagram of the G1

the chassis (after 1952 there were two). Because the group possessed only one oscilloscope, my first task while learning about bits and bytes was to build an oscilloscope for myself.

Implementation step by step

The construction of the computer was performed step by step using the HAR layout as a guide. Because we didn't have any electronic test equipment (other than the two *scopes*), the sequence of implementation was dictated by the requirement that parts of the machine which had already been completed had to serve as a sufficient testbed for the next device. While the master clock pulse generator, other timing circuits, and the read and write amplifiers were being assembled and put into operation, I scribbled with a pencil on old paper sheets the wiring diagram of the distributor register. All in all I used 50 vacuum tubes. The technician had a hard job deciphering my scribbles, but he did a fine job!

Once the chassis had been assembled, wired, mounted on the rack and connected to the amplifiers, pulse and power sources, it could be tested statically. This was done by applying constant voltages, resulting in a switching of the gates. During these tests the corresponding waveform generator chassis was assembled and wired. It was then mounted on the rack, connected to the other devices, and after plugging 29 vacuum tubes the distributor could be tested dynamically.

The last step of that design phase (in the spring of 1951) involved connecting the typewriter (modified for electrical input/output by the application of contacts and magnetic coils to specific keys) via a set of relays. Now the input and output from the typewriter to the memory, including decimal to binary conversions and all memory functions, were operational. We finally had a test-bed for the subsequent design and construction of the arithmetic unit which was again performed step by step as follows: first the accumulator (ACC) register, and the ACC waveform generator which controlled addition and subtraction, had to be designed, constructed and tested. Subsequently the multiplicand (MD) and multiplier (MR) registers were built, using a single chassis. Afterwards the corresponding waveform generators which control multiplication and division were added.

The G1 distributor

Fig. 4 shows a more detailed diagram of the distributor, which Billing called the "central part of this computer." The output from the reading head R2 is amplified and shaped by A2 and fed to a set of 3 gates. G0 transmits the pulses to the inputs of the adder without delay, G1 and G3 transmit with delays of 1 and 3 bits, respectively. In standard mode, only G0 is open. This establishes a circulation period of exactly one double word. If G1 and G3 are

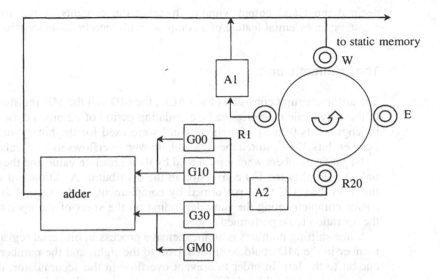

Figure 4: G1 distributor

opened, the circulating number will be multiplied by 10 at each turn. When a number is fetched, the gate GM connected to the reading amplifier of the static memory is opened, and the pulses are fed to the other input of the adder. If A2 is switched off and A1 switched on, the circulation period corresponds to a single word. According to the timing, this copies the bits in the distributor register from the low word positions to the high word positions, or vice versa.

To execute a cyclic permutation of the 4 numbers stored in one of the tracks a, b, c, d, both amplifiers A1 and A2, as well as the erasing head E are switched off. Then, all 4 sectors of the selected track are copied through GM onto the distributor track. In the next step, the writing head W is switched off, and A1 is opened for writing back the distributor track, delayed by one sector, to the static memory. Thus, only 2 revolutions of the drum are needed for the interchange of the 4 numbers.

In the even sectors preceding the memory locations 0 to 9, the binary equivalents of the numbers $0....9 \times 10^{-10}$ are stored permanently. However, the value of these numbers refers to the binary point in the subsequent high word. For a decimal input the low word according to the input digit is read through GM and added to the contents of the distributor. In the next register cycle the content of the distributor is multiplied by 10, as described above.

For the decimal output, the 4 bits to the left of the binary point are latched onto a set of 4 flip-flops and are transmitted to the I/O hardware, before being cleared in the register. The number is again multiplied by 10, thus bringing the next 4 bits into the leftmost position. This enables the G1 to perform decimal input and output without changing the contents of the arithmetic registers: an essential feature of a computer with merely 26 address locations!

The arithmetic unit

The arithmetic unit consisted of the ACC, the MD and the MR registers. Each of these dynamic memories had a circulating period of 72 bits, i.e. two words in length. Bits 0 to 31 of a single word were used for the binary value of a number, bits 32+33 stored the sign and allowed overflows to be checked.

Negative numbers were represented by their absolute value and the sign, in order to simplify the I/O conversion in the distributor. Addition and subtraction into the ACC were performed by complementing the second argument and/or complementing the sum, depending on the signs of the operators and the operation to be performed.

Since shifting numbers is an inexpensive process in bit-serial registers, the number in the MD could be shifted 1 bit to the right, and the number in MR one bit to the left. In order to prevent overflow in the accumulator, the contents could be halved by shifting the number one bit to the right on command.

The multiplication is performed top-down, i.e. according to the uppermost bit in MR, the contents of MD are sent to the *external* input of the adder of the ACC. Simultaneously, these numbers are shifted one bit. Since the contents of the ACC are neither shifted nor cleared, all bits of a 32×32-bit multiplication can be added using the accumulator. Thus, during summing up of simple products, rounding errors could not occur. Only when being stored in memory or used as a multiplicand, the two words were transferred from ACC to DIS and rounded to a single word, before being brought to the memory or into the MD.

To speed up the machine, non-restoring division was implemented by adding a circuit that compared the values in ACC at the output of the adder/inverter and the values in MD. Thus the contents of MD and MR were shifted in the same way as in multiplication. After 32 steps (or 16 revolutions of the drum) the last partial remainder was cleared in the ACC, and the quotient was copied from MR to ACC.

These operations were implemented and tested until the end of 1951. In the meantime the designers considered the idea of including the square root operation in hardware because it was the most needed function. In the algorithm published by Zuse in 1951[9], binary square-root extraction can be regarded as a specific type of division. Presumably, it is the same algorithm Zuse used for the implementation of that function in his Z3 and Z4 computers: the divisor is treated as a number, to be extended by a single bit during the division process. This could be implemented just adding a few tubes to the waveform generator which controlled multiplication and division.

---

[9] Heinz Rutishauser, Ambros P. Speiser and Eduard Stiefel, *Programmgesteuerte digitale Rechengeräte* (Basel, 1951), 36.

By January 1952 all the parts had been assembled and all these commands and operations had been tested successfully. The members of Biermann's astrophysics division could start working on the G1.

The punched tape system

In the autumn of 1951 the idea of connecting a punched tape reader to facilitate the implementation of frequently needed small programs was put forward, e.g. for the calculation of trigonometric or exponential functions. Discussions on that topic resulted in the decision to add 4 punched tape readers to the G1 system for general control purposes. The only readers available at that time were the 5-hole readers of the standard CCIR-TELEX systems, which operated at 7 characters/sec. They were rebuilt for parallel output on 5 signal lines. As all other parts of the G1 were ready and working, the output from the readers had to simulate an input from the contacts of the typewriter. This was done by a tree of relays acting as a 5 to 32 binary decoder. The output contacts were connected in parallel to the appropriate input lines coming from the typewriter.

Additionally, the circuitry for the control of the readers and the circuitry for the control of a tape punch had to be designed and built. The latter was to be used for punching all input typed into the typewriter (thus a program tape could be punched in parallel to the execution of the input by the computer), for copying tapes from reader to punch, and for punching results in parallel to their printing.

It was Easter 1952 before the tape system was assembled. It worked correctly but was unreliable: when it was tested in continuous operation at 7 characters/sec the unreliability of all our relays, bought cheaply from wartime surplus shops, became noticeable. Luckily, Zuse came to our aid: he sent us approximately 120 relays, the standard type used in his Z5 computer. After replacing the relay boxes at the end of May 1952, all 476 vacuum valves and all 101 relays of the G1 worked perfectly.

The instruction set

The commands available for handling operators are shown in the first two groups of the list of instructions (Table 1): each of the commands 1 to 8 exists in two versions. The *a* version is provided for immediate execution of operations on numbers which have been input directly. Otherwise version *b* is active: the execution of the command is suspended until, by inputting the next one or two characters, the memory location is specified from where the operand has to be fetched into the distributor. Having loaded the distributor with the operand, the suspended command is continued.

Table 1: Table of instructions for the tape-controlled computer G1. The operations 3a and 3b must be followed by operation 4 or 5!

	No.	Operation	A	M	z
Operations with entered numbers	1a	Z+	\<A> + Z	-	-
	2a	Z -	\<A> - Z	-	-
	3a	Z m	-	Z	-
	4a	Z ×	\<A> + Z×\<M>	0	-
	5a	Z $\underline{\times}$	\<A> - Z×\<M>	0	-
	6a	Z :	\<A> : Z	0	-
	7a	Z → z	-	-	Z
	8a	Z D *(Print the number Z)*	-	-	-
	8a	Z L *(Print and punch the number Z)*	-	-	-
Operations with stored numbers	1b	+ z	\<A> + \<z>	-	-
	2b	- z	\<A> - \<z>	-	-
	3b	m z	-	\<z>	-
	4a	× z	\<A> + \<z>×\<M>	0	-
	5b	$\underline{\times}$ z	\<A> - \<z>×\<M>	0	-
	6b	: z	\<A> : \<z>	0	-
	7b	→ z	0	-	\<A>
	8b	D z *(Print the number \<z>)*	-	-	-
	8b	L z *(Print + punch the number \<z>)*	-	-	-
Other Operations	10	]	0	\<A>	
	11	√	√\<A>	0	-
	12	H	\<A> : 2	-	-
	13	A b, c, d, accordingly	Cyclical permutation	\<a0> \<a1> \<a2> \<a3>	a1 a2 a3 a0
Tape operations	14	Sa *(Start tape reader a (b,c,d accordingly))*	-	-	-
	15	F *(Stop tape reader)*	-	-	-
	16	SI *(Start conditional branch)*	-	-	-
	17	I *(End conditional branch)*	-	-	-

The table uses the letters A, M, and z to denote the accumulator, multiplicand register and a memory location, respectively. It shows how the contents of each register or memory location is changed by an instruction. \<R> denotes the contents of a register, a zero, that the register has been cleared. Z denotes a decimal number entered from the keyboard or the punched tape.

Command 10 transfers the contents of the accumulator to the multiplicand register for the calculation of multi-factorial products. Command 13 is provided in order to overcome the disadvantages of a fixed addressing system, at least for iterative algorithms, e.g. for the step by step numerical integration of differential equations.

The last group of commands in Table 1 were added to run the tape system: Sa (b,c,d) starts the appropriate tape reader. If this command is started from another reader, that *calling reader* is stopped simultaneously. Command F stops the readers. By use of rotary switches on the operator panel an *embedded* tape command may be started: when the input termination symbol (p or, n) is sensed after reading a number from any tape, that tape is stopped and the selected tape reader is started.

The only conditional command of the G1, implemented a year later, was the tape command *SI*, which performed as follows: if SI is read from the tape and if the ACC is not negative, then a *tape skip* is performed, i.e. the following instructions are read but no operation is executed. If the computer is executing a tape skip, it resumes normal operation after finding the code for I (if not immediately preceded by an S).

Thus, a piece of calculation or the start of another tape reader may be skipped. A nested set of SIs may also be used, however the terminator for all skips can remain the same *I*.

Figure 5: The G1 (H. Billing and W. Hopmann (sitting), May 1952)

Years of regular operation

On June 6, 1952, Billing presented a paper on the G1 at the GAMM conference in Brunswick.[10] The participants traveled to Göttingen the following day to see the computer in operation. Fig. 5 shows the G1 at the laboratory in May 1952.

In September 1952, the G1 was moved to the Max Planck Institute of Physics where it ran up to 24 hours a day, seven days a week during the first years of operation. When it was permanently taken out of operation on June 6, 1958, it had served for 5 years and 8 months, or a total of 49,560 hours. The G1 was switched on for 33,946 of those hours, an average of 16.5 hours a day; 82% of that time was used for computations, a mere 5% was lost due to unexpected crashes (this includes maintenance), the rest was spent on preventive maintenance and other tests. The magnetic drum is the only surviving part of the G1, now kept at the *Deutsches Museum* in Munich.

Keep in mind that the G1 was the only electronic computer in Germany to be used for scientific computations from September 1952. It was not until January 1955 that there was a second such computer, the G2, followed in May 1956 by the PERM in Munich.

Learning by doing

The proper use of the few memory cells and the commands (Table 1) were so simple that they could be learned in an hour. Programming of the G1 and the use of the cyclic permutation for iterative procedures could be mastered by experimenting with the machine. With the command list to their left and the algorithms to be programmed to their right, the scientists could sit down at the typewriter and start practicing the *art* of programming the G1. The computer responded immediately by typing the commands and the results. Nearly a decade passed before the same ease of programming reappeared in interactive BASIC or FOCAL interpreters.

A summary of the tasks performed on the G1 (and the G2) up until the autumn of 1955 is given in Table 2.[11]

Many of our in-house colleagues and guests from elsewhere made their first acquaintance with programmable computers in interactive sessions with the G1. Some of them later became directors of physics or astrophysics institutes or heads of industrial R&D computing centers . Thus, in addition to

---

[10] For the most detailed description of the G1 (in German), including two programming examples, see Heinz Billing, Wilhelm Hopmann and Arnulf Schlüter, "Die Göttinger bandgesteuerte Rechenmaschine G1," *Zeitschrift für angewandte Mathematik und Mechanik* 33 (1953): 50–60.

[11] Ludwig Biermann, "Überblick über die Göttinger Entwicklungen, insbesondere die Anwendung der Maschinen G1 und G2," in *Elektronische Rechenmaschinen und Informationsverarbeitung,* ed. Johannes Wosnik (Brunswick, 1956), 36-39.

producing valuable results, the G1 helped to introduce computers into science and industry over the next half decade. It was also responsible for the increased use of small computers in laboratories for the gathering and pre-evaluation of experimental data.

3. The G2 computer

In December 1954, Herbert Öhlmann completed the construction of the G2. The magnetic drum *static* memory contained 2048 58-bit locations, arranged on 64 tracks of 32 sectors each. A sector could store either a single number or two command words. The 4 dynamic memories or registers with separate heads for writing and reading were allocated to a separate track. Manchester phase encoding was used for both types of memory.

Since the drum rotated at a speed of 50 rps, a master clock frequency of approximately 92.8 kHz resulted. Thus the G2 was not a very fast computer, compared with other bit-serial machines, which had clock rates in the MHz range through the use of fast delay-line or electrostatic tube memories. The mean operation speed of the G2 was approximately 30 Ops/sec. The chassis, containing a total of approximately 1200 vacuum tubes, were mounted onto 9 racks.

Table 2: Subdivision of computing time at G1 and G2

Mathematical structure of the problems	Description of the problem	G1	G2
Tabulation of integrals with algebraic sums	Quantum mechanics of molecules	2200	
Systems of ordinary differential equations	Quantum mechanics of electron shells	3700	
	Störmer trajectories	1300	291
	Planetary orbits		97
Initial value problems of partial differential equations	Quantum mechanics of atom nuclei	1300	
	Hydrodynamic problems from atomic physics	1600	
	Hydrodynamic problems from aerodynamics	1300	
Algebraic tasks	Response of quadrupole filters	700	
	Shock waves in magnetic field		74
	Matrix inversion		300
Others	Quantum mechanics of metals		24
Computing time (hours)		12100	786

The 1951 layout was a relatively standard one for early computers. In addition to the 3 registers of the arithmetic unit (ACC, MD, MR) a fourth register, the index register IR, served to index the address in a command: when a command was called up from the memory, it went through an adder in the IR to the command control. If the index flag of that command word was raised, then the content of the index register was added to the address. The index value itself could be exchanged with the ACC.

Conditional branching of the command sequence was controlled by the 5 different conditional commands: two commands sensed either sign of the ACC, the third initiated a branch if the last number read from the memory contained a raised flag bit. That command was very useful, e.g. when matrix operations had to be programmed: if the numbers in the last column of a matrix were *flagged*, the programming of column-counting loops could be omitted.

The last two conditional commands sensed the position of 2 switches on the operator panel. They were used as program debugging tools: if, for example, a program crashed or went into an endless loop, these commands could be used to initiate a *stop on demand* in order to allow the printing of results, counts, etc., using the keyboard.

Between January 1, 1955 and June 30, 1961, the G2 served in the Max Planck Institute of Physics for a total of 56,952 hours.[12] The G2 was switched

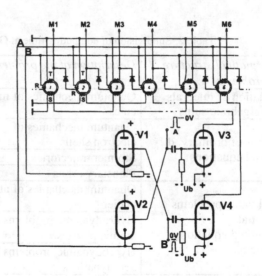

Figure 6: Magnetostatic delay line

[12] Heinz Billing, "Die Göttinger Rechenmaschinen G1, G2 und G3," *MPG-Spiegel* 1982, No. 4, 41-49.

on for 36,076 hours, an average of 15.2 hours daily. 78% of that time could be used for computing, losses of only 5% were encountered due to hardware crashes and corrective maintenance. The remaining 17% was spent on preventive maintenance and development. Billing's G2 was the first computer in Germany to use memory stored program control.

4. New technologies

After the G1 and the G2 had been completed and were operating regularly, Billing's team tried using new computer components. Apart from the long-life tubes and the crystal diodes, the first ferrite cores, which arrived in the spring of 1952, attracted the most interest.

Using the 2.7 mm type, Billing started the design of a core matrix memory with 160 elements. That matrix was used as a test-bed to investigate different addressing methods. The tests began in September 1952. The aim was to build a core memory with a capacity of 4096 words with a length of 43 bits, to be used as the memory of a fast bit-parallel computer, the G3.

At that time, I evaluated the suitability of 9-mm ferrite cores for application in *Magnetostatic Delay Lines* following Wang and Woo[13] in the USA. It turned out that these lines could not replace Billing's *dynamic* memories because, at shifting rates beyond 20 kHz, the cores became too hot. But they were possibly suitable for other applications.

In order to reduce the number of cores, I redesigned the original setup and reduced the number of cores used per stage from 3 to 2 by adding two vacuum tubes generating alternatively *bias* voltages (Fig. 6)[14]. Setting up 120 cores and crystal diodes to form a loop containing only a single "1," a ring counter was obtained for generating 120 individual timing pulses. The source impedance of the 50 V pulses generated by the output coils was so low that the pulses could be used for triggering flip-flops directly and without any other amplification.

The paper written by Maurice V. Wilkes et al. on microprogramming[15] pointed to another application of this type of magnetostatic delay line: a command can be executed by a specific sequence of micro-operations, e.g. the shifting of a number or its transfer between registers. The sequence of trigger-pulses releasing the micro-operations is generated by the output from the cores of a command specific magnetostatic delay line. Because command

---

[13] An Wang and Way Dong Woo, "Static Magnetic Storage and Delay Line," *Journal of Applied Physics* 21 (1950), 49.

[14] Heinz Billing and Wilhelm Hopmann, "Mikroprogramm-Steuerwerk," *Elektronische Rundschau* 9 (1955), 349-353.

[15] Maurice V. Wilkes and J.B. Stringer, "Micro-Programming and the Design of the Control Circuits in an Electronic Digital Computer," *Proceedings of the Cambridge Philosophical Society* 49 (1953), 230-238.

execution may need conditional branching or looping within the sequence of micro-operations, the required parts have been added according to Fig. 7: output and input coils of cores 1 to 6 in the upper row are connected as shown in Fig. 8 (the shift pulses and coils are omitted for simplicity). M1 to M6 denote connections to the corresponding micro-operations. However, if, during resetting of core number 2, the bus line C is pulsed instead of the biasing bus A, then the micro-operations M3', M4' and M5' will be released.

Fig. 8 shows the *head* of a *micro-sequencer* for several instructions. The operation-code of a command, split onto 2 predecoders, magnetizes one of the cores of a matrix decoder using the I/2 current adding method. Each of the matrix cores acts as the *head core* of the associated magnetostatic chain.

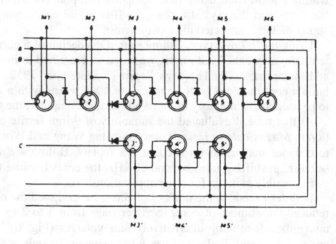

Figure 7: Branch in a magnetostatic delay line

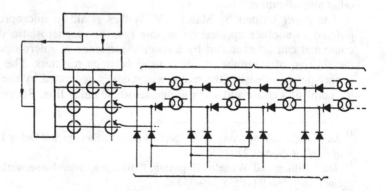

Figure 8: Microprogram control using magnetostatic delay lines

That system of micro-control was implemented in the G1a, using subunits of 10 cores and diodes, mounted on a small board. The bit parallel processing computer G3 featured the same method of micro-sequencing (see below).

Moreover, due to the cooperation and exchange of information between Billing's group and Zuse, an improved version of the two-phase magnetostatic chain went into the Z22. Lorenz Hanewinkel, who constructed that bit-serial computer, used these lines as current drives for the bit sequential I/O from and to the cores of the fast registers of that computer. These registers had a word length of 38 bits and up to 32 registers could be used.[16]

5. Other members of the Göttingen computer family

The G1a computer

The design of the G1a[17] aimed at a computer to be used like the G1: direct input of commands for immediate execution from a typewriter and output of results to that device. Floating-point arithmetic was used in order to avoid the problems of keeping numbers within a fixed range. The complex set of waveform generators as used in the G1 was simplified by using magnetostatic delay lines as ring counters and for microprogramming.

The magnetic drum stored 1800 words on 30 tracks with 60 words of 60 bits each. Due to this memory size, it was possible to omit a separate distributor register for decimal I/O. The communication between ACC and memory, including the cyclical permutation, was handled by the MD register. There were 10 track switching relays connected to the rotor arms of 10 selectors, each of these connecting in turn to one of the 30 tracks. Through programmed positioning of the selectors an address interpretation akin to bank switching was performed. Moreover, it was possible to replace wildcards in the address positions of a command by a decimal number output from the ACC.

Program control was performed by a set of 10 mechano-optical tape readers of our own design at a speed of up to 180 chars/sec. Because the input and the output control acted autonomously, the next command could be prefetched in parallel to the execution of the current one. After conversion in the

---

[16] It is remarkable, that due to this design, the Z22 became the only bit-serial computer in which the fast circulating registers used neither ultrasonic delay-lines, nor Billing's circulating registers, nor Williams tubes, but bit-serial ferrite core matrices.

[17] Wilhelm Hopmann, "Zur Entwicklung der G1a," in *Elektronische Rechenmaschinen und Informationsverarbeitung,* ed. Johannes Wosnik (Brunswick, 1956), 92-98.

ACC, a result could be printed or punched out from a separate output register while the next commands of the program were executed.

Two G1a's were built at Göttingen. One was put into operation at the Max Planck Institute for Hydrodynamics, the other went to the Institute for Plasma Physics in Jülich in 1958 and was moved to a high school at Neuss in 1961, from where it was later recovered and finally moved to the *Deutsches Museum* in Munich. A third G1a was built at the University in Helsinki using our diagrams and mechanical parts from our workshop. That computer, called ESKO, can be seen now at the Museum of Technology in Helsinki.

Table 3: The Göttingen computer family

Computer	G1	G2	G1a	G3
Operation mode	serially, 2 ops/sec	serially, 30 ops/sec	serially, 20 flops/sec	parallel, 5000 flops/sec
Memory	26 words of 32 bits	2048 words of 50+1 bits	1840 words of 60 bits	4096 words of 42+1 bits
Drum?	magnetic drum	magnetic drum	magnetic drum	ferrite core matrix
Paper tape	Yes	No	Yes	No
Number representation	Fixed-point $z < 8$	Fixed-point $z < 8$	Floating-point 43 bit mantissa 8 bit exponent 1 bit marker	Floating-point 33 bit mantissa 9 bit exponent 1 bit marker
Program control	punched tape 4 readers typewriter	memory (2 commands/ word)	punched tape 10 readers 180 char/sec typewriter	memory (2 commands/ word)
Commands	16	32	22	64
Micro coding			magnetostatic delay lines	magnetostatic delay lines
Address modification	no, but cyclic permutation 4 x 4	yes, 1 index register, conditional branching	involved, but cyclic permutation 30 x 2...60 words, 30 x 10 bank switching	yes, 7 index registers, bracket handling, 16 word hardware stack
Lifetime	1.11.52 30.6.58 = 49560 hrs.	1.1.55 - 30.6.61 = 56.952 hrs	a)1958 - 1963 b)1960 - 1970 c)1958-1961/68 (88)	1.1.61 - 9.11.72 = 104200 hrs
Operation time	33946 hrs. (16.5 hrs./day)	36076 (15.2 hrs./day)	no specific statistics	57300 hrs. (13.2 hrs./day)
Efficiency	82%	78%	app. 80 %	85.9%
Hardware crashes and corrective maintenance	5%	5%	a)AVA Göttingen b)Mathematical Institute Helsinki c)IPP Jülich/ Neuss / *Deutsches Museum*, Munich	1.1 %

The bit-parallel computer G3

Using the latest technologies, Billing planned the G3, a succesor to the G2. The computer became one of the *big* machines of that time. It was the first computer to have a hardware stack pointer and stack registers. In comparison to other contemporary parallel computers with a word length of approx. 40 bits, the number of tubes used in the G3 (1500) is relatively small. This is due to the extensive use of magnetostatic delay lines for microprogramming of all operations. Circa 600 cores were used in those lines. The arithmetic unit of the G3 could also calculate the addresses in a command, thus saving ca. 150 tubes otherwise needed for a separate address unit. Because the core and diode subunits were assembled as pluggable groups, the last computer of the Göttingen family presented no problems for the revision of old or the implementation of new macro commands. Thus a digital tape recorder and a graphic display were attached to the G3 in the last years of operation.

A single plane of the core memory and a flip-flop plug-in unit can be seen in the Heinz Nixdorf MuseumsForum at Paderborn, probably the only relics of that last member of the Göttingen Computer Family.

WILHELM HOPMANN made his hobby, electronics, his vocation while working on his physics thesis. He transformed military radar systems into scientific measuring devices useful for ionospheric research at the Max Planck Institute near Göttingen. In January 1951, he joined the *Working Group on Numerical Calculators* headed by Dr. Heinz Billing, associated with the Max Planck Institute for Physics directed by Werner Heisenberg. Working under Billing's direction, Hopmann implemented the G1. Thereafter, he designed and constructed the G1a. After leaving the computer field in 1959, Hopmann headed the Electronics Laboratory at the Institute for Plasma Physics of the Nuclear Research Center in Jülich. Here he was responsible for the development and construction of electronic devices for the control and timing of large fusion experiments. With the development of the ADC for fast discharge experiments, Hopmann went back to the application of computers. He contributed to the design of computerized control and data acquisition systems for fusion experiments e.g. the JET experiment at Culham, UK.

Konrad Zuse and Industrial Manufacturing of Electronic Computers in Germany

Hartmut Petzold

Abstract. Although Konrad Zuse is widely recognized as one of the pioneers of the invention of the computer, his role as entrepreneur in the 1950s and 1960s has been overlooked – quite unjustly. In fact, Zuse started the industrial production and distribution of electronic computers in Germany. He was the founder and director of the first German computer company, not considering IBM. For several years, Zuse was able to compensate his company's modest development capacity by incorporating some of the most up-to-date results of projects conducted by research institutes in Germany, France, the Netherlands, Switzerland and Austria.

1. Before the electronic machine

When Konrad Zuse decided to build a program controlled automatic calculator in 1935, his main objective was to construct a prototype which would validate the operational concept. However, this could only be a first step. His vision of the computer and a computerized society was based on the successful commercialization of his new machine. Zuse was not able to realize his vision until after the war, during the early years of the young Federal Republic of Germany. In 1955/56 his company built the model Z11, a special machine for surveying and optical calculations.[1] Although his commitment to obtaining a patent was immense,[2] his priority was always the construction of

---

[1] This story has been repeatedly told and analyzed. Konrad Zuse, *Der Computer. Mein Lebenswerk* (Berlin, 1984) (revised version of the first edition with the same title, Munich 1970). Hartmut Petzold, *Rechnende Maschinen. Eine historische Untersuchung ihrer Herstellung und Anwendung vom Kaiserreich bis zur Bundesrepublik* (Düsseldorf, 1985), 291–372. Rolf Zellmer, *Die Entstehung der deutschen Computerindustrie. Von den Pionierleistungen Konrad Zuses und Gerhard Dirks' bis zu den ersten Serienprodukten der 50er und 60er Jahre* (Diss. University of Cologne, 1990), pass.

[2] Friedrich L. Bauer, "Konrad Zuse – Fakten und Legenden," in *Die Rechenmaschinen von Konrad Zuse*, ed. Raúl Rojas (Berlin, 1998), 5–21; Hartmut Petzold,

a functioning machine. Zuse often told the story of how he managed to do the extensive work which was necessary to build his first machines, before he founded his company in 1942.[3] It seems that he possessed a touch of genius when it came to organizing development and manufacturing.

While Zuse's role as a pioneer of the invention of the computer in Germany is uncontested, it runs the risk of veiling the complexity of the invention itself. Recognition of his role as an entrepreneur during the 1950s and 1960s has been largely overshadowed by his failure in 1967, when his company was taken over by Siemens.[4] However, there can be no doubt that he also was a pioneer of the *electronic* computer in Germany, which coincided with the role he played as the founder and spirit of the first German computer company.[5] This paper discusses some of the problems which Zuse faced when organizing the development, manufacture, and distribution of his first electronic computers, the Z22 and the Z23, between 1956 and 1963.

2. The historical role of the Z22 and Z23 computers

Zuse was the first to embark on the industrial production and distribution of electronic computers in Germany, independently of IBM. The Z22, the first model produced in series and introduced in 1958/59, was, in his words, "a good commercial success". This product enabled the Zuse KG to establish itself as an electronic computer company and even to pose a challenge to IBM in Germany. The company enjoyed the esteem of the scientific and engineering community, which meant that Zuse could compensate for his company's modest development capacities by incorporating some of the latest findings of several projects from scientific institutes in Germany, France, the Netherlands, Switzerland and Austria. This was an important entrepreneurial achievement.

During the first half of the 1960s the Zuse computers, Z22 and Z23, and the specific ideas and experience connected with them, had a strong influence on the introduction of the computer into many fields of science and engineering in the Federal Republic of Germany. They paved the way for new

---

"Die Mühlen des Patentamts," in *Die Rechenmaschinen von Konrad Zuse*, ed. Raúl Rojas (Berlin, 1998), 63–108.

[3] Zuse (n. 1 above), pass.; Karl-Heinz Czauderna, *Konrad Zuse. Der Weg zu seinem Computer Z3* (Munich, 1979).

[4] This also supports the completely unjustified German cliché of the genial but naive engineer, totally lacking in business acumen.

[5] Paul Ceruzzi has pointed out the pioneering role of the entrepreneur Zuse in the international computer business. Paul E. Ceruzzi, "Die frühen Arbeiten von Konrad Zuse im Kontext der Erfindung des digitalen Computers. 1935–1950," in *Deutsches Museum. Wissenschaftliches Jahrbuch 1992/93* (Munich, 1993), 170–186.

concepts, like the "Minima"[6] machine and the propagation of the programming language ALGOL, which were both born in Germany and neighboring countries. During the 1960s the spectrum of ideas in the young computer science community broadened and the voice of the European protagonists could be heard slightly better in the international scene. In Germany, the Z22 and Z23 models played an important role in opening up and preparing a new computer market alongside the traditional IBM markets. The Z22 and Z23 made their imprint on the new engineering discipline, which came to be called "Informatik" in Germany.

The credo of the young Federal Republic was a liberalized "social market economy." At the same time the strategic doctrine of obtaining maximum control over the new electronic computer technology prevailed. This could not be done within the framework of military projects, as was the case in the UK and USA. Therefore, the first computer development projects were started as research projects at scientific institutes and were usually financed by the Deutsche Forschungs Gemeinschaft (DFG). Projects were started at the Max-Planck-Institute for Astrophysics, Göttingen (Ludwig Biermann, Heinz Billing and their team) in 1947, without DFG financing, and at the Technical Universities in Darmstadt (Alwin Walther and his team) and Munich (Robert Sauer, Hans Piloty and their team) between 1950 and 1955.

After 1955, a growing number of scientists and engineers tried to obtain computers for their institutes or universities. It was becoming clear that computers would be indispensable to those wanting to compete internationally in the fields of sciences and engineering.

3. Architectural decisions

Zuse's experience and knowledge in the new field of computer science and technology was reflected in his entrepreneurial decisions during the time spent developing the vacuum tube machine Z22 (shipped since 1958) and the transistorized Z23 (shipped since 1961). He had to take into consideration the technological situation: although in 1936 Helmut Schreyer had already suggested building the new machine with electronic tubes, instead of electromechanical relays, Zuse hesitated and waited for about 18 years before he decided to build an electronic computer. In computers, tubes could not compete with relays in terms of reliability until 1952, when the manufacturers started producing tubes for digital circuits.

---

[6] The "Minima" architecture is the subject of the chapter "STC ZEBRA" in Gerrit A. Blaauw and Frederick P. Brooks, *Computer Architecture: Concepts and Evolution* (Reading, Mass., 1997), 724–739. The term "Minima," which was introduced by Theodor Fromme, and which became well known in Germany, is not used in that book.

Figure 1: Assembly of Z22 in the Zuse KG factory in Bad Hersfeld (Germany)

Zuse mastered the electromechanical relay technology better than others. His technique of switching without current made his computers superior to most other relay machines, and an improved relay, manufactured by the Zettler company, gave his machines maximum reliability. Zuse's relay machines could work for many hours without operator supervision.

Zuse was able to acquire know-how about transistor circuits and computers in 1954. He knew at that time that four years later he would be able to hire the well-trained engineer Richard Bodo, from the Institute of Low Frequency Technology at the Technical University of Vienna. Heinz Zemanek developed and built a computer between 1956 and 1958 in Vienna – at about the same time the Z22 came into being. Zuse knew that Zemanek's machine, with the Viennese name "Mailüfterl," would be a "Minima" with transistor circuits.[7]

While developing the Z22 and Z23, Zuse had access to the rare expertise in computer technology that was scattered over Germany and Europe, and he incorporated it into his small company. He adopted tube circuits for the model Z22, which had been developed and tested by the Max Planck Institute

[7] Personal communication by Richard Bodo. Heinz Zemanek, *Weltmacht Computer: Weltreich der Information* (Esslingen, 1991), 229. Richard Bodo, *Über den optimalen Entwurf von Transistorrechenmaschinen* (Diss., Technical University of Vienna, 1958).

for Physics and Astrophysics in Göttingen, and he obtained the necessary know-how on transistor circuits for a "Minima" machine from Zemanek's institute. He did not need to develop the architecture of the electronic machine in his company. He simply adopted the "Minima" architecture from Amsterdam, Freiburg and St. Louis, France[8]. In each of these cases he made competent decisions.

Zuse was initially motivated by a request to manufacture the G1a machine, which was a further development of the first successful electronic computer in Germany, the G1, built by the Max Planck Institute (MPI) in Göttingen.[9] The MPI and the DFG were looking for an industrial company which could manufacture the G1a in small series. The G1a had not yet been completed when the MPI agreed contractually, in January/February of 1955, to make all the know-how they had acquired with the G1 and G2 computers available to Zuse. The G1 (since 1952) and the G2 (since December 1954) were the only electronic computers running in Germany at that time. The contract with the MPI was liberal and forward-looking and did not limit Zuse to manufacturing only the punched tape controlled model G1a, but explicitly allowed him to manufacture the "Minima."[10]

A further impulse came from the inventors of the "Minima" concept who were looking for some way to build the machine. Willem Louis van der Poel, the inventor of the principle, could not build a "Minima" with the Netherlands Post, Telephon, and Telegraph (PTT). He had completed the PTERA at that institution and had already experimented, in 1952, with a minimal design, later called ZERO. This model ran for a few weeks and served as a test model for the ZEBRA. The ZEBRA became commercially available through Standard Telephone and Cables (STC) in 1959.[11]

Theodor Fromme, who was a mathematician at Professor Henry Goertler's mathematical institute at the University of Freiburg and who was also enrolled at the French institute for armament research at St.Louis, ensured that the "Minima" would be developed and built at the institute in Freiburg.[12]

---

[8] St. Louis is a suburb of the Swiss town Basle, but is situated in France and not very far from Freiburg in Germany.

[9] Compare in this volume: Wilhelm Hopmann, *The G1 and the Göttingen Family of Digital Computers.*

[10] This contract is part of the Billing Papers, NL 106/050 in the Archives of the Deutsches Museum, Munich.

[11] Blaauw and Brooks (n. 6 above), 724. Van der Poel's theoretical one-instruction machine was part of his Ph.D. thesis: Willem Louis van der Poel, *The Logical Principles of Some_Simple Computers (Ph.D. thesis University of Amsterdam 1956)* (Excelsior, 1962). The story of W.L. van der Poel and the Netherlands PTT is told by Eda Kranakis, "Early computers in the Netherlands," *Centrum voor Wiskunde en Informatica CWI Quarterly* 1:4 (December 1988), 61–84.

[12] There are several references in Billing Papers, NL 106/051, Archives of the Deutsches Museum, Munich.

While the Z22 was being developed by Zuse's company, Fritz-Rudolf Güntsch, an assistant from the Mathematical Institute of the Technical University of Berlin, introduced some new instructions to the machine – without being asked to do so by Zuse, and in conflict with Zuse's vision of the pure "Minima." This event illustrates some of the enthusiasm many young people had for the new machine.[13]

Conditions for financing the project were favorable and Zuse knew how to take advantage of this situation.

At the time, Zuse's entrepreneurial decisions and concepts were accepted and backed by the DFG and, therefore, by the federal government. The computers procuring committee, initiated by the *Gesellschaft für angewandte Mathematik und Mechanik* (GAMM), and installed by the DFG in 1952, tried to get financial support from government agencies for Zuse's work on the Z22.[14] In 1956, the West German government and the Bundestag decided to finance the acquisition of twelve electronic computers for twelve university institutes – three from Siemens, three from Standard Electric Lorenz (SEL) (which was an amalgamation of several traditional German companies under the roof of ITT), three from the Zuse KG, and also three model 650 computers from IBM. Distribution was organized by the DFG and the procuring committee. At the time only IBM's computers had actually been built. The other companies could use half of the funds for development. Because Zuse's planned Z22 was smaller than the Siemens and SEL machines, he received less funding than the others. Therefore while this decision helped the Zuse KG, it also stimulated the competitors.[15]

A year later, the DFG asked the minister for economic affairs to organize an additional credit for the Zuse KG, which was needed to go into series production. Because of his experience and knowledge, Zuse was considered to be the ideal chief executive for a German company specializing in the manufacture and distribution of computers.[16]

4. Programming

Though Zuse had invented the "Plankalkül"[17] and had become an expert in programming computers much earlier than others, and though the "Minima" was based on the principle that the hardware should be minimal and the pro-

---

[13]Zuse (n. 1 above), 121 and several personal communications.

[14]There are several references in Billing Papers, NL 106/078, Archives of the Deutsches Museum, Munich.

[15]Petzold, Rechnende Maschinen (n. 1 above), 402–416.

[16]There are several references in Billing Papers, NL 106/050,051,078, Archives of the Deutsches Museum, Munich.

[17]Compare in this volume: Friedrich L. Bauer, "The Plankalkül of Konrad Zuse – Revisited."

gramming possibilities maximal, Zuse was not able to establish the Plankalkül. In 1955/56, users enjoyed writing their own programs even at the machine language level. Demand for a high level language grew only in the following years, but the Zuse KG lacked the capacity to give the "Plankalkül" a more practical form. In 1959/60 the users of the Z22 and Z23, and also the Zuse KG, placed their stakes on the new ALGOL 60. The company did not employ a programmer and was not able to write a compiler for ALGOL. They obtained it from the Institute of Applied Mathematics of the University of Mainz (Friedrich L. Bauer, Klaus Samelson). Zuse's company, which was one of the early members of the ALCOR-group,[18] promoted a user organization that supported the exchange of programs.

Unlike FORTRAN, ALGOL had been initiated in Europe by university institutes which also actively promoted it.[19] The non-IBM industry adopted it for some years, but did not try to develop it further.

5. The competitors

At approximately the same time as the Z22 was "transistorized" by Bodo (to become the Z23), Zuse's company developed another machine with transistors, the Z31 model. The big electrotechnical companies, Siemens & Halske, Telefunken, and SEL, had been developing and delivering electronic digital computers in Germany since 1959. SEL ceased producing computers immediately after it had delivered its first ER 56 models. The middle-sized company, Schoppe und Faeser, manufactured the American models LGP 30 and Libratol 500 under license. IBM Deutschland, which had existed in Germany since 1910 as a very efficient company under the name Deutsche Hollerith Maschinen-Gesellschaft (Dehomag), manufactured the models IBM 650 (since 1956) and IBM 305 RAMAC (since 1958) in Sindelfingen. Both machines had been developed in the USA.

By the middle of 1960, the Zuse KG had delivered 41 Z22 machines, Siemens ten 2002 machines, and SEL four of the ER56 machines. Although the design was still unfinished, Telefunken had four orders for the TR 4 model. At that time, 52 IBM 650 models, 14 IBM 305 RAMAC, one IBM 704, one IBM 705, one IBM 7050, 24 UNIVAC UCT from Remington Rand,

---

[18] In 1959 several institutions had founded the ALCOR-group with the goal of making the ALGOL compilers (ALgol COnverteR) uniform and to exchange experiences with ALGOL. Richard Baumann (Ed.), "ALGOL-Manual der ALCOR-Gruppe, Part 1," *Elektronische Rechenanlagen* 3 (1961), 206–212.

[19] Friedrich L. Bauer, "Die ALGOL-Verschwörung," in *Leitbilder der Informatik– und Computer-Entwicklung,* ed. Hans Dieter Hellige (Bremen, 1994), 42–55.

6 X1 models from the Netherlands company N.V. Electrologica, and 6 LGP 30 models were in use in Germany.[20]

Although Zuse KG was very active in the West German market, it also had to hold its own in a sector that was oriented to the world market, a task it achieved successfully between 1958 and 1965. Zuse KG was not short of good ideas, but they could not to amass a sufficient capital base to consolidate the company.[21] The failure or takeover of a small or medium computer company was not a unique event. Even the large companies all over the world had significant problems making strategic decisions in the computer field. We also know what an important role small high tech companies could play – at least in the USA. The computer was no longer a toy for individuals, it was the lever which would revolutionize technology and science.

HARTMUT PETZOLD holds a degree in electrical engineering and a Ph.D. in history. He was editor of the German journal "Technikgeschichte." He has been the curator for computer science and time measurement at the Deutsches Museum in Munich since 1988. He has published the books *Rechnende Maschinen. Eine historische Untersuchung ihrer Herstellung und Anwendung vom Kaiserreich bis zur Bundesrepublik*, Düsseldorf 1985 and *Moderne Rechenkünstler*, Munich 1992. Together with William Aspray and Oskar Blumtritt he is the editor *of Tracking the history of Radar*, Piscataway, N.J. 1994.

---

[20] All figures were published by the Deutsche Arbeitsgemeinschaft für Rechenanlagen (DARA), *Stand des elektronischen Rechnens und der elektronischen Datenverarbeitung in Deutschland* (Darmstadt, 1961), 16 and 23.

[21] Looking back, Heinz Zemanek's determined judgement was: "Es ist eher für Deutschland schade – fast möchte man sagen: eine Schande – , daß die Zuse KG nicht zum Kern einer deutschen Rechnerindustrie geworden ist." Zemanek, *Weltmacht Computer* (n. 7 above), 230.

Helmut Hoelzer – Inventor of the Electronic Analog Computer

Thomas Lange

Abstract. During World War II, a young German engineer, Helmut Hoelzer, studied the application of electronic analog circuits for the guidance and control system of liquid-propellant rockets. He developed a special purpose analog computer, the "Mischgerät," and integrated it into the rocket. The development of the fully electronic, general purpose, analog computer was a spin-off of this work. It was used to simulate ballistic paths by solving the equations of motion. At the time, Hoelzer did not use the word "computer" but referred to "electronic modeling" or "transformation of equations into hardware."

1. Introduction

In 1941, Konrad Zuse built his computing machine Z3. It has long been forgotten, however, that in the same year a young electronical engineer developed a machine which was, at that time, probably equally important. His name was Helmut Hoelzer (1912-1996), and his invention the program-controlled, fully electronic analog computer. During World War II, Hoelzer and Zuse had no contact whatsoever. It was only after the end of the war that they heard about each other.[1] Curiously, their computers found the same eventual destination, when in 1945 Zuse moved his Z4 to safety, taking it from Berlin to a mining tunnel in the Harz mountains. The A4 rockets were being produced in the same tunnel and were guided to their targets by Hoelzer's analog device, the "Mischgerät."

With the help of Wernher von Braun's staff, Zuse was eventually able to bring his Z4 to safety in the Allgäu region. It was also Wernher von Braun who brought Hoelzer to the army test center in Peenemünde in 1939.

---

[1] Helmut Hoelzer, "V2-Simulator, Interview on 2 Cassette Tapes, 24 June 1983," in Special Collections MS87-8, ed. James E. Tomayko (Wichita, 1983); Konrad Zuse, "Greetings at the unveiling of a commemorative stone for Helmut Hoelzer" (Hünfeld, October 1995).

The results of Hoelzer's research and development work in Peenemünde can be summarized in three points:

- His contribution to the development of a radio beam ("Leitstrahlverfahren") for rockets
- The integration of electronic analog circuits, working in real time, into the control system of the rocket
- The invention of the fully electronic, program-controlled analog computer and its use as simulator

The first task assigned to Hoelzer in Peenemünde was the development of a guidance system based on a radio beam. This had already been done for aircraft.[2] Since the rocket was unmanned, the correction of a deviation had to be made automatically, using both built-in controls and a control by radio link. Modifying an idea he had in 1935, Hoelzer used a capacitor to build differentiation and integration devices and applied them to the guidance and control system of the rocket. He called the whole circuit the "Mischgerät."

Hoelzer noticed that the electronic circuits could also be used to solve the equation of motion of the rocket (Hoelzer called it "transformation of equations into hardware"). For this purpose, he developed electronic circuits for multiplication, division, square root, and various other functions, in addition to his differentiation and integration devices. A large class of differential equations could be solved using his methods.

Hoelzer described the results of his research in a Ph.D. thesis submitted in 1941. However, the thesis was soon declared secret and could not be published. During the massive air raid on Peenemünde on August 18, 1943, this first version was destroyed. Later, in 1946, Hoelzer submitted a new version to the Technische Hochschule Darmstadt (TH Darmstadt). His supervisor was the professor of applied mathematics Alwin Walther. The American Military Government ordered him to divide the dissertation into two parts, a civil and a military section, and the latter could not be published.

Hoelzer did not achieve recognition for a long time, so the invention of the fully electronic analog computer was not associated with his name. This is also clearly due to the fact that the analog computer was superseded by the digital computer. It was not until the beginning of the 1980s that Hartmut Petzold[3] and James E. Tomayko[4] reassessed Hoelzer's contribution. The correspondence conducted between Tomayko and Hoelzer in 1983 has preserved many technical details for historical research and is now kept in the

---

[2] Fritz Trenkle, "Die deutschen Funklenkverfahren bis 1945," in AEG-Telefunken-Anlagentechnik, Geschäftsbereich Hochfrequenztechnik (Ulm 1982): 113-118.

[3] Hartmut Petzold, *Rechnende Maschinen. Eine historische Untersuchung ihrer Herstellung und Anwendung vom Kaiserreich bis zur Bundesrepublik* (Düsseldorf, VDI 1985): 85-88.

[4] James E. Tomayko, "Helmut Hoelzer's Fully Electronic Analog Computer," in *Annals of the History of Computing* 7 (July 1985): 227-240.

archives of the Charles Babbage Institute.[5] In 1993, Hoelzer began to build a replica of his analog computer of 1941. The company ADS in Huntsville, Alabama, placed an office at his disposal, in which the reconstruction could be coordinated and carried out[6]. The machine is now part of the collection of the *Deutsches Technikmuseum* in Berlin[7].

An analog computer works using *analog units*, for example, segment lengths in mechanical analog computers, and voltages and currents in electronic analog machines. At that time, analog computers were much faster than digital computers. For this reason, they could be used to solve real time problems, such as the guidance and control of a rocket. They were also capable of parallel computation – a feat that digital computers only performed years later.[8]

I describe here the electronic analog computer and the "Mischgerät" based on the existing sources. I explain the architecture and functionality of both devices using some circuit diagrams I have redrawn.

2. Electronic modeling of differentiation and integration

Hoelzer started working on the analog modeling of differential expressions, which he had often came across in control engineering, at a rather early stage. Analog modeling could be implemented mechanically or electronically. Through Professor Walther at the TH Darmstadt, Hoelzer was already familiar with mechanical analog computers. But electronic modeling was less expensive for production in series and better suited to the requirements of aircraft.

Differential equations consist of several differential expressions. It is possible to arrange the devices for modeling the differential expressions in such a way, that the total circuit corresponds to the differential equation to be solved. It is also necessary to implement some algebraic calculations in order to describe nonlinear combinations of the differential expressions. If, in addi-

---

5 Helmut Hoelzer, "Helmut Hoelzer Papers 1946-1983," in Charles Babbage Institute, CBI 33, University of Minnesota (Minneapolis).

6 Beth Boone, "History Rediscovers Computer Pioneer," in ADS Environmental Services Inc. Publication (Huntsville/AL, 1993): 5; Beth Boone, Angely Koons, "Computer Pioneer is in Residence at ADS," in Axel Johnson Inc. Publication 13 (Stamford/CT 1993): 8-9; James Mc Williams, "Pioneer rebuilding Computer," in The Huntsville Times (Huntsville, 5 August 1993): A1-A2.

7 Helmut Hoelzer, "Reconstruction of the Analog Computer," organized by Hadwig Dorsch, in Deutsches Museum für Verkehr und Technik (Berlin, 1993).

8 Gunter Schwarze, "Speech at the unveiling of a commemorative stone for Helmut Hoelzer," Manuscript of the speech (Usedom, 27 October 1995).

tion, it is possible to set up the initial conditions, one obtains a machine capable of solving differential equations.

From his work on control engineering, Hoelzer was particularly familiar with differential expressions of time-dependent variables. In his impressive dissertation[9], he describes how time derivatives and time integrals can be reduced to two physical electric phenomena:

- Inductivity (coil inductor): The first phenomenon is Faraday's Induction law. A variable magnetic field produces a voltage in a conductor surrounding the field. The device in which the phenomenon is used is the coil illustrated in Figure 1. In the ideal case, the induced voltage u is directly proportional to the time derivative of the current i. The proportionality factor is the inductivity L. But in practice, the variable magnetic field can only be generated by a current whose field must be amplified by a material with high permeability, for example iron. But then the nonlinear characteristic curve of iron mediates between the magnetizing current and the induced voltage, producing large errors. Furthermore, an "intrinsic" current is needed, that is, one not influenced by opposite voltages across impedances in the circuit. This can only be guaranteed by a voltage source with very high internal resistance, or by a regulated current.

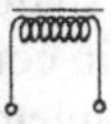

$$u = L \frac{di}{dt}$$

Figure 1: Coil

- Capacitance (capacitor): The second physical phenomenon is the fact that a variable electrical field produces a compensation current. In electrostatics no material exists whose dielectric constant can compete with the permeability of iron. In contrast to iron, the materials required to increase the field strength have linear characteristic curves. As in the first case, a variable electric field must be produced with an intrinsic voltage. This could be done by a voltage source with very low internal resistance, or by a regulated voltage. The device which uses the second physical phenomenon is the capacitor illustrated in Fig. 2. In the ideal case, the loading current is directly proportional to the time derivative of the voltage u applied to the capacitor. The proportionality factor is the capacitance C.

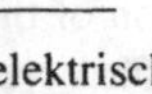

$$i = C \frac{du}{dt}$$

Figure 2: Capacitor

---

[9] Helmut Hoelzer, "Anwendung elektrischer Netzwerke zur Lösung von Differentialgleichungen und zur Stabilisierung von Regelvorgängen," Dissertation D.87, Technische Universität Darmstadt (Darmstadt, 1946).

3. The capacitor as differentiator and integrator

In principle, differentiators and integrators can be built using capacitors or coils. Since 1935, while studying at TH Darmstadt, Hoelzer decided to use the capacitor because of its advantages. He was, like many scientists in Peenemünde, an enthusiastic glider pilot and, also in 1935, he designed a device that could determine the ground speed, i.e. the true speed of an aircraft over the ground[10]. He used an electronic integrator which was able to calculate (simulate) the absolute velocity $\bar{v}_{ground}$ from the acceleration $\bar{a}$. The acceleration could be measured by a special device (a mass-spring damping system) in all three axes[11].

In Fig. 3, the measured value u_1 controls a voltage source, and the loading current of the capacitor is proportional to the time derivative of the measured value. This is only true when the internal resistance of the voltage source is $R_i = 0$.

In Fig. 4, the capacitor is connected to a controllable current source. The voltage over the capacitor is the integral of the measured value u_1. It is important to set the internal resistance of the current source to $R_i = \infty$.

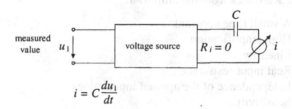

$$i = C\frac{du_1}{dt}$$

Figure 3: Capacitor as differentiator

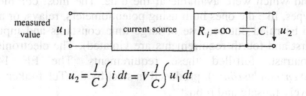

$$u_2 = \frac{1}{C}\int i\,dt = V\frac{1}{C}\int u_1\,dt$$

Figure 4: Capacitor as integrator

---

[10] Helmut Hoelzer, "Oral History Interview, 10 November 1989," in Peenemünde Interviews, National Air and Space Museum, ed. Michael Neufeld (Washington 1989); Gunter Schwarze, "Obituary of Dr.-Ing. Helmut Hoelzer," in RZ-Mitteilungen 13, Rechenzentrum der Humboldt-Universität Berlin (Berlin, January 1997): 43-44.

[11] James E. Tomayko (n. 4 above); Klaus Biener, "Helmut Hoelzer, Wegbereiter der Informatik," in *Mikroprozessortechnik* 5 (1991): 1.

Pure voltage and current sources are not easily available. However, it is not too complicated to obtain intrinsic voltages and currents by a simple regulation mechanism. A voltage is applied to the electric circuit to compensate the sum of voltages across the resistances at any given moment. The resulting voltage is measured at the additional resistance and is returned to the electric circuit, after amplification, with the sign reversed. Then the final current is only determined by the capacitor. Intrinsic voltages can be produced in the same way.

At that time, paper-winded capacitors with a capacitance of about 30 µF could be used. If the measured value is an oscillation of low frequency (lower than 0.1 Hz), the maximum loading current can become smaller than 1mA. Such low currents must be amplified.

4. Amplification

Amplifying elements are needed to produce intrinsic currents and voltages, as said before. Very low voltages and loading currents at the capacitor require the use of an amplifier. In his dissertation, Hoelzer demands the following characteristics from the amplifier:

1. A small time constant
2. High amplification
3. Linearity
4. Real input resistance
5. Independence of the applied input values
6. Sensitivity
7. Zero point safety

Hoelzer discusses the types of amplifier that could satisfy these requirements and which were available at the time. The most commonly used amplifier types, like the ones built using potentiometers, relays, or a magnetic coil, had to be ruled out because of large time constants and input inductivities (the first and fourth requirements are violated). The electronic tube amplifier, in contrast, fulfilled these requirements. The EF 14 electronic tube (*Wehrmachtsröhre*), produced in steel by the Telefunken company, was relatively failsafe and robust[12].

However, the electronic tube cannot be driven as a DC amplifier (direct current), because this does not guarantee the independence of the applied measured values. The fifth requirement is not fulfilled. Due to the known

---

[12] Helmut Hoelzer, "50 Jahre Analogcomputer," in *Computer als Medium*, ed. N. Bolz, F.A. Kittler, C. Tholen (München, Wilhelm Fink Verlag, 1991): 69-90; Klaus Biener, "Computerpionier zu Gast in Berlin," in *Mikroprozessortechnik* 8 (1992): 40-41.

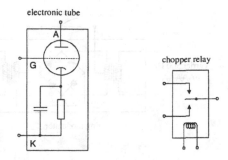

Figure 5: Elements used for a chopper amplifier

drift of the DC amplifier, zero point safety is not guaranteed and requirement seven is violated. All the other points are fulfilled satisfactorily.

Reacting to these problems, Hoelzer came up with the idea of driving the electronic tube as an AC amplifier (alternating current). The applied measured value was transformed into an alternating voltage by chopping up, or modulating, with a suppressed carrier. The amplification is zero point safe, because the measured voltage is completely separated from the operational DC voltage of the electronic tube. The open gate of the secondary side guarantees complete independence of the applied measured values.

The elements shown in Fig. 5 are used for the electronic tube chopper amplifier. The measured value at the input is chopped up with a relay, triggered by an alternating current. After amplification and separation from the anode DC voltage, the AC-voltage is rectified by a synchronously oscillating relay (mechanical rectification). In this case, the amplified output voltage is directly proportional to the input voltage. In addition, one is able to keep the amplification factor of the electronic tube constant.

The electronic tube amplifier with modulation needs the elements illustrated in Fig. 6. Instead of a relay, a ring modulator is used, which is well known from the field of telecommunications and which chops up the input signal, not rectangularly, but sinusoidally. When the measured voltage changes sign, the modulated AC voltage changes phase by 180°. At the output of the amplifier, a rectification takes place in a phase bridge with the addition of the AC voltage as carrier, so that the output voltage is directly proportional to the input voltage.

In both the ring modulator and the phase bridge rectifier, diodes of copper-oxide (Cu_2O) were used. These proved to be very stable and zero point safe.[13] The electronic tube amplifier with modulation is illustrated in Fig. 7.

---

[13] Klaus Biener (n. 12 above).

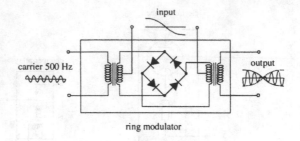

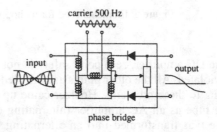

Figure 6: Elements used for a modulation amplifier

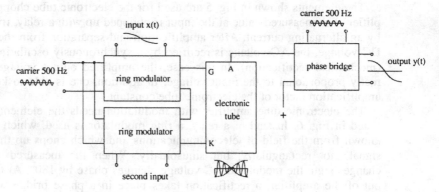

Figure 7: Electronic tube amplifier with modulation

5. Differentiators and integrators

Hoelzer built the devices for differentiation and integration using one *RC* circuit and two amplifier stages in each case. The *RC* circuits are shown in Fig. 8.

RC circuit for differentiation   RC circuit for integration

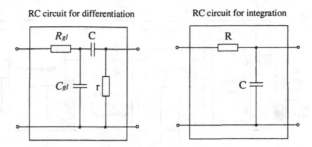

Figure 8: RC circuits for differentiation and integration

A smoothing device is added in front of the differentiation element to eliminate the harmonic series which occur due to rectification. The influence of the resistance R_{gl} on the differentiation process is eliminated by the feedback. The capacitor C_{gl} is small compared to C ($\approx 1/1000$); for that reason, the error contribution is small.

One could combine these RC elements using either a chopper amplifier or a modulation amplifier. The measured values at the input had to be fed in modulated form. Only for differentiation and integration there was an intermediate transformation to direct current. The feedback was also done on the alternating current side[14].

In Fig. 9, this is illustrated using the example of a chopper amplifier. The switching frequency of the chopper relay is 200 Hz.

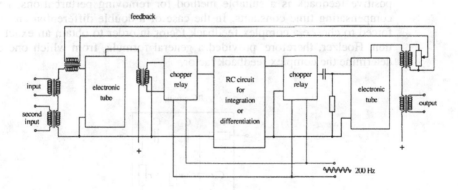

Figure 9: Solution with a chopper amplifier

[14] Helmut Hoelzer, "Guidance and Control Symposium," *in The Eagle has returned, Proceedings of the Dedication Conference of the International Space Hall of Fame* 43, Space and Technology, ed. Ernst A. Steinhoff (Alamogordo/NM, 5-9 October 1976): 301-316.

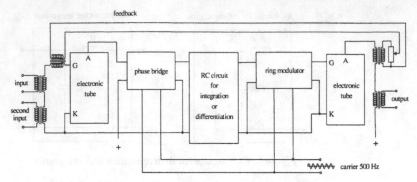

Figure 10: Solution with modulation amplifier

Instead of the chopper amplifier, one could use a modulation amplifier, as illustrated in Fig. 10. The relay is substituted by a ring modulator and a phase bridge. Using a modulating frequency of 500 Hz, the influence of the smoothing device is even smaller in the case of differentiation.

The differentiators and integrators described previously can be combined at random. In this way, it is possible to represent high-order differential and integral expressions. In the case of second-order expressions, Hoelzer adopted a different approach for his analog computer. He combined two *RC* elements immediately behind each other without the use of intermediate amplifiers. This is illustrated in Fig. 11 for a double differentiation.

In his dissertation of 1941, Hoelzer provides a theoretical justification. The positive feedback is a suitable method for removing perturbations, i.e. for compensating time constants. In the case of a double differentiation, one is forced to choose a complex feedback factor in order to obtain an exact solution. Hoelzer, therefore, provided a general formula, from which one could determine the complex feedback factor.

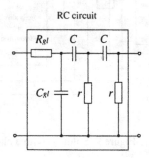

Figure 11: RC circuit for double differentiation

6. Electronic modeling of differential equations

Using the integrators and differentiators we have described, it is possible to solve differential equations. In his dissertation, Hoelzer demonstrated this using a fourth-order linear differential equation.

$$\frac{d^4 y}{dt^4} + a\frac{d^3 y}{dt^3} + b\frac{d^2 y}{dt^2} + c\frac{dy}{dt} + d \cdot y = F(t) \tag{1}$$

This differential equation can be modeled in various ways. First, we consider using integrators only. To this end, the differential equation can be transformed in the following way:

$$\frac{d^4 y}{dt^4} = F(t) - a\frac{d^3 y}{dt^3} - b\frac{d^2 y}{dt^2} - c\frac{dy}{dt} - d \cdot y \tag{2}$$

The composition of integrators illustrated in Fig. 12 emulates this expression.

The fourth derivative of y is integrated until y is obtained. The solution function $y(t)$ appears as a curve in the oscilloscope. The signals at the output of the integrators are multiplied by the coefficients a, b, c, and d and added to $F(t)$ at the input of the first integrator, taking the sign into consideration. Multiplication with constants is carried out by applying a voltage divider or an amplifier, if necessary. At the input of the integrators, potentiometers are used to adjust the initial conditions. The differential equation can also be solved using differentiators only. It has to be transformed in the following way:

$$y = \frac{F(t)}{d} - \frac{c}{d}\frac{dy}{dt} - \frac{b}{d}\frac{d^2 y}{dt^2} - \frac{a}{d}\frac{d^3 y}{dt^3} - \frac{1}{d}\frac{d^4 y}{dt^4} \tag{3}$$

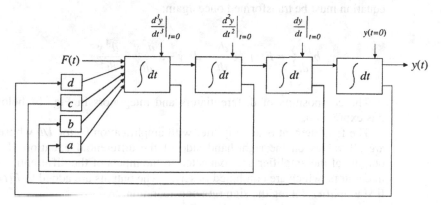

Figure 12: Solution of the differential equation using integrators

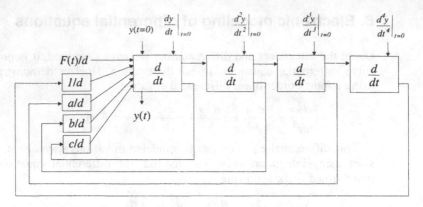

Figure 13: Solution of the differential equation using differentiators

The composition of differentiators illustrated in Fig. 13 belongs to this expression.

The quantity y is obtained by adding certain differential expressions gained by continuous differentiation of y to the disturbance function $F(t)/d$. The output signals of the differentiators are multiplied by the coefficients c/d, b/d, a/d, and $1/d$, and are then added to $F(t)/d$ at the first differentiator, taking the sign into consideration. Potentiometers at the input of the differentiators are used to adjust the initial conditions of the differential equation. The variable y is transmitted to an oscilloscope before the first differentiation.

The advantage of the machine with differentiators is that a deviation in the zero point calibration of the devices will not have a serious effect. Constant errors at the input of the differentiators will not propagate to the output. On the other hand, the output signal will appear noisy when the input signal is not sufficiently continuous. To minimize the disadvantages, Hoelzer adopted a compromise with two integrators and two differentiators. The differential equation must be transformed once again:

$$b\frac{d^2y}{dt^2} = F(t) - c\frac{dy}{dt} - d \cdot y - a\frac{d^3y}{dt^3} - \frac{d^4y}{dt^4} \qquad (4)$$

The composition of differentiators and integrators in Fig. 14 belongs to this expression.

The first element is an amplifier with amplification factor $1/b$ whose input are all values on the right-hand side of the differential equation. The two outputs of the amplifier are connected to the inputs of the differentiators and integrators, which are combined in series. The outputs are added to $F(t)$ at the first amplifier, taking the sign into consideration.

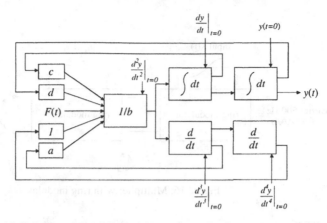

Figure 14: Solution of the differential equation using integrators and differentiators

In his dissertation, Hoelzer also formulated detailed procedures for the calibration of the three sets of analog devices. The solution of a system of linear second order differential equations presented him with an even greater challenge. This problem often appears in physics, for example, in a mechanical system with three independent dimensions (coordinates x, y, z) and with acceleration, velocity and distance coupling. To handle this problem, Hoelzer proposed a circuit where a series connection of two integrators with a mixing amplifier in front is used for each coordinate x, y, z, i.e., three-dimensional dynamic problems were being solved in Peenemünde[15] using analog devices.

While only sums and differences of the differential expressions appear in

Figure 15: The final form of the analog computer in Peenemünde, 1943

---

[15] Gunter Schwarze, letter to the author (Berlin, Humboldt University, 5 June 1998).

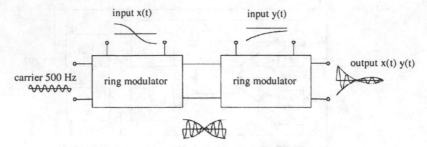

Figure 16: Multiplier with ring modulators

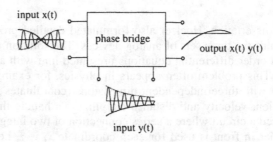

Figure 17: Multiplier with modulated input values

linear differential equations, we find products, quotients, powers and various nonlinear functions in the case of nonlinear differential equations. This was why Hoelzer worked on the electronic modeling of these functions, and it led him to develop a fully electronic analog computer. Fig. 15 shows the machine in its final form. The bottom panel contains the power supply unit. In addition to the differentiators and integrators, the top panel contains the calculating devices which are considered in the following sections.[16]

7. Multiplier, square, and cube devices

The device which Hoelzer used to create multipliers, square and cube units was the ring modulator. Fig. 16 shows how multiplication is performed with two ring modulators connected in series. Multiplication is carried out with the modulation of $x(t)$ and $y(t)$ with a constant AC voltage. This circuit can be

---

[16] Klaus Biener (n. 11 above).

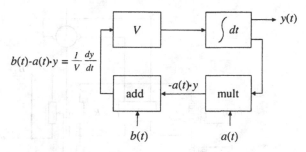

Figure 18: Division by solving a differential equation

used when both factors are given in demodulated form and the result is presented in modulated form. If both carriers have been modulated, a phase bridge has to be used, as shown in Fig. 17. However, the result appears in demodulated form and has to be transformed for further processing.

8. Divider

Hoelzer used a special differential equation to build a divider:

$$\frac{1}{V}\frac{dy}{dt} + ay = b \tag{5}$$

The differential equation has a homogeneous solution y_h for the compensation process and a particular solution, $y_p = b/a$, after infinite time ($t = \infty$).

The device shown in Fig. 18 solves this differential equation and supplies

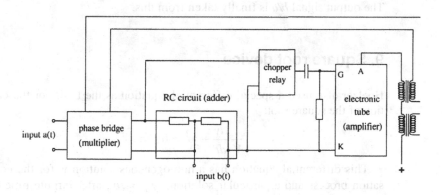

Figure 19: Circuit for division with negative feedback

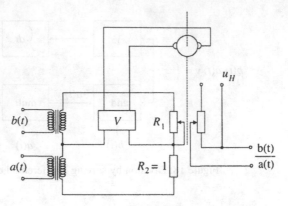

Figure 20: Division by a self-balancing bridge

the quotient b/a. If there is a fast compensation process, a and b are functions of time. This is achieved by a high amplification V. One can also eliminate the integrator in Fig. 18, if the amplification chosen is high enough. The new circuit is shown in Fig. 19. It could be described as a negative feedback, which has the advantage that the time constants are absolute.

Another method proposed by Hoelzer was the use of the equivalence conditions of a balanced bridge for modeling the quotient. This circuit is illustrated in Fig. 20.

The input signals are given in modulated form and are fed through transformers into the bridge circuit. The other branch of the bridge is composed of a fixed resistance R_2 and a controllable resistance R_1. The voltage in the zero branch of the bridge adjusts the resistance R_1 via an amplifier and a small motor until the bridge is balanced. Then one can obtain $R_1=(b/a)R_2$. Another potentiometer, supplied by voltage U_H, is adjusted simultaneously with R_1. The output signal b/a is finally taken from this.

9. Square root device

Hoelzer also used a special differential equation as the basis for the calculation of the square root:

$$\frac{1}{V}\frac{dy}{dt} + y^2 = b \tag{6}$$

This differential equation has a homogeneous solution y_h for the compensation process and a particular solution, $y_p = \sqrt{b}$, after infinite time $(t=\infty)$. For this reason, the device illustrated in Fig. 21 is suitable for producing the square root of a given value.

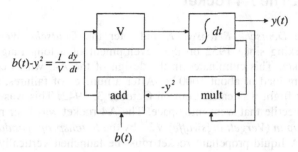

$$b(t)-y^2 = \frac{1}{V}\frac{dy}{dt}$$

Figure 21: Producing the square root by solving a differential equation

If b is a function of time, the compensation process must be accelerated by a high amplification. Here, as well, the integrator can be spared in practice (negative feedback) if the amplification factors are high enough.

10. Device for modeling arbitrary functions

There was no method for modeling arbitrary functions electronically in Hoelzer's time. Therefore, he used a mechanical representation using a curve disk. The voltage which occurs at a mechanical comparator describes the sampled function. The mode of operation is shown in Fig. 22.

The value $y(t)$ is applied to the coil of a motor and is compensated by the feedback coil after adjustment of the motor. In this way the plane $f(y)$ is adjusted and the associated function value $f(y(t))$ is taken from the potentiometer.

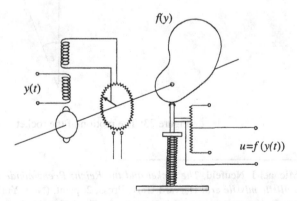

Figure 22: Modeling the function $f(y(t))$

11. The A4 rocket

The *Deutsche Reichswehr*, and later the *Deutsche Wehrmacht*, had been working since 1932 on the development of a long range liquid propellant rocket. This culminated in the design of the A4, which could carry a maximum load of about 1000 kg. After a number of failures, the first successful test flight was carried out on October 3, 1942. This was the first controlled projectile that went into space. The A4 rocket was later renamed *vengeance weapon* (*Vergeltungswaffe*) V2[17] by the *Reichspropagandaministerium*.

A liquid-propellant rocket must be launched vertically. At a certain altitude, it rotates 45° in the direction of launch and, at a defined time, the rocket engine is stopped. The rocket continues on a ballistic path towards the target. The final path can differ from the calculated one due to lateral deviations or because the rocket does not reach the intended distance. The latter could result from faulty adjustment of the angle of attack or of the cutoff time of the rocket engine. The lateral deviations had to be corrected both by remote-control (radio beam guidance) and by on-board control (inertial guidance system).

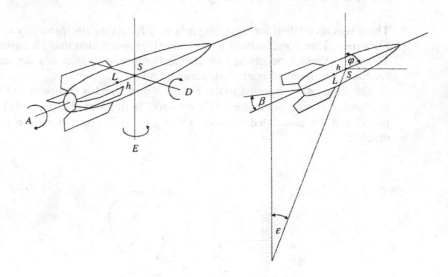

Figure 23: The motion of the rocket

[17] Michael J. Neufeld, *The rocket and the Reich: Peenemünde and the coming of the ballistic missile era*, Harvard Univ. Press, 2. print. (New York, Mass, 1995); Gregory P. Kennedy, *Vengeance weapon 2: The V-2 guided missile*, National Air and Space Museum (Washington D.C., 1983).

Aerodynamic arrow stabilization was provided to the A4 rocket by fins at the rear end. Because of this construction, the attack point L of aerodynamic forces lies behind the center of gravity S. The mass reduction due to fuel consumption causes a shift in the center of gravity. But arrow stabilization does not guarantee that the desired path will be kept. Due to external forces and torques, a turn of thrust force $\vec{P}$ occurs, which is rigidly coupled with the axis of the rocket. The effect of the turning of the thrust force is much larger than the direct effect of the forces and torques. A guidance and control system has to compensate the deviations due to these forces and torques, and has to suppress them when they arise. The opposite forces and torques required for this are generated by thrust and air rudder. They were driven by servomotors, which received commands from the guidance and control system[18].

12. Equations of motion and control of the rocket

In order to build the control device, one has to determine which physical quantities are relevant for rocket motion. This is illustrated in Fig. 23.
The motion of the rocket is described by two differential equations. The first is obtained by the equivalence of forces and the second by the equivalence of torques applied to the rocket. In the air attack point L wind force is applied, which produces a wind torque in relation to the center of gravity S. The forces and torques due to air resistance also have to be taken into account. A third differential equation specifies the control system of the rocket. It describes how the rudders have to be adjusted when a deviation ε is given:

$$m_1 \frac{d\beta}{dt} + m_0 \beta = E(\varepsilon) \qquad (7)$$

To determine whether or not a disturbance could be stabilized, Hoelzer used the Nyquist method, which is well known in electronics. In his dissertation, he describes the following Gedankenexperiment: If the rudder of the rocket is perturbed with sinusoidal frequency (angle β_{ein}), the guidance and control system will try to compensate this motion with an oscillation of another phase and amplitude (angle β_{aus}). This will be repeated for all frequencies ω between 0 and ∞. When the oscillation at the axis of the rudder is

---

[18] Helmut Hoelzer, "Anwendung elektrischer Netzwerke zur Stabilisierung von Regelvorgängen und zur Lösung von Differentialgleichungen, gezeigt an der Stabilisierung des Fluges einer selbst- bzw. ferngesteuerten Großrakete (Application of Electrical Networks to the Stabilization of Regulating Processes and to the Solution of Differential Equations, Demonstrated on the Stabilization of the Flight of an Automatically Controlled, Respectively Guided Large Scale Rocket)," included in Helmut Hoelzer Papers CBI33 (Darmstadt 1946).

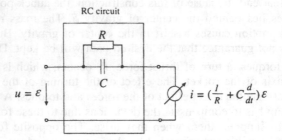

Figure 24: Phase modifier

represented in the complex plane, the form of the local curve according to the Nyquist method provides stability. The point (1;0) must always be on the left side of the curve, in the direction of increasing ω. Remote control by radio link and on-board inertial guidance were used for the A4 rocket.

13. Remote control of the rocket

A pure amplifier cannot be used for the control device $E(\varepsilon)$. Every small disturbance would result in an amplification of the rocket motion. For that reason, a phase modifier had to be used too, as is illustrated in Fig. 24. But this would fail in a very slow drift of the rocket ($\omega \approx 0$), because the servomotors of the rudder would not react. Hoelzer determined that only a series combination of four phase modifiers could be successful.

However, in the course of measurement, he noticed that the value from the HF-receiver did not have an ideal smooth curve. Due to noise in radio communication, the curve itself is noisy, making it impossible to carry out four differentiations of ε. Only one differentiation is possible. Hoelzer, therefore, supplemented remote-control with on-board control.

However, the errors due to the faulty adjustment of gyroscopes, amplifiers, and the rocket engine and the variance of m_0 caused by the uncontrollable burning down of the rudder are not taken into account. In order to do this, we have to add a term where ε is integrated over time into the control equation. Finally, we obtain the following expression for $E(\varepsilon)$, which has to be superimposed by the rocket's own control system:

$$E_{-1}\int \varepsilon dt + E_0 \varepsilon + E_1 \frac{d\varepsilon}{dt} \tag{8}$$

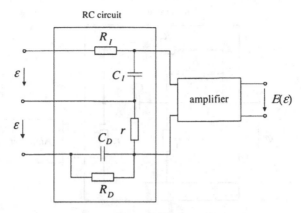

Figure 25: Network for remote-control

In the ideal case, the coefficients for remote-control are determined as $E_{-1} =$ 75, $E_0 = 750$, and $E_1 = 2250$. When Hoelzer considered modeling this expression, he noticed the similarity of a series combination of capacitance, resistance and inductivity to Ohm's law. At that time, it was only possible to create capacities of at most $C = 60\,\mu F$. If one adopts the ratio 1:10:30 for the coefficients, the required resistance ($R = 167\,k\Omega$) and inductivity ($L = 667000$ H) can be computed. But such an inductivity could hardly be created in practice.

Hoelzer avoided the inductivity L by the union of serial and parallel combination, i.e. he used a serial combination of R_I and C_I for integration, and a parallel combination of R_D and C_D for differentiation. This circuit is illustrated in Fig. 25.

The radio beam on-board device has galvanic separated outputs, so that ε is applied to the series and to the parallel circuit. After that, an amplification is made with the electronic tube driven as an AC-amplifier. However, the measured value has to be applied as voltage at the input. A small resistance has to be connected to the parallel circuit for this. Hoelzer called the circuit the "Isodrome Network." It was not intended to produce exact values for the integrals and differentials, as is important in the case of the solution of equations, but it was intended to correct the deviations from the calculated path due to disturbances.

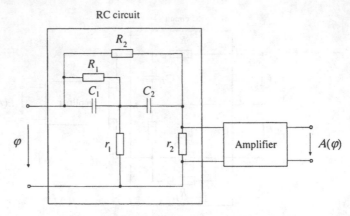

Figure 26: Network for on-board control

14. Automatic guidance of the rocket

The quantity which has the same phase position as the second derivation of ε is the position angle φ. This angle is delivered by the inertial navigation system and could be used to model the higher order derivations of ε. Hoelzer obtained the following differential expression $A(\varphi)$, which has to be added to $E(\varepsilon)$:

$$A_0\varphi + A_1 \frac{d\varphi}{dt} + A_2 \frac{d^2\varphi}{dt^2} \tag{9}$$

The differential expression for on-board control was also modeled through a parallel combination of R and C. Because a double differentiation is required, two RC elements were combined immediately behind each other, as shown in Fig. 26.

The introduction of feedback to compensate the time constants, as was done in the analog computer, does not yield a significant advantage. The coefficients of on-board control were determined in the ideal case as $A_0=10$, $A_1=4$ and $A_2=3$.

The compensation of disturbances due to external forces occurs more slowly by replacing the second derivative of ε with the position angle φ.

15. The "Mischgerät"

The networks for remote-control and on-board control, including the amplifying elements, were put together by Hoelzer in one apparatus which he called the "Mischgerät." Later on, in letters to James E. Tomayko,[19] he translated the name as "mixing computer." The "Mischgerät" was the world's first on-board computer. Fig. 27 shows the original circuit diagram given in Hoelzer's dissertation.[20]

The detailed diagram of the "Mischgerät" is shown in Fig. 28. The deviation positions of the three axes D, E, and A are read from the gyroscopes's potentiometers ("Kreiselhorizont" and "Kreiselvertikant")[21] and fed into the control device. The phase shift was carried out with the RC elements according to the equation selected for the stabilization calculation. The voltage at

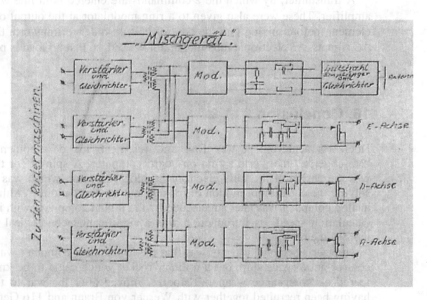

Figure 27: Original circuit plan from Hoelzer's dissertation

[19] James E. Tomayko (n. 4 above)

[20] Helmut Hoelzer (n. 18 above).

[21] Stefan Karner, "Die Steuerung der V2: Zum Anteil der Firma Siemens an der Entwicklung der ersten selbstgesteuerten Grossrakete," in *Technikgeschichte* 46 (1979): 45-66; Archive of the Heeresanstalt Peenemünde, Series 86 (Steering), 87 (Flight Mechanics), 73, 79 (Navigation), 115, 119 (Inertiale Navigation), in

the output of the *RC* elements has to be amplified. Because the electronic tubes are driven as *AC* amplifiers, the voltage has to be modulated with an *AC* voltage in a ring modulator, whereby the carrier is suppressed. At the output of the amplifier a rectification is carried out with the addition of the carrier. The output *DC* current then controls oil-hydraulic servomotors (Askania Lrm5), which move the rudder. The commands from the *D* and *E* axes are conducted in each case to two amplifiers with the same phase; the commands from the *A*-axis (spin) are given with opposite phase to the two amplifiers of the *E*-axis. Rudders 1 and 3 will control the *E*-axis if they move in phase, and the *A*-axis if they move in opposite phase. Rudders 2 and 4 are prevented from moving in opposite phases (i.e. they are synchronized), so that an over determination of the spin control does not occur. This is done by a potentiometer construction at the rudder, whose commands are given in opposite phase to the amplifiers of the *D*-axis.

A transmitter, by which the ε-commands are entered is in line with the *E* amplifier. These were also given to a ring modulator at the output of the *RC* element, before mixing into the control unit, in order to transform them into *AC* signals. An electronic tube modulator is used, so that no load is put on the capacitor.

16. Conclusion

Hoelzer's achievement was first and foremost, the development of the "Mischgerät," the world's first on-board computer. A spin-off of this work was the design of the fully electronic analog computer, which was used for solving the equation of motion of rockets and for simulations.[22] Although the analog computer was later superseded by the digital computer, with its higher calculation speed, its introduction was a significant technological development at the time.

Hoelzer's personal tragedy was that the first analog computer was used to control not a space ship, but a military rocket which came to be known and feared as "vengeance weapon." Hoelzer moved after the war to the USA, having been recruited together with Werner von Braun and 116 German scientists as part of the so-called "Operation Paperclip." He became director of the Computation Laboratory at the Marshall Space Flight Center and contributed to the development of the Saturn V rocket.

Deutsches Museum, Abteilung Archive, Sondersammlungen und Dokumentationen (München).

[22] Eric A. Weiss, "Helmut Hoelzer, 1912-1996," *Annals of the History of Computing* 20, No.2 (Los Alamitos/CA, April-June 1998): 62-63.

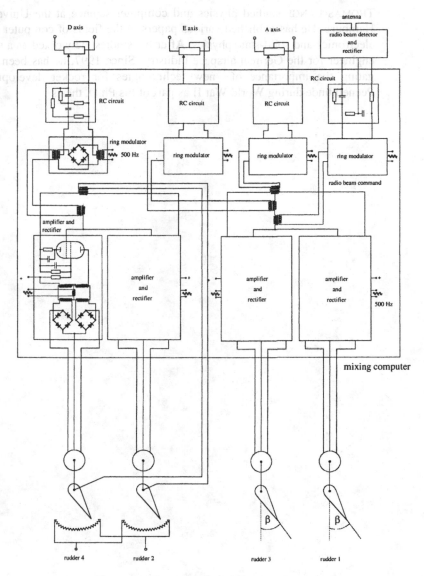

Figure 28: The "Mischgerät" (Mixing Computer)

THOMAS LANGE studied physics and computer science at the University of Hamburg. He has published various papers in the fields of computer science, electronics and solid-state physics. After his studies, he worked as a software engineer for the German airspace industry. Since 1997, he has been investigating the importance of new technologies for rocket development in Peenemünde during World War II as part of his Ph.D. thesis.

Part IV: The British Scene

EDSAC
(Cambridge)

Mark Computers
(Manchester)

Colossus
(Bletchley Park)

The Colossus of Bletchley Park – The German Cipher System

Anthony E. Sale

Abstract. Colossus was designed and built at the Post Office Research Laboratories at Dollis Hill in North London in 1943 to help Bletchley Park in decoding intercepted German telegraphic messages. The telegrams were enciphered using the Lorenz SZ42 cipher machine. Colossus was the world's first large electronic valve programmable logic calculator. Not just one but ten of them were built and were operational in Bletchley Park, home of Allied World War II code-breaking.

This paper describes the German codes and the methods used by Colossus to decipher messages. Some of the components of Colossus are reviewed in more detail.

1. The German cipher system

The German Army High Command asked the Lorenz Company to produce for them a high security teleprinter cipher machine to enable them to communicate by radio in complete secrecy. The Lorenz Company designed the SZ42 cipher machine based on the additive method for enciphering teleprinter messages invented in 1918 by Gilbert Vernam in America.

The Vernam system enciphered the message text by adding to it, character by character, a set of obscuring characters thus producing the enciphered text which was transmitted to the intended recipient. The simplicity of Vernam's system was that if the obscuring characters were added in a rather special way (known as modulo 2 addition) then exactly the same obscuring characters added in the same way to the received enciphered message, cancelled out the obscuring characters and retrieved the original message.

Vernam proposed that the obscuring characters should be completely random and pre-punched onto paper tape to be consumed character by character in synchronism with the input message characters. Such a cipher system using purely random obscuring characters is unbreakable.

The difficulty was, in a hot war situation, to make sure that the same random character tapes were available at each end of a communications link and that they were both set to the same start position. The Lorenz Company decided that it would be operationally easier to construct a machine to generate the obscuring character sequence. Because it was a machine it could not generate a completely random sequence of characters. It generates what is known as a pseudo random sequence. Unfortunately for the German Army it was more "pseudo" than random and that was how it was broken.

The amazing thing about Lorenz is that the code breakers in Bletchley Park never saw an actual Lorenz machine until right at the end of the war, but they had been breaking the Lorenz cipher for two and a half years.

The first intercepts

The teleprinter signals being transmitted by the Germans enciphered using Lorenz machines were first heard in early 1940 by a group of policemen on the South Coast who were listening out for possible German spy transmissions from inside the UK.

Brigadier John Tiltman was one of the top code breakers in Bletchley Park and he took a particular interest in these enciphered teleprinter messages. They were given the code name "Fish" and the messages which, as was later found out, were enciphered using the Lorenz machine were known as "Tunny." Tiltman knew of the Vernam system and soon identified these messages as being enciphered in the Vernam manner. Because the Vernam system depended on addition of characters, Tiltman reasoned, if the operators had made a mistake and used the same Lorenz machine starts for two messages, (a Depth), then by adding the two cipher texts together character by character, the obscuring character sequence would disappear. He would then be left with a sequence of characters each of which represented the addition of the two characters in the original German message texts. For two completely different messages it is virtually impossible to assign the correct characters to each message. Just small sections at the start could be derived but not complete messages.

The German mistake

As the number of intercepts increased, now being made at Knockholt in Kent, a section was formed in Bletchley Park headed by Major Ralph Tester and known as the Testery. A number of Depths were intercepted but not much headway had been made into breaking the cipher until the Germans made one horrendous mistake. It was on 30th August 1941 and a German operator had a long message of nearly 4,000 characters to be sent from one part of the German Army High command to another, probably Athens to Vienna. He correctly set up his Lorenz machine and then sent a twelve-letter indicator, using the German names, to the operator at the receiving end. This operator then set his Lorenz machine and asked the operator at the sending end to start sending his message. After nearly 4,000 characters had been keyed in at the sending end, by hand, the operator at the receiving end sent back by radio the equivalent, in German, of "didn't get that – send it again."

They now both put their Lorenz machines back to the same start position. Absolutely forbidden, but they did it. The operator at the sending end then began to key in the message again, by hand. If he had been an automaton, and used exactly the same keystrokes as the first time, then all the interceptors would have got would have been two identical copies of the cipher text. Input the same – machines generating the same obscuring characters – same cipher text. But being only human and being thoroughly disgusted at having to key it all again, the sending operator began to introduce differences in the second message compared to the first.

The message began with that well-known German phrase SPRUCHNUMMER, "message number" in English. The first time the operator keyed in S P R U C H N U M M E R. The second time he keyed in S P R U C H N R and then the rest of the message text. Now NR means the same as NUMMER what's the difference? It meant that immediately following the N the two texts were different but the machines were generating the same obscuring sequence, therefore the cipher texts were different from that point on.

The interceptors at Knockholt realized the possible importance of these two messages because the twelve letter indicators were the same. They were sent post haste to John Tiltman at Bletchley Park. Tiltman applied the same additive technique to this pair as he had to previous Depths. But this time he was able to get much further with working out the actual message texts because when he tried SPRUCHNUMMER at the start he immediately spotted that the second message was nearly identical to the first. Thus the combined errors of having the machines back to the same start position and the text being re-keyed with just slight differences enabled Tiltman to recover completely both texts. The second one was about 500 characters shorter than the first, where the German operator had been saving his fingers. This fact also allowed Tiltman to assign the correct message to its original cipher text. Now Tiltman could add together character by character, the corresponding cipher and message texts revealing for the first time a long stretch of the obscuring character sequence being generated by this German cipher machine. He did not know how the machine did it, but this was what it was generating!

The denouement

John Tiltman then gave this long stretch of obscuring characters to a young chemistry graduate, Bill Tutte who had recently arrived to Bletchley Park from Cambridge.

Bill Tutte started to write out the bit patterns from each of the five channels in the teleprinter form of the string of obscuring characters at various repetition periods. Remember this was BC, "Before Computers," so he had to write out vast sequences by hand. When he wrote out the bit patterns from

354 Anthony E. Sale

channel one on a repetition of 41, various patterns began to emerge which were more than random. This showed that a repetition period of 41 had some significance in the way the cipher was generated. Then over the next two months Tutte and other members of the Research section worked out the complete logical structure of the cipher machine which we now know as Lorenz. This was a fantastic *tour de force* and at the beginning of 1942 the Post Office Research Labs at Dollis Hill were asked to produce an implementation of the logic worked out by Bill Tutte and colleagues. Frank Morrell produced a rack of uniselectors and relays, which emulated the logic. It was called "Tunny." So now when the manual code breakers in the Testery had laboriously worked out the settings used for a particular message, these settings could be plugged up on Tunny and the cipher text read in. If the code breakers had got it right, out came German. But it was taking four to six weeks to work out the settings. This meant that although they had proved that technically they could break Tunny, by the time the messages were decoded the information in them was too stale to be operationally useful.

The machine age

The mathematician Max Newman now came on the scene. He thought that it would be possible to automate some parts of finding the settings used for each message. He approached TRE at Malvern to design an electronic machine to implement *the double delta method* of finding wheel start positions that Bill Tutte had devised. The machine was built at Dollis Hill and was known as Heath Robinson after the cartoonist designer of fantastic machines.

There were problems with Heath Robinson keeping two paper tapes synchronized at 1,000 characters per second. One tape had punched onto it the pure Lorenz wheel patterns that the manual code breakers had laboriously worked out. The other tape was the intercepted enciphered message tape. The double delta cross correlation measurement was then made for the whole length of the message tape. The relative positions then moved one character and the correlation measurement was repeated. The code breaker was looking for the relative position which gave the highest cross correlation score, which hopefully corresponded with the correct Lorenz wheel start position.

Heath Robinson worked well enough to show that Max Newman's concept was correct. Newman then went to Dollis Hill where he was put in touch with Tommy Flowers who was the brilliant Post Office electronics engineer who designed and built Colossus to meet Max Newman's requirements for a machine to speed up the breaking of the Lorenz cipher.

Tommy Flower's major contribution was to propose that the wheel patterns be generated electronically in ring circuits thus doing away with one paper tape and completely eliminating the synchronization problem. This required a vast number of electronic valves, but Tommy Flowers was confident it could be made to work. He had, before the war, designed Post Office repeaters using valves. He knew that valves were reliable provided that they were never switched on and off. Nobody else believed him!

Colossus design started in March 1943. By December 1943 all the various circuits were working and the 1,500 valve Mk1 Colossus was dismantled shipped up to Bletchley Park and assembled in F Block over Christmas 1943. The Mk1 was operational in January 1944 and successful on its first test against a real enciphered message tape.

The contribution to D Day

Colossus reduced the time to break Lorenz messages from weeks to hours and just in time for messages to be deciphered which gave vital information to Eisenhower and Montgomery prior to D Day. These deciphered Lorenz messages showed that Hitler had swallowed the deception campaigns, the phantom army in the South of England, the phantom convoys moving east along the channel, that Hitler was convinced that the attacks were coming across the Pas de Calais and that he was keeping Panzer divisions in Belgium. After D Day the French resistance and the British and American Air Forces bombed and strafed all the telephone and teleprinter land lines in Northern France, forced the Germans to use radio communications and suddenly the volume of intercepted messages went up enormously. The Mk1 had been rapidly succeeded by the Mk2 Colossus in June 1944, and eight more were quickly built to handle the increase in messages. The Mk1 was upgraded to a Mk2, and there were thus ten Mk2 Colossi in the Park by the end of the war. By the end of hostilities 63 million characters of high grade German messages had been decrypted, an absolutely staggering output from just 550 people at Bletchley Park, plus of course the considerable number of interceptors at Knockholt, with backups at Shaftsbury and Coupar in Scotland.

2. The Colossus computer

Each of the ten Colossi occupied a large room in F Block or H Block in Bletchley Park. The racks were 90 inches high (2.3 m) of varying widths. There were eight racks (Fig. 1) arranged in two bays about 16ft (5.5 m) long plus the paper tape reader and tape handler (known as the bedstead).

The front bay of racks, spaced 5ft (1.6m) from the rear bay, comprised from right to left, the J rack holding the master control panel, the plugboard, some cathode followers and the AND gates. Next came the K rack which contained the very large main switch panel together with the very distinctive sloping panel at the front which was a duplicate patch panel for the thyratron rings. Next came the S rack which held the relays used for buffering counter output and making up the typewriter drive logic. The left-hand rack at the

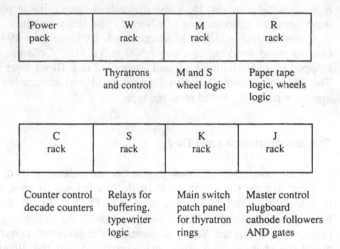

Power pack	W rack	M rack	R rack
	Thyratrons and control	M and S wheel logic	Paper tape logic, wheels logic

C rack	S rack	K rack	J rack
Counter control decade counters	Relays for buffering, typewriter logic	Main switch patch panel for thyratron rings	Master control plugboard cathode followers AND gates

Figure 1: The racks of Colossus

front was the C rack which held the counter control logic on the front and the decade counters on the back

The rear bay of Colossus contained four racks, the R rack holding the staticizer and delta boards for the paper tape reader output and the K and S wheel thyratron ring outputs, the M rack for the M wheel staticizers and S wheel motion logic. The very large W rack held, on one side all the thyratrons making up the wheel rings, 501 in all, and on the other side the 12 thyratron ring control panels. Also on the W rack were the link boards for the wheel patterns and the uniselectors for setting wheel start positions. The end rack of the back bay held the power packs. These were 50 volt Westat units stacked up in series to give +200 volts to –150 volts. The total power consumption was about 5 Kilowatts most of which was to the heaters of the valves.

The circuit layout was all surface mounting on metal plates bolted to the racks. The valve holders were surface mounting with tag strips for the components. This form of construction had much to commend it, firstly both sides of a rack could be used, secondly wiring and maintenance were very easy and lastly cooling of the valves was expedited by them being horizontal.

How Colossus worked

The following information summarizes the structure and capabilities of Colossus:

- Input: Cipher text punched onto 5-hole paper tape,
                    read at 5,000 cps.
- Output: Buffered onto relays,
                    typewriter.
- Processor: Memory of 5 characters of 5-bits held in a shift register,
                    pluggable logic gates,
                    20 decade counters arranged as 5 by 4 decades.
- Clock speed: 5 KHz, derived from sprocket holes in the input tape.
- Valves: 2500.

Colossus read teleprinter characters, in the international Baudot code, at 5,000 characters per second from a paper tape. These characters were usually the intercepted cipher text which had been transmitted by radio. The paper tape was joined into a loop with special punched holes at the beginning and end of the text.

The broad principle of Colossus was to count throughout the length of the text the number of times that some complicated Boolean function between the text and the generated wheel patterns had either a true or false result. At the end of text the count left on the counter circuits was dumped onto relays before being printed on the typewriter during the next read through the text, an early form of double buffering.

Colossus had two cycles of operation. The first one was controlled by the optical reading of the sprocket holes punched between tracks 2 and 3 on the paper tape. The sprocket signal was standardized to 40 microseconds wide. The optical data from the paper tape was sampled on the back edge of the standardized sprocket pulse as were the outputs from the rings of thyratrons representing the Lorenz wheel patterns. The result of the logical calculation was sampled on the leading edge for feeding into the counter circuits.

The second cycle of operations occurred at the beginning and end of the text punched onto the paper tape. The paper tape was joined into a loop and special holes were punched just before the start of text between channels three and four (called the start) and just after the end of text between channels four and five (called the stop). This long cycle of operations began with the electrical signal from the photocell reading the stop hole on the tape. This stop pulse set a bistable circuit which stayed set until the optical signal from the start hole was read. The setting of this bistable thus lasted for the duration of the blank tape where the text was joined into a loop, typically about 100 milliseconds. The first operation after the stop pulse was to release any settings on the relays from the previous count. Next the new count was read onto the relays. Then the counters and the thyratron rings were cleared and then the thyratron rings were struck at the next start point to be tried. When the bistable was reset by the start pulse, sprocket pulses were released to precess the thyratron rings, to sample the data read from the paper tape and to sample the calculation output to go to the counters.

The various components of Colossus were the optical reader system, the master control panel, the thyratron rings and their driver circuits, the optical data staticizers and delta calculators, the shift registers, the logic gates, the counters and their control circuits, the span counters, the relay buffer store and printer logic.[1]

The optical reader system

In order to break the Lorenz codes in a reasonable time the cipher text had to be repeatedly scanned at very high speed. This meant at least 5,000 characters per second and in 1942 this implied hard vacuum photocells to optically read the holes in the paper tape. The smallest photocells available were some developed for proximity fuses in anti aircraft shells. Six of these in a row meant an optical projection system to enlarge the image of the paper tape about 10 times. Dr Arnold Lynch designed the paper tape reader and used slits cut into black card to form a mask in front of the photocells.

The output from the data channels went to the staticizer and delta circuits.

The master control panel

This was where the start and stop pulses from the optical reader set and reset the bistable. Monostable delay circuits generated the voltage waveforms for releasing the relays, for staticizing the counters, for resetting the counters and thyratron rings, and for striking the rings. Gate circuits controlled the flow of sprocket pulses.

The thyratron rings and their driver circuits

These circuits were the most complex on Colossus. Thyratrons are gas filled triodes which strike a discharge arc between anode and cathode when the grid voltage is raised to allow electrons to flow. This discharge, when struck, continues quite independent of the grid voltage. Thus the thyratron acts as a one-bit store. It can only be switched off by driving both the anode and the grid negative with respect to the cathode. To construct a shift register with thyratrons requires that the striking of the next thyratron in the ring also quenches the previous thyratron. This leads to a biphase circuit with anodes of alternate thyratrons connected together and the grid voltage partially biased by the cathode voltage of the previous thyratron.

---

[1]   Thomas H. Flowers, "The Design of Colossus," *Annals of the History of Computing*, 5:3 (1983), 240–252.

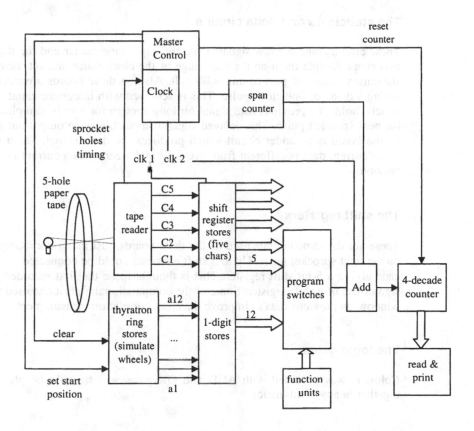

Figure 2: Block diagram of Colossus

The complication arises when a Lorenz wheel contains an odd number of setting lugs. The thyratron ring controller for this requires a complete set of circuits to handle just the odd thyratron in order to get back to the biphase circuits for the rest of the ring. The thyratrons in a ring conduct sequentially stepped round by the sprocket pulses. Each thyratron cathode is brought out to a patch panel which allows the cathode pulse to be connected to a common output line when a link is plugged into the patchboard. Thus as the ring precesses round, a sequence of pulses appears on the common output line. By selecting the link positions this sequence can replicate the mechanical lugs set on the Lorenz wheel. Alongside the patch panel is a Uniselector which selects the thyratron cathode to which the ring strike pulse goes. This is the start position of the ring when sprocket pulses come in at the start of text.

The common line output went to the staticizer (one-digit stores) and delta circuits (program switches).

The staticizers and delta circuits

These circuits take the raw signals from the paper tape reader and the thyratron rings, sample them on the back edge of the clock pulse and set them to
standard voltages of plus or minus 80 volt. Also on these boards are circuits
giving a delay of one clock pulse. This is achieved with integrator capacitors
which "hold" the previous data signal for long enough for it to be sampled on
the next sprocket pulse. This delayed signal is available as an output but also
on the board is an adder circuit which produces the delta signal, i.e., a one
when current data is different from previous and a zero when current equals
previous.

The shift registers

These are the same circuits used on the delta boards, just integrators sampled
on the next sprocket pulse. Up to 5 shift elements could be connected in cas-
cade giving a 5-bit shift register. This is thought to be the first recorded design or use of a shift register. Some of the computational algorithms used this
window on previous data to improve the cross correlation measurement.

The logic gates

Colossus was provided with AND and OR gates which could be plugged
together in any combination.

The counter and counter control circuits

The decade counter circuits were based on a pre-war design by Wynn-
Williams. They used a divide by two circuit followed by a ring of five pento-
des. Four decades were required for each of the five counters used and each
control circuit covered four decades of counters. The inputs to the control
circuits were the output from the logic gates, the sprocket pulse for strobing
and the reset pulse from the master control panel. Also on the control panels
were comparison circuits between the outputs of the decade counters and
switches on another panel. These switches could be set to any number in the
range 0 to 9999. The output of the comparison could be included in the logic
calculations thus for instance suppressing printing of scores below a set
value.

The span counters

These were the same design of counters and counter control circuits with switches on another panel which could be set in the range 0 to 9999. The purpose of the span counters was to be able to ignore sections of the cipher text which were corrupted, possibly due to fading radio signals. The comparison output was used to gate the sprocket pulses which went to the main counter controllers, cutting off these pulses stopped the sampling of the logic calculation and thus ignored the section of text covered by the span counters.

The relay buffer store and printer logic

Latching relays held the ending count on the decade counters. The start positions of the thyratron rings and the count for the previous run through the text are clocked out sequentially onto the typewriter by the printer relay and uniselector logic.

3. Programming Colossus

Programming of the cross correlation algorithm was achieved by a combination of telephone jack plugs and cords and switches. The main plug panel was on the rack nearest to the paper tape reader. The direct and delta signals from the paper tape reader and the K wheel thyratron rings were on this panel. The changeover from direct to delta could also be achieved by switches. Also on the main plug panel were the input and output sockets for the AND gates and the so-called "Q" sockets which took the calculated output to the main switch panel on the next rack to the left. This very large switch panel allowed signals to be combined through further logic gates and the results switched to any of the five result counters.

As an example take the simple double delta algorithm as devised by Bill Tutte. This requires two wheels to be run simultaneously so take K4 and K5.

First the delta outputs from channel 5 from the paper tape reader is combined in an AND gate with the delta output of the K5 thyratron ring, then this result is ANDed with the AND output of delta channel 4 and the delta output of the K4 thyratron ring. This result is plugged to Q1 and on the switch panel Q1 is switched to counter 1. The output can be negated before being counted so that the count can represent either the number of times the double delta calculation equals one or zero.

The end of Colossus

After Victory Day, suddenly it was all over. Eight of the ten Colossi were dismantled in Bletchley Park. Two went to Eastcote in North London and then to GCHQ at Cheltenham. These last two were dismantled in about 1960 and in 1960 all the drawings of Colossus were burnt, and of course its very existence was kept secret. In the 1970s information began to emerge about Colossus. Professor Brian Randell of Newcastle University started researching the machine. Dr. Tommy Flowers and some of the other design engineers wrote papers in the 1980s describing Colossus in fairly general terms.

4. The Colossus rebuild project

When some colleagues and I started, in 1991, the campaign to save Bletchley Park from demolition by property developers, I was working at the Science Museum in London restoring some early British computers. I believed it would be possible to rebuild Colossus. Nobody believed me. In 1993 I gathered together all the information available. This amounted to the eight 1945 wartime photographs taken of Colossus plus some fragments of circuit diagrams which some engineers had kept quite illegally, as engineers always do!

I spent 9 months poring over the wartime photographs using a sophisticated modern CAD system on my PC to recreate the machine drawings of the racks. I found that sufficient wartime valves were still available as were various pieces of Post Office equipment used in the original construction.

In July 1994 His Royal Highness the Duke of Kent opened the Museums in Bletchley Park and inaugurated the Colossus Rebuild Project. At that time I had not managed to obtain any sponsorship for the project and my wife Margaret and I decided to put our own money into it to get it started. We both felt that if the effort was not made immediately there would be nobody still alive to help us with memories of Colossus. Over the next few years various private sponsors came to our aid and some current and ex Post Office and radio engineers formed the team that helped me in the rebuild.

The switch-on

By 1995 the optical paper tape reader was working, (helped by the memories of Dr Arnold Lynch who designed it in 1942) and the basic circuitry of Colossus had been recreated. Colossus first worked at two-bit level (out of the five bit channels from the paper tape). HRH The Duke of Kent returned to the Park on 6th June 1996 to switch-on the basic working Colossus. This was a marvelous occasion with Dr Tommy Flowers present and a large number of the people who worked at Knockholt, in the Testery and the Newmanry during the war.

One reason for wanting to get Colossus working in 1996 was that for far too long the Americans have got away with the myth that their ENIAC computer was the first in the world. It was not, but they got away with it because Colossus was kept secret until the 1970s. As 1996 was the 50th anniversary of the switch-on of ENIAC, I made sure that Colossus was rebuilt and working in Bletchley Park, just as it was in 1944. There has been a stunned silence from across the water!

The American information

One ironic twist to the Colossus story is that most of the information about how Colossus was used has come from America. In 1995 the American National Security Agency (equivalent to GCHQ) was forced by application of the Freedom of Information Act to release about 5,000 World War II documents into the National Archive. The listing of these documents was put onto the Internet and I quickly obtained a copy of the list. When I scanned this I was amazed to see titles like "The Cryptographic Attack on FISH." I managed to get copies of these documents only to find that they were reports written by American service men seconded to Bletchley Park when America entered the war. The most important one was written by Albert Small and is a complete description of Colossus code breaking. Having this report has enabled us to work out the function of many more of the circuits and program switches on Colossus. We have now, we think, incorporated nearly all the circuits and although there may still be some parts which cannot be worked out, we think we have about 90% of Colossus correct and working.

The performance of Colossus

Colossus is not a stored program computer. It is hard wired and switch programmed, just like ENIAC. Because of its parallel nature it is very fast, even by today's standards. The intercepted message punched onto ordinary teleprinter paper tape is read at 5,000 characters per second. The sprocket holes down the middle of the tape are read to form the clock for the whole machine. This avoids any synchronization problems, whatever the speed of the tape, that's the speed of Colossus. Tommy Flowers once wound up the paper tape drive motor to see what happened. At 9,600 characters per second the tape burst and flew all over the room at 60 mph! It was decided that 5,000 cps was a safe speed.

At 5,000 cps the interval between sprocket holes is 200 microseconds. In this time Colossus will do up to 100 Boolean calculations simultaneously on each of the five tape channels and across a five character matrix. The gate delay time is 1.2 microseconds which is quite remarkable for very ordinary valves. It demonstrates the design skills of Tommy Flowers and Allen Coombs who re-engineered most of the Mk2 Colossus.

Colossus is so fast and parallel that a modern Pentium PC programmed to do the same code breaking task takes twice as long as Colossus to achieve a result!

Acknowledgments

The rebuilt and working Colossus can now be seen in the Museums at Bletchley Park which are open to the public every other weekend throughout the year. It is marvelous tribute to Tommy Flowers, Allen Coombs and all the engineers at Dollis Hill and a great tribute to Bill Tutte, Max Newman, Ralph Tester and all the code breakers involved at Bletchley Park. Not forgetting all the WRNS who operated and supported Colossus and the interceptors at Knockholt without whom there would have been no messages to break.

The financial sponsors are: A E & M D Sale, Mr. Frank Morrell, the Mrs. L D Rope Third Charitable Settlement, Mr. Keith Thrower OBE. Contributions by special low prices: Charles Head (Blacksmiths), Billington Exports Ltd., Claude Lyons Ltd. Thanks to the Bletchley Park Trust which has allowed free use of the room in H Block where Colossus has been rebuilt and to the many hundreds of individuals who have searched their garages and lofts and sent valves for Colossus.

Lastly I would like to thank my wife Margaret for agreeing to the use of our own money to start up the project and for her continuing support and encouragement.

The Colossus team

The regulars: Cliff Horrocks, David Stanley, Paul Bruton (deceased), John Lloyd, Bob Alexander. Every other weekend: John Pether, Don Skeggs, Adrian Cole, Ron Clayton. Intermittently: Don Grieg, Philip Hopkins, Richard Watson, Derek Turton, Mark Hyman. By parcel post to Wales: Gil Hayward, the original designer of Mk2 Tunny who turned some 600 pattern plugs and modified over 1,000 octal valve bases.

ANTHONY E. SALE is Museum's Director to the Bletchley Park Trust and has established in Bletchley Park a Museum to commemorate the World War II codebreaking work which significantly helped the allied war effort. His varied careers include electronics, intelligence with MI5, running a computer software company for 12 years and working at the Science Museum in London for 4 years restoring early computers. All this experience has come together to enable him to rebuild the wartime Colossus computer and to research and present to the public the outstanding codebreaking technologies of Bletchley Park. Tony Sale is an Honorary Fellow of the British Computer Society and was the Society's Technical Director from 1986 to 1989.

The Manchester Mark 1 Computers[1]

R. B. E. Napper

Abstract. This paper provides a brief history of the four related computers that were designed and built in Manchester from 1948 to 1951, and a summary specification of their architecture. Each machine has a strong claim to be a "first" in the history of stored-program computers, specifically all-electronic computers with an electronic store. The SSEM (June 1948) was the first such machine to work, thus realizing and proving the von Neumann ideal. The Manchester Mark 1 (Intermediate Version, April 1949) was the first full-sized computer available for use. The completed Manchester Mark 1 (October 1949), with a fast random access magnetic drum, was the first computer with a classic two-level store. The Ferranti Mark 1 (February 1951) was the first production computer delivered by a manufacturer.

1. Introduction

By 1946 F.C. Williams had been working at the Telecommunications Research Establishment (TRE) for seven years, and was an electronics engineer of international reputation. He had led for some years a small team that specialized in solving problems for other groups working in radar and airborne electronics. On a visit to the U.S. in July 1946, Williams saw that electronic digital storage was the problem holding up the development of electronic computers. So on return he looked into the problem of storing digital information on a conventional radar cathode ray tube (CRT). By November 1946 a single bit could be stored.

In December Williams moved to the University of Manchester as Professor in the Electrical Engineering Department (originally Electrotechnics). He arranged for Tom Kilburn (a member of his group) and another person to be seconded from TRE, with the authority to order parts from TRE, so that they could continue their CRT storage research.

---

[1] This paper was written with the assistance of Prof. Tom Kilburn, who was unable to attend the Conference.

2. The CRT store

By October 1947 the team had developed a system for storing 2048 bits on a standard CRT. The mechanism used a 64×32 array of phosphor charges on the screen, each planted in one of two different ways, representing a 0 or a 1. But the charge would decay within 0.2 seconds; so a detector was placed in front of the CRT, and a mechanism devised so that, as the array was swept again by an electron beam, the type of charge at each position could be detected and refreshed to the same type of charge before the beam moved on to the next position. With the surface of the tube being refreshed at regular intervals, before the charge could decay significantly, the CRT could hold a pattern of 0s and 1s indefinitely. Resetting of particular bits and reading values could be interleaved with the refresh mechanism.

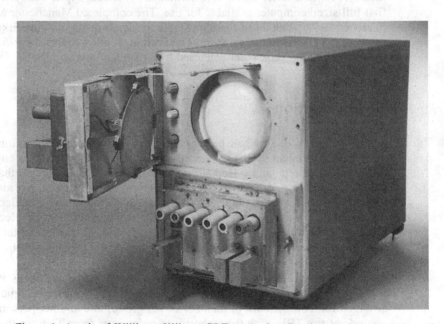

Figure 1: A pair of Williams-Kilburn CRT memories. Reprinted with permission of the Department of Computer Science, University of Manchester.

Tom Kilburn submitted a report to the TRE management, *A Storage System for use with Binary Digital Computing Machines* (Progress Report, 1st December 1947),[2] which explained in detail the research carried out in 1947, and explained how the CRT store could be used inside a stored-program computer. The report was circulated widely and was influential in several American and Russian organizations adopting the Williams-Kilburn CRT storage system.

3. The Small Scale Experimental Machine

By June 1948 the team had built a Small Scale Experimental Machine to Tom Kilburn's design, the SSEM, or just "The Baby"; this was to test that the CRT store would still work effectively at the fast speeds required in a computer. The SSEM only had a 32×32-bit word main store, using one CRT. It had a second CRT just holding a 32-bit accumulator *A*, and a third holding the address of the current instruction *C*, and the instruction itself (*PI*). A fourth CRT, without any storage mechanism, was placed on the console and could be switched to show a copy of the current bit pattern on any of the storage tubes. This was used as the output device, and the input device was a keyboard of 32 buttons plus manual switches; these could be used to set any bit pattern in any word.

The SSEM used a whole 32-bit word for an instruction, using bits 13–15 for the function code and bits 0–12 for the store address (though as only one CRT was fitted for the main store, which could hold 32 words, only five bits were used). There were just seven instructions initially, i.e., where *S* represents the *contents* of the given store address:

1)	A = –S	Load S negated into the accumulator
2)	A = A – S	Subtract S from accumulator
3)	S = A	Reset S to the value in the accumulator
4)	C = S	Reset C to (the address in) S
5)	C = C + S	Add (the address in) S to C
6)	If A < 0 then C = C + 1	Skip next instruction if accumulator < 0
7)	Halt.	

Note that both relative (5) and absolute (4) unconditional jumps were provided, and they were both indirect (the more general case) rather than direct. *C* had to be set to the instruction *before* the next instruction to be obeyed, since *C* was always incremented at the *start* of each instruction. The branch instruction (6) simply consisted of testing the accumulator and skipping an

---

[2] Department of Computer Science, University of Manchester, *The Computer that Changed the World* CD-ROM June 1998. This contains copies of early papers including the Technical Report of 1947, video interviews and explanatory material

instruction if it was negative. The awkward use of negative operations (1 and 2) was simply to avoid having to build a full 32-bit adder as well as a 32-bit subtractor before the SSEM could be tested – of course X + Y can be computed as X – (0 – Y). Following instructions were added a few months later:

- A = S
- A = A + S
- A = A & S

A 32-bit line on the main store CRT could be read, written or refreshed in just over 300 microseconds. Refreshing scans, cycling in turn through each line in the store, were interleaved with "action" scans of the same length, so that the regular rhythm of obeying an instruction was as follows:

1. Refresh the next line in turn; add 1 to the control address *C*.
2. Read the line given by *C* into *PI*.
3. Refresh the next line in turn; decode *PI*.

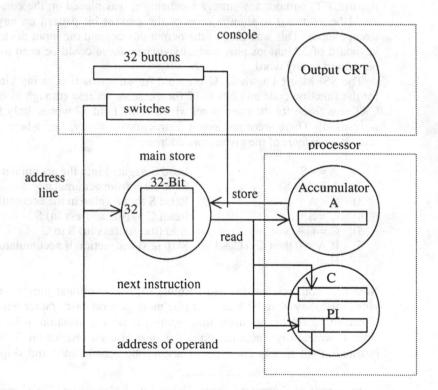

Figure 2: Schematics of the SSEM architecture

4. Read/write any line *S* as required and complete the instruction. (Note that any addition/subtraction involved was done serially bit by bit, and was overlapped with the serial read operation.)

Instructions were therefore executed in around 1.2 milliseconds, and the main store CRT was refreshed every 16 instructions.

The SSEM immediately proved the suitability of the CRT storage device and the effectiveness of the stored-program computer. Within days of the first successful run on June 21st 1948, a run of two million instructions in under an hour had been achieved. New programs (albeit short!) could be loaded in minutes.

4. The Manchester Mark 1

By October 1948 a decision had been made to build a full-sized computer in the Electrical Engineering department, the basic architecture had been decided, the team had been expanded, and the government had contracted Ferranti Ltd. to build a commercial production computer based on it. Tom Kilburn was again responsible for the design. The other full time members of the team, with their specialities, were D.B.G. (Dai) Edwards (CRT store and central processor), A.A. (Alec) Robinson (multiplier, starting in June 1947), G.E. (Tommy) Thomas (drum) and G.C. (Geoff) Tootill (general, joined June 1947, seconded from TRE).

The full-sized machine, the Manchester Mark 1, included two significant advances in computer design over the SSEM and the basic von Neumann model: a fast two-level store and address modification registers (B-lines).

It was the general opinion that a computer would need a few thousand words of main store to make it useful for general applications. But it was decided that, although there was no particular architectural problem in providing this amount of memory, the cost would be significant. So it was decided to attach a fast magnetic drum store. The drum would rotate at a speed synchronized with the refresh scans, so that a read transfer resetting a CRT would be completed in a single revolution.

The second innovation was the introduction of the B-tube, containing two special 20-bit registers (more usefully eight in the Ferranti machines), called B-lines. Each instruction contained a field which specified one of the B-lines, and as soon as the next instruction was fetched from store, the contents of the B-line were added to the instruction before it was fully decoded and executed. The contents of B0 were by convention always kept at 0, so that B0 was used wherever a modification was not required. The ability to modify a whole instruction was a dangerous facility, that was rarely required. The real usefulness was in modifying instruction *addresses*; this soon became a fundamental capability of most computers, with modification being restricted to the address of the store operand in an instruction. This address would now in general have two components, the fixed part F in the store address field of the instruction plus a variable part V, being the current contents of a specified modification register. This classically facilitates access to general data, for example (a) accessing an element of a fixed-position array, where F gives the array base and V the index, and (b) accessing a scalar in a record or stack, where V gives the record/stack base and F the relative address of the scalar. Of course as store sizes increase, the B-lines enable access to store addresses outside the range that can be specified by the address field of an instruction. Therefore, the B-lines constitute, as they were used, what we would call today "index registers."

The Manchester Mark 1 was fully operational by around October 1949, with 128 40-bit words as main store, and with a magnetic drum capacity that continued to be increased up to about 3,000 words. However an *Intermediate Version* was available for general use by other departments and Ferranti from April 1949. It was recorded as completing a successful 9-hour overnight run on Mersenne Primes on June 16–17, 1949. It was still without paper tape I/O and programmable drum transfers. The magnetic drum was working, but the circuitry to incorporate transfers into the instruction set had not yet been completed; so transfers were by human intervention with the program halted. However, it was invaluable for holding previously prepared input data and programs, output, useful routines, and intermediary calculations for long program runs.

From the summer of 1949 the effort moved increasingly from developing the Manchester Mark 1 to transferring the design, and enhancing it, for the Ferranti Mark 1. The first machine was delivered to the University in February 1951. This had essentially the same architecture, but the number of instructions, the CRT and drum stores, and the machine speed, were all increased. The engineering of the machine was considerably improved, the multiplier was redesigned and the drum re-engineered. The main store now contained 256 40-bit words and the maximum drum capacity was about 16,000 words.

Specification of the Manchester Mark 1

The Manchester Mark 1 evolved as a physical extension of the SSEM, with sections being added, revised or replaced as appropriate over a period of time. It had two specially manufactured CRTs for its main store, each of which held two arrays of 32×40-bit words. Each 40-bit word could contain two 20-bit instructions. With a multiplier unit now available, the accumulator was extended to 80-bits.

The general specification of the Manchester Mark 1 was as follows, using square brackets to indicate comparable figures for the Ferranti Mark 1:

- Store organized in 40-bit addressable "words," a word containing a number or two 20-bit instructions,
- Double-length accumulator, with serial 80-bit and 40-bit integer arithmetic; hardware addition, subtraction, multiplication, and logical operations,
- Two [8] 20-bit modifier registers – "B-lines," for modifying instructions,
- Four [8] pages of random access main store, 32×40-bit words per "page," using two CRTs,
- 128 [512] page capacity drum backing store, two pages per track, about 30 milliseconds revolution time; random access to any track,
- Single-address format order code (plus a field for the B-line),
- Standard instruction time: 1.8 [1.2] milliseconds, multiplication costing roughly an extra half instruction time for every 1 in the multiplicand,
- Peripheral instructions: read and punch a line of 5-hole paper tape; transfer a given page (or track) on drum to/from a given page (or page pair) in store.

Instruction Set

The full Ferranti Mark 1 instruction format had three fields, bits 0–9 giving the CRT store address, bits 10-12 the B-line and bits 14–19 the instruction code. The Manchester Mark 1 only used subsets of the fields.

The order code allowed the user to interpret a 40-bit number in four ways, as either a signed number or an unsigned number, and as either an integer or a binary fraction. A double-length accumulator of 80 bits was provided, especially for use with multiplication and double-length or multi-length operation. The most significant 40 bits were referred to as Am and the least significant as Al. It was therefore necessary to provide different instructions in some places where a 40-bit Store word S was interacting with the accumulator A, to distinguish a number being used as an unsigned number (extend to 80 bits with 0s) or signed (sign-extend). Unsigned orders are indicated by using Su instead of S in the instruction codes below.

An extra register D was used to hold the multiplicand in multiplication; this could be set directly from main store, signed or unsigned, and would hold its value until reset. Multiplication was between two 40-bit numbers, in D and S, yielding an 80-bit answer that was added to the accumulator. Remember that S refers to the contents of the address coded in the bits 0–9 of the instruction.

Control Orders (same as for the SSEM)

Stop
C = S
C = C + S
If A< 0, C = C + 1

Accumulator Orders

A = Su	A = A neq S
A = S	D = Su
A = –S	D = S
S = A	A = A + D × Su
S = Am	A = A + D × S
A = A + S	A = 0
A = A + Su	S = Am, Am = 0
A = A – S	swap Am and Al
A = A AND S	S = Al, Al = Am, Am = 0
A = A OR S	Am = Am + S

Other Orders

B0 = S
B1 = S
Peripheral transfer using code word in S (not in the Intermediate Version)

Peripheral code word

The peripheral operations, for transfers between magnetic drum and CRT store, and for paper tape input/output, had a large number of variations, the extent and detail of which were not clear when the basic architecture was designed in October 1948. So it was decided to provide separate decoding and execution hardware for them, using a code word in store, and to have just one instruction in the main instruction set, to carry out a peripheral transfer according to a specified value S, i.e., the contents of a memory address. The Manchester Mark 1 used a sparsely coded 40-bit word and had a complicated paper tape system. The specification of the Ferranti Mark 1, which is better documented, is given below. This provided the same functionality neatly specified in a 20-bit line.

Field	Meaning
Bits 0-9	Track address on drum
Bit 10	Even/Odd page on track
Bit 11	One/two page transfer, swapped in CRT store if bit 10 = 1
Bit 12	Transfer without/with full check
Bit 13	Read/write from/to drum
Bit 14	Drum/paper-tape transfer (see below)
Bits 16-18	CRT page address in store

In the case of paper tape I/O the following conventions were used: B14 is set and only bits 10 to 14 are used; bit 13 set means "read next character to A"; bits 10-13 = 0 means "punch next character from A"; bit 13 = 0 and bits 10-12 non-zero punches various special characters.

5. The Ferranti Mark 1

The first Ferranti Mark 1 was delivered to the University of Manchester in February 1951. It had the same basic architecture as the Manchester Mark 1, but it included many improvements, most obviously industrial-standard engineering and increased store sizes. To improve main store reliability, a CRT now held only one page. Also the page layout was changed so that there were 64 20-bit lines on a page, rather than 32 40-bit words. Each line could hold one instruction, and a 40-bit number could start at any line (odd or even) and would consist of that line and its successor. Alec Robinson made a major contribution to increased speed in numerical computation by redesigning the multiplier using extensive parallelism. Multiplication now took a fixed time (2.16 milliseconds), about five times faster than the average time in practice on the Manchester Mark 1.

In the order code a few instructions were added to the accumulator orders, and a number of miscellaneous orders were also introduced. The B-line concept was shown to be so useful that eight were provided and the two instructions to set B0 and B1 replaced by eight new orders, including simple arithmetic. Use of B-lines was now extended from their role as an index modifier to the related, but more general, role as a loop control counter. In the classic case of a simple loop counting down to zero and accumulating a total in A, four instructions could be saved (save A, load counter, save counter, load A). Where the loop was processing a vector, the same B-line could be used for both indexing and counting (in steps of 2).

The program input mechanism and run-time management system for the new machine was devised by Alan Turing and Cicely M. Popplewell, using their experience with the Manchester Mark 1. This was known subsequently as Scheme A. It was heavily based on a 32-digit numbering system, with each digit d being represented by the character corresponding to the 5-bit binary

pattern d on the paper tape equipment. It did not provide any program translation, for example of function code mnemonics or of symbolic names or
labels; binary 10-bit addresses, 20-bit instructions and 40-bit integers were
written as 2, 4 and 8 character groups respectively, using the base-32 character set. But there were comprehensive facilities for handling the two-level
store. There were obviously many library routines, that could for example
carry out division and floating point arithmetic, and read and print numbers,
and these were held on the drum store, as were the routines of a large program. Scheme A was augmented by an improved alternative Scheme B by
R.A. (Tony) Brooker soon after he replaced Turing in 1951. In 1954 Brooker
produced the Mark 1 Autocode, which provided users with a simple algorithmic language based on interpreted floating-point variables V1 to V5000
and integer indices N1 to N18, with indirect references of the form VNi.
However an autocode program would run much more slowly than a machine
code program.

Some further enhancements were made after the first two machines were
sold, and it was now called the Mark 1*. In all, nine machines were sold
publicly, three abroad (in Canada, Holland and Italy), but then in 1957 Ferranti replaced the Mark 1* with the Mercury, again based on a prototype
(MEG) designed and built at the University of Manchester under Tom Kilburn.

Appendix A: The Ferranti Mark 1 Instruction Set

The following additions or changes were made in the Ferranti Mark 1 relative to the instruction set of the Manchester Mark 1.

Accumulator Orders (in addition to the 20 orders in section 4)

$A = A - D \times S$ $S = Al, A = 0$
$A = A - D \times Su$ $A = 2 \times S$

B-line Orders and Control Orders (replacing those in section 4)

$B = S$	Stop
$S = B$	$C = S$
$B = B - S$	$C = C + S$
$B =\# S$	If $A >= 0, C = S$
$S =\# B$	If $A >= 0, C = C + S$
$B =\# B - S$	If $Bx >= 0, C = S$
	If $Bx >= 0, C = C + S$

Note that two versions were provided of the orders to load, store and decrement a B-line. One version (=#) suppressed the usual B-line modification by B before an instruction was executed; this was the most common requirement.

Note that the conditional test of the Manchester Mark 1 (If $A < 0$, $C = C + 1$) was replaced by two conditional tests on the sign of the accumulator and two on the sign of the last B-line set or stored (Bx). This provided both relative and absolute jumps if the test on the accumulator or last B-line used was satisfied. The jumps were however all indirect, as before. It could be argued that the opportunity was missed here of changing to direct jumps, e.g. $C = Sa$, where Sa is the address in the address field of the instruction itself. This would have made life much easier for the programmer, and the less frequent indirect jumps could have been programmed simply and clearly using B-modification, but using two instructions, e.g. $B7 = S$; $C = 0\{7\}$, where Sa was 0, but modified by B7.

Miscellaneous instructions

There was a set of miscellaneous instructions, some referring to the 20-bit number currently set on the console hand switches (H):

Peripheral transfer with code word in S   Send Pulse to Hooter
Peripheral transfer with code word in H $S = H$

Al = random number (in bits 0-19)   Halt if console switch /G is set
Am = Am + number of 1's in S   Halt if console switch /L is set
Am = Am + position of the most
            significant 1 in S

Appendix B: The first program

The first program to run on the SSEM was a program to find the highest
factor of a number N. The version given below is a reconstruction of it made
by Tom Kilburn and Geoff Tootill in 1996, based on the observations made
by them in Tootill's notebook on the day of the first run and a revised version
of the program written down there a few weeks later.

The method used was to try each number J in turn from $N - 1$ down and
see if it divided into N by setting the accumulator A to N and repeatedly subtracting J from it until it became 0 (in which case J was the answer) or negative.

The instructions are written using the notation of the order code in Section
3, but with e.g. [23] replacing S as the contents of a store line, i.e. meaning
the contents of line 23 in store. Remember that the control address C is always the address of the *Current* instruction, and is incremented at the *start* of
each new instruction.

Line	Instruction	Effect
1	A = – [18]	Clear Accumulator (not necessary; done as "good practice")
2	A = – [19]	Start of OUTER LOOP: A = N (line 19 holds –N)
3	A = A – [20]	Start of INNER LOOP: A = A – J (line 20 holds J)
4	If A<0, C=C+1	If A is negative, skip 1 instruction, so exit inner loop
5	C = C + [21]	A still >= 0, so jump back to line 3 by C =C + –3 = 5 – 3 = 2
6	A = A – [22]	A = A + J; add back J (line 22 holds –J), so A = remainder
7	[24] = A	Store remainder in [24]
8	A = – [22]	A = J (line 22 holds –J)
9	A = A – [23]	A = A – 1
10	[20] = A	Reset J, now = old J – 1
11	A = – [20]	A = –J
12	[22] = A	Reset –J, now = – (old J – 1)
13	A = – [24]	A = negated remainder
14	If A<0, C=C+1	If A negative, i.e. if remainder is not 0, skip an instruction
15	C = [25]	Otherwise, we have a factor, so C = 16; jump to end program
16	C = [23]	C = 1; jump back to start of outer loop (line 2) to test next J
17	Halt	End the program (this could have been done at line 15)
18	0	Initialized to 0 (used in line 1)
19	–N	Initialized to –N

20	N – 1	Initialized to N – 1, the initial value of J
21	–3	Initialized to –3
22	– (N–1)	Initialized to –(N–1), the initial value of –J
23	1	Initialized to 1
24	–	Not initialized; used to hold the remainder of N/J
25	16	Initialized to 16

Lines 0 and 26 to 31 were not used, and may or may not have been cleared to 0 before loading the program. The program is entered at line 1 (by setting C=0 before initiating the run). At the end of the program you get the answer by reading the value in line 20 (now the successful J – 1) and adding 1 to it – no problem if you were testing a prime!

If you want to immediately rerun the program with a new N, then lines 19, 20 and 22 have to be reset to –N, N–1 and – (N–1), and the program reentered at line 1. Note that once one of the values has been worked out in binary the other two are easily formed (in particular –N is the one's complement of N–1). If you want to rerun with the same N, then lines 20 and 22 still have to be reset, as they are used to hold J and –J. The revised program allowed you to rerun with the same value of N without resetting, and showed the answer J (not J – 1) in the appropriate line.

R.B.E. NAPPER has been at the University of Manchester working in Computer Science as research student and lecturer since 1960. His first program produced some of the wiring schedules for the manufacture of the first Atlas computer, using the Ferranti Mercury. His main research was to develop the ideas of the Brooker-Morris Compiler-Compiler into a high-level, user-extensible, systems programming language (RCC). After his official retirement in October 1997, he contributed to the 50th Anniversary celebration of the Mark 1, in particular developing the www.computer50.org web-site and helping to produce a CD-ROM, both telling the story of the Mark 1 computers. This was done in regular contact with Tom Kilburn and Dai Edwards.

Rebuilding the First Manchester Computer

Christopher P. Burton

Abstract. The University of Manchester Small-Scale Experimental Machine, the "Baby," first ran a stored program on June 21, 1948, thus claiming to be the first operational general-purpose computer. To celebrate the fiftieth anniversary of that event, a fully authentic replica has been built by the Computer Conservation Society. The paper describes some of the deductions made about the detailed design of the original machine, and how the actual reconstruction work was carried out.

1. A stored-program computer

A historic event took place at 11:15 on Monday, June 21, 1948. For the first time ever in the world, a program stored in an electronic computing machine successfully ran and produced the expected answer. Today, almost half a century later, countless millions of people routinely carry out a range of tasks on their computers, from word-processing, through accounting to games. They would recognize that event in 1948 as "like" what they do all the time. And no earlier computer, not Zuse's Z3 in Germany, nor Colossus at Bletchley Park in Britain, nor ENIAC in the United States, was in that way "like" a modern computer. Loading a program into a computer (which can modify its own program) makes it a "Universal Machine," capable of performing any task within its capacity. The computing machine which first achieved that breakthrough was built at the University of Manchester by Professor F.C. Williams, Tom Kilburn and G.C. Tootill, three unsung heroes of the Second Industrial Revolution, the Information Age.[1]

The reason why the Manchester team got there first has been suggested by Tootill. All three men had been working on radar development during the war, where they were used to the idea of a "crash program" to get things designed and built in a hurry. With their knowledge of the radar circuit techniques which they had perfected, Williams and Kilburn moved to the Univer-

---

[1] S.II.Lavington, "A History of Manchester Computers," National Computing Centre, 1975. Second edition, British Computing Society 1998. S. Lavington, *Early British Computers*. Manchester University Press 1980 and Digital Press, 1980

sity of Manchester in 1946 to carry out research into using cathode ray tubes as a memory system for a hypothetical computing machine.[2] By the end of 1947 they had made such a data storage system hold 2048 binary digits, and decided to test it by building a computing machine around it. This was done in the remarkably short time of nine months, thereby beating other teams in the world who were striving to build a computer. They worked long hours, improvising as they went along, in the spirit of a crash program, to get it done.

2. The Computer Conservation Society

We now jump forward to 1989, when the Computer Conservation Society (the CCS) was founded to promote the study of the history of the computer industry, and to help preserve and restore old computing machines. The CCS is constituted as part of the British Computer Society and operates in association with the Science Museums in London and Manchester and at Bletchley Park.[3] The first secretary of the Society, Tony Sale, has recently completed a working reconstruction of Colossus, a specialized electronic computer which was built in 1943 and which contributed to decoding intercepted secret messages between Hitler and his generals. That reconstruction helped to inspire me to try to celebrate the approaching fiftieth anniversary of the running of the first program by a similar project. At the end of 1994, I proposed that we should build a replica of the Manchester computer, the Small-Scale Experimental Machine, or "Baby" as it was familiarly called, and re-run the first program on the anniversary day, Sunday June 21, 1998.

Why bother to go to all that trouble? And what happened to the original machines? Firstly, the pioneers in the late 1940s started a revolution which touches everyone in the world, yet they are largely unknown. Their names are not as familiar as those of James Watt and Thomas Edison. Secondly, it is important to signal a triumph of British innovation – it is not true that computers originated in beige desktop boxes from anywhere but Britain. And the technology the pioneers used has long been superseded by transistors and extraordinary modern technology – but the old valve technology ought to be a source of amazement to succeeding generations. Colossus and Baby and similar machines filled a room, weighed a ton, and were scrapped a few years after they were built; Colossus because of its secrecy – nothing was publicly known until 1974, and Baby to make way for bigger and better computers. In both cases, negligible detailed information has come down to us about how they were built.

---

[2] F.C.Williams and T. Kilburn, "A Storage System for use with Binary-Digital Computing Machines," *Proc. IEE*, Vol. 96, Part III, No. 40, pp. 81–100, March 1949.
[3] See World Wide Web location http://www.cs.man.ac.uk/CCS/.

My proposal and plan to rebuild Baby was enthusiastically taken up by the CCS and by the University of Manchester, who offered to provide space for this project only 100 yards from where the original was built. If we were successful, then the rebuilt machine would make a fine focal point for the celebrations to take place in Manchester in June 1998. The then chairman of our Society, Peter Hall OBE, former Ferranti and ICL director, was able to persuade ICL, Europe's leading systems company, that this was an important opportunity. After a presentation by me to Tom Hinchliffe, managing director of ICL High Performance Systems, he agreed to sponsor the substantial cost involved. This was a happy arrangement, since ICL in West Gorton is the descendant of the old Ferranti company, who collaborated with the University in 1949 to produce the world's first commercially-sold and delivered computer, based on the design of Baby. So the ground was set to see if it could be done.

3. What was the "Baby" machine like?

1995 was the year of finding out. The design was known in general terms from the published learned papers,[4] and a few photographs have survived. The outline schematic diagram in Fig. 1 was published in *Nature*, and shows how simple the machine was. The main store for program and data held 32 words of 32 bits. The Program Counter and Present Instruction were held in the Control Store, from where they could select a location in the Main Store and set up function decoding. The only arithmetic function was subtraction, the result being held in the accumulator. The sign of the number in the accumulator could be used to enable a conditional jump. All the elements of a universal computing machine were present.

There were never any engineering drawings – there was no need, it was an experimental computing machine. At the time, the only documentation was the set of circuit diagrams in the laboratory, which have long since disappeared. Fortunately, two of the pioneers, DBG Edwards and AA Robinson, made copies of those diagrams in their notebooks, which they have kept, and they kindly provided photocopies of the relevant pages. But these diagrams date from many months after our target date of June 1948, so their interpretation has to be treated with great care to take into account the many changes which took place in those months. For example, the order code was expanded from seven to nearly twenty types of instruction, the storage capacity was increased, and the B-tube or index register was invented and added.[5] By

---

[4] F.C.Williams, T. Kilburn and G.C.Tootill, "Universal High-Speed Digital Computers: A Small-Scale Experimental Machine," *Proc. IEE*, Vol. 98, Part II, No. 61, pp. 13–28, February 1951.

[5] A fuller description of many of these innovations can be found in the companion paper in this volume, "The Manchester Mark 1 Computers," by R.B.E. Napper.

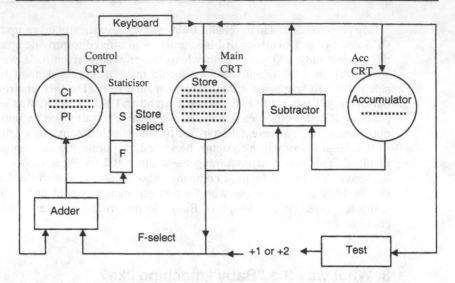

Figure 1: Simplified schematic of Small-Scale Experimental Machine. (After: Kilburn, *Nature*, Vol. 64, 1949, p. 684).

carefully comparing the various circuit diagrams and talking to the pioneers who did it, it was possible to arrive at a plausibly accurate picture of the functionality. Inevitably of course, there may have been vestigial details still in the machine in June but which were now unused and had been removed from the diagrams. We have had to ignore this possibility. We then created our own set of circuit diagrams using a CAD system on a PC, so that we had a reference set representing our best view of the circuits as at June 21, 1948. This tells us the abstract functionality, but not the physical description of the machine.

For the physical picture, the surviving photographs are an extremely valuable resource. The best-known one, shown here as Fig. 2, was actually taken in about March or April 1949, after significant modifications to the machine. Indeed it is the Mark 1 prototype rather than the SSEM. I have an original copy of that photo which has superb resolution – the markings on some of the valves are clearly visible under magnification. One example of its use is that careful examination of the push-button "typewriter" switches reveals the dimensions of the fixing holes and pitch of the buttons. By a stroke of serendipity, I recognized these as being identical with some switches in my junkbox. Furthermore, I still had the catalog of war surplus items for sale dating from 1953 when I bought the switches. This identified the item by its wartime reference number and it turned out that the switches were part of the control unit for the VHF radios in fighter aircraft. Turning to the Operator's Manual for the Spitfire and the Beaufighter reveals the control box in the photograph of the left side of the cockpit.

Figure 2: The Manchester University Mark 1 prototype. Reprinted with permission of the Department of Computer Science, University of Manchester.

Another well-known photograph is in the popular magazine *Illustrated London News* for June 25, 1949.[6] This is a two-page-spread wide-angle picture with many annotations. When discussing this picture with Geoff Tootill, he explained that it was actually a composite picture made from about twenty separate photographs taken by Alec Robinson. When I went to talk to Alec Robinson, he kindly loaned me a set of most of the original prints, and he had also kept nearly all the negatives, which are now in the University of Manchester archive. The importance of these photographs is that they are the earliest known of the SSEM, having been taken on December 15, 1948, as recorded in Robinson's notebook. Furthermore they show excellent detail. We have scanned the images so that, on the screen of a PC, we can make measurements of the dimensions of the various chassis, where the holes are, where the components are placed, what kind of components were used and so on. From this information, we produced CAD engineering drawings of the physical construction. Again, the six-month interval between June and December 1948 meant that we had to be very careful in our interpretation of the images.

Yet another interesting deduction had to be made to correlate a given circuit with its physical location in the computer. We were helped by the anno-

---

[6] "A Marvel of Our Time – The Memory Machine which can Solve the most Complex Mathematical Problems," *The Illustrated London News*, June 25, 1949.

tations on the *Illustrated London News* photo, but many of the decisions had
to be made by, for example, counting valves and other components on a
photo of a chassis and then comparing with possible circuits. A few circuits
just could not be identified, and, in such cases, we would try to put ourselves
in the shoes of the pioneers and do what we thought they would have done.
Geoff Tootill had emphasized that the machine was experimental, and they
did not hesitate to move chassis about as the development proceeded, so that
the actual layout was to some extent fluid.

Finally, one of the most precious surviving documents is the notebook
kept by Geoff Tootill at the time.[7] This remarkable book has notes discussing
the designers' thoughts and decisions on a number of logic design and circuit
design issues. It also has a wealth of information about early programs and
fragments of routines, and trial runs of those routines. The book is preserved
on loan to the Museum of Science and Industry in Manchester.

4. Building the replica

Interesting though deducing the details was, it would not lead to a replica
unless the appropriate old parts could be found. Electronic technology has
moved forward at such a pace, that half a century ago is almost pre-history!
The key items were the thermionic valves which have long been replaced by
transistors, and the cathode ray tubes. Amazingly, there are dealers who still
have stocks of these items at reasonable prices. Some valves are new, in
original boxes marked as packed by RCA in the United States in 1943, so
they will have been shipped as part of the war effort, dodging the submarines
across the Atlantic. Other items will have been made in the 1950s, but are
precisely identical with the originals. The Spitfire switches were available
from a dealer in vintage aircraft parts, and I cleaned out his stock. I spread the
word about and obtained masses of relevant components from individuals all
over the country. One kind person even donated two of the seven-foot high
steel racks, which he was using to hold his garden back from falling into the
River Severn! The remaining racks were prepared by ICL from some donated
by Tony Sale from Bletchley Park.

A team of volunteer friends from the Computer Conservation Society was
formed at the beginning of 1996 to do the detailed design and construction.
Many were former colleagues from the old days when we worked on valve
technology, and so knew what to do. They include Charlie Portman, Keith
Wood, Ken Turner and George Roylance, all connected with ICL, Bill Purvis
of Daresbury Labs, Adrian Cornforth, our web designer, and Suzanne
Walker, our photographer. Each engineer took a part of the machine and

---

[7] G.C. Tootill, "Digital Computer- Notes on Design and Operation," National Archive for the History of Computing, University of Manchester on loan to the Archive of the Museum of Science and Industry, Manchester.

created engineering drawings using modern Computer Aided Design software. The drawings would be taken to the workshop of ICL in Manchester, where the metal chassis would be made. Then the chassis would be taken home by the team member, with a bag full of components, where he assembled and wired up that unit.

The power supplies to provide the high voltages, +300V, +200V and −150V at several amps, do not appear on any surviving documents. We believe they were modular 50V 10A Post Office units, located in an adjacent room. We have made no attempt at an authentic replication here. Fairly modern, extelephone exchange switch-mode units have been modified to provide the supplies. Six modules in series are mounted in a rack behind the replica machine.

As chassis were completed, so they were brought to the university, where we built up the machine. By the end of 1996, it was almost complete − seven feet high, eighteen feet long, weighing nearly a ton, and looking most impressive!

Now we could start trying to make it work. Throughout 1997, the team would gather every Tuesday and gradually get parts operational. It was a time to learn a lot − where our detailed design was not quite right, why things had been done in certain ways, how to operate the machine and so on. The most difficult part, the cathode ray tube stores, took many months of patient work, but what a relief when we got them working as the pioneers had, almost half a century before! It was a great pleasure to have Tom Kilburn and Dai Edwards come to sit with us in front of an oscilloscope to help at this stage.

Figure 3: The Small-Scale Experimental Machine replica in May 1998. Photo courtesy of the Museum of Science and Industry in Manchester.

By the end of 1997, the computing machine was complete, apart from cosmetic details, and we could run programs; in particular, the world's first program. This had been written by Kilburn and it calculated the highest factor of a number, using a rather laborious method, as a proof of concept. A copy of an amended version has survived in Tootill's notebook, and is well known to historians. However, Kilburn and Tootill have now deduced what the unamended program was like. The programming work was aided by use of a simulator written by a student at the university. At this stage also, we launched a worldwide programming competition to see who could write the most interesting program for the Baby, with a chance for the winner to run their entry on the replica machine. About 130 entries were received from 19 different countries.

Early in 1998, the computer was dismantled and moved from the university to the Museum of Science and Industry in Manchester, see Fig. 3. This was done with very great care by the project team and the operatives from ICL, so that within a few days it was operational again. It is located in the 1830 Warehouse, the world's first railway goods warehouse, in a setting very reminiscent of the laboratory where the original machine was built. The reliability was gradually improved in the succeeding months, though the new environment seems to have produced some difficulties. In the middle of June 1998, one of the storage tubes became temperamental such that during the week of celebrations commencing June 17 we sometimes had to bypass it. However, the running of the first program at 11:15 on Sunday June 21, 1998 was achieved to the great pleasure of participants and audience– a fitting tribute to those remarkable men and women who pioneered our industry.

Acknowledgments

The rebuilding of the Baby has been possible due to the generosity of many organizations and individuals. Vital support has come from the University of Manchester, the Museum of Science and Industry, and the far-sighted sponsorship by ICL. The technicians there, the volunteer members of the project team, the pioneers themselves and numerous donors, have all made it possible with their time, skills and enthusiasm.

CHRISTOPHER P. BURTON, FIEE, FBCS graduated in electrical engineering from the University of Birmingham in 1955, and was awarded an honorary MSc by the University of Manchester for his work on the replica SSEM. He worked on computer developments with Ferranti Ltd and then ICT and ICL from 1957 until his retirement in 1989. A member of the Computer Conservation Society and Project Manager of the Small-Scale Experimental Machine Rebuild Project, he lives in Shropshire in England, and he can be contacted at chris@envex.demon.co.uk.

The Atlas Computer

Frank H. Sumner

Abstract. The Atlas computer was designed and built in the Department of Computer Science at the University of Manchester with the collaboration of ICL. When it was completed, in December 1962, it was the most powerful computer in the world. It included many innovative design features of which the most important were the implementation of virtual addressing and the one-level store. This paper gives a brief summary of the structure of the Atlas and of the effect of extensive instruction overlapping on the final performance. The main part of the paper describes the implementation of the one-level store including the learning program which efficiently managed the movement of pages between the two physical levels of store.

1. Introduction

After the "Baby" ran its first program in June 1946, Kilburn and Williams expanded the design of the Mark 1. In collaboration with Ferranti Ltd., a commercial version was constructed and installed at Manchester University in February 1951.

Nine other Mark 1 systems were built and delivered to customers, mainly in government research laboratories. In parallel with the production of the Mark 1, work continued at the University under the leadership of Tom Kilburn on two projects: the MEG, a faster extension of the Mark 1 with engineered floating-point arithmetic, and two transistor based computers.

The Meg design was adopted by Ferranti and commercialized as the Mercury after changing the memory from cathode ray tube stores to core stores. Meanwhile, the team at the University started a project with the grand aim of building the most powerful computer in the world. The machine, named the Atlas, was built in collaboration with Ferranti and was operational in December 1962. It outperformed the IBM Stretch and maintained its premier position until the arrival of the CDC 6600. The most significant aspects of the design were the introduction of virtual addressing, paging and a one-level store, features which have since become the norm in computer designs.

The starting and completion dates of the first five Manchester computers are listed in Table 1.

Table 1: Summary of Manchester University Computer Systems 1946 to 1962

Name	Started	Operational	Commercial Production by	First Delivery	No. Made
Baby	1946	June 21st 1948			
Mk 1	1948	April 1949	Ferranti Mk 1	Feb. 1951	10
Meg	1951	May 1954	Ferranti Mercury	Aug. 1957	19
Transistor	1952	April 1953			
Transistor	1953	April 1955	Metropolitan Vickers MV950	1956	6
Muse	1956		ICL Atlas	Dec. 1962	3

2. Instruction formats

All Manchester computers had single-address instructions; arithmetical operations used an accumulator for one of the inputs and for the result.

The Mark 1 and the Mercury used five-hole paper tape for input and had a character size of five bits. The word length was 40 bits or 8 characters. Full and half-word addressing were provided and instructions occupied a half-word. The instruction format for both computers was:

Function	Modifier	Address
7 bits	3 bits	10 bits

The machines could address only 1024 words and both had drum stores with the capability of transferring fixed sized pages of data between the two levels of store.

The Atlas

For the Atlas the character size was increased to 6 bits with an 8 character or 48 bit word (there was a parity bit with each half-word giving a total word length of 50 bits).

The number of index registers was expanded to 128. The motivation for this large increase was to use them as fixed-point arithmetic registers as well as index registers. There were two index, or modifier, fields in an instruction to permit indexed addressing of operands for fixed-point and logical operations. This allowed double address modification for floating-point operations. In the final design, some of the index registers were allocated to be copies of other registers, for example, the three instruction address registers and the exponent field of the floating-point accumulator.

With a 48-bit word the approach of the earlier designs would have given 24 bit instructions. With two seven-bit index fields this was insufficient so the decision was made to have 48 bit instructions with 24 bits allocated to the operand address. This huge address space, far greater than any store feasible at the time, was probably the most important decision that was made from the point of view of the overall design. The final instruction format was

Function	Modifier	Modifier	Address
10 bits	7 bits	7 bits	24 bits

Full and half-word addressing were again implemented in hardware using the top 21 or 22 bits of the address respectively. The bottom two bits could, and with hindsight should, have been used to give hardware addressing of characters. In practice, the bottom 2 bits of the address were used by some programmers as a character address with any character manipulation being implemented by software.

The maximum address space of 2^{21} words was considerably greater than any direct access store that was feasible at the time, e.g. the 2.5 Mbit store built at M.I.T was very expensive and only a fortieth of this address space.

The machine order code was of the single address type, with a comprehensive range of basic functions provided by normal engineering methods.

Also available to the programmer were a number of extra functions termed "extracodes" which gave automatic access to and subsequent return from a large number of built in subroutines. These routines provided:

- A number of orders which would have been expensive to provide in the machine both in terms of equipment and also time because of the extra loading on certain circuits. An example of this is the order: "Shift accumulator contents n places left."
- The more complex mathematical operations, e.g. sin x, log x, etc.
- Control orders for peripheral equipment, card readers, parallel printers etc.
- Input-output conversion routines.

Special programs concerned with storage allocation to different programs being run simultaneously, monitoring routines for fault finding and costing purposes, and the detailed organization of drum and tape transfers.

All this information was permanently required and hence was kept in part of the private store called the fixed store which operated on a read-only basis.

3. Multiple stores

The Atlas was designed to have up to five distinct direct access stores together with magnetic drums and magnetic tape backing stores. The desired direct access store was selected by the value of the top three bits of the address.

000 to 011   Main Store (maximum size: 2^{20} words of 48 bits)
100   Fixed Store
101   Not used
110   V-Store
111   Subsidiary Store

The fixed store

This store consisted of a woven wire mesh into which patterns of small ferrite and copper rods were inserted to represent the desired digital information. The information content could only be changed manually. The store was arranged in two units each of 4096 words, a unit consisting of 16 columns of 256 words, each word being 50 bits. The access time to a word was 0.4 microseconds which made it the fastest store available at that time. The fixed store operated in conjunction with the subsidiary store which programs in the fixed store used as a data store.

The subsidiary store

This was a private system store not accessible to normal programs. It had 1024 words with a cycle time of 1.8 microseconds. It was used by the supervisor program and as working space for routines in the fixed store.

The V-store

In the earlier Manchester computers the peripherals were restricted to simple I/O devices and were controlled by a small subset of the instructions. This was not suitable for the Atlas and a different method had to be found. The solution was named the V-Store. It was certainly not a store in the conventional sense, rather an address space with words of up to 24 bits. The peripheral could be controlled and information transferred by writing or reading the appropriate bit patterns to or from the address or addresses associated with the peripheral.

As well as controlling the standard peripherals this technique was very useful in other applications, for example, the angular position of any of the drums could be determined by reading the appropriate V-Store address.

A very important application was in determining the source of interrupts and in the case of multiple interrupts identifying the one of highest priority. Up to 512 sources of interrupt were permitted, arranged in order of priority and stored in 64 groups of eight in the V-store. A second level merged the 64 groups into eight groups, a third level into one and finally into the interrupt signal. One of the index registers was modified so that if it was loaded with an eight bit pattern and then read, the output was the position of the most significant digit. This enabled the highest priority interrupt to be identified in five instructions.

New peripherals could be easily added, including ones which required complex control systems, for example one of the peripherals on the Manchester Atlas was an X-ray crystallographic unit. At a later date an input/output interface was provided for speech recognition and synthesis.

The main core store

On the Manchester Atlas the core store was 16K words, each of 50 bits arranged as 4 stacks of 4K words in two pairs with interleaved addressing. This means that consecutive addresses were on different stacks and therefore that a pair of instructions could be read at the same time.

The drum store

The Atlas had four magnetic drums, each with a capacity of 24K words. Later versions of the Atlas had more drums and larger core stores.

4. Performance

In the early stage of the development of the system it was realized that the time from initiation to completion of a single floating-point addition instruction would be in the region of six microseconds even though the actual addition time was only 1.2 microseconds. This clearly showed the need for ensuring that as many operations as possible could take place at the same time as others.

Whilst the steps of an individual instruction must generally be executed sequentially it is possible to arrange for different parts of the computer to be executing parts of different instructions at the same time. Atlas made extensive use of this feature of "overlapping." The effect of overlapping together with the ability to read pairs of instructions was to reduce the time for a sequence of floating-point add operations to 1.6 microseconds per instruction. This corresponds to a continuous overlapping of between three and four instructions.

Total overlapping is possible in some cases e.g. a sequence of floating-point multiplication instructions in which the multiplier is always busy and all the other operations can be overlapped with those of multiplication.

5. The one-level store

On the Mark 1 the CRT store was in pages, each page corresponding to one CRT and the drum was divided into two page blocks with the ability to transfer either one or a pair of pages between the two stores.

On the Atlas both the core store and the drum store were divided into pages of 512 . This gave 32 pages in the core store and 192 pages on the drum store. These 224 pages were only about 10% of the address space. The total address space was called the virtual address space and addresses in this space could be translated into real addresses in either the core store or the drum store.

For the core store 32 Page Address Registers held the virtual page addresses of the corresponding real page. There was always at least one empty page, how this was arranged will be described in section 6. When an address was presented to the store the virtual page address was compared in parallel with all 32 PAR's and if one was the same, the 5 bits of the real core store page address were concatenated with the rest of the presented address and used to access the core store.

If there was no match, a "non-equivalence" signal caused an interrupt and entry into the Supervisor, which used tables in the subsidiary store to locate the address on the drum store of the required page and then initiated the transfer of this page from the drum to an empty page in the core store. When the transfer was completed the PAR was set to the new page address and the location on the drum from which the page had been transferred was declared to be empty. Pages only existed in one place, either in the main store or on the drum.

6. The drum transfer learning program

It was stated above that there was always an empty page in the core store. In order to achieve this, as soon as the drum to core store transfer was initiated there was a check that there would still be an empty page in the core store after the transfer had been completed, which would be required when the next nonequivalence occurred. If this was not the case, the system chose a page in the core store to be transferred to the drum. This transfer, which took place as soon as the drum to core store transfer was completed, was to the first available empty block position on any of the four drums, i.e., the empty block with the shortest access time.

Many computer users believed that the choice of the page to be transferred to the drum could only be made by the programmer, as was the case in the Mark 1 and the Mercury. They were convinced that an automated system would lead to many more unnecessary transfers between the two levels of store with a consequent decrease in performance.

The selection of the page to be transferred could be made at random; this could easily result in many additional transfers occurring, as the page selected could be one of those in current use or one required in the near future. The ideal selection, which would minimize the total number of transfers, could only be made by the programmer. To make this ideal selection the programmer would have to know

- precisely how his program operated, which was not always the case;
- the precise amount of core store available to his program at any instant.

This latter information was not generally available as the core store could be shared by other central machine programs, and almost certainly by some fixed store program organizing the input and output of information from slow peripheral equipment. The amount of core store required by this fixed store program was continuously varying. The only way the ideal pattern of transfers could be approached was for the transfer program to monitor the behavior of the main program and in so doing attempt to select the most appropriate page to be transferred to the drum. The techniques used for monitoring were subject to the condition that they must not slow down the operation of the program to such an extent that they offset any reduction in the number of transfers required. The method described occupied less than 1 per cent of the operating time, and the reduction in the number of transfers was more than sufficient to compensate for this.

That part of the transfer program which organized the selection of the page to be transferred was called The Drum Transfer Learning Program. In order for this program to have some data on which to operate, the machine was designed to supply information about the use made of the different pages of the core store by the program being monitored.

With each page of the core store there was associated a "use" digit which was set to "1" whenever any line in that page was accessed. The 32 "use" digits existed in two lines of the V-store and could be read by the learning program, the reading automatically resetting them to zero. The frequency with which these digits were read was governed by a clock which measured not real time but the number of instructions obeyed in the operation of the main program. This clock caused the learning program to copy the "use" digits to a list in the Subsidiary store every 1024 instructions. The use of an instruction counter rather than a normal clock to measure "time" for the learning program was due to the fact that the operations of the main program may be interrupted at random for random lengths of time by the operation of peripheral equipment. With an instruction counter the temporal pattern of the blocks used will be the same on successive runs through the same part of the program. This was essential for the learning program to make use of this pattern to minimize the number of transfers.

When a non-equivalence occurred and after the transfer of the required block had been arranged, the learning program again added the current values of the "use" digits to the list and then used this list to bring up to date two sets of times also kept in the subsidiary store. These sets consisted of 32 values of t and T, one of each for each page of the core store. The value of t was the length of time since the block in that page had been used. The value of T was the length of the last period of inactivity of this block. The accuracy of the values of t and T was governed by the frequency with which the "use" digits were inspected.

The page to be written to the drum was selected by the application in turn of three simple tests to the values of t and T.

- Any page for which $t > T+1$,
- That page with t not equal to zero and the maximum value of $(T - t)$,
- If all t were equal to zero, then that page with the maximum value of T.

The first rule selected any page which has been currently out of use for longer than its last period of inactivity. Such a page had probably ceased to be used by the program and was therefore an ideal one to be transferred to the drum. The second rule ignored all pages with $t = 0$ as they were in current use, and then selected the one which, if the pattern of use was maintained, would not be required by the program for the longest time. If the first two rules failed to select a page the third ensured that if the page finally selected was wrong, in that it was immediately required again, then, as in this case, T would become zero and the same mistake would not be repeated.

In order to make its decision the learning program had only to update two short lists and apply at the most three simple rules; this could easily be done during the 2 msec transfer time of the block required as a result of the non-equivalence. As the learning program used only fixed and subsidiary store addresses it was not slowed down during the period of the drum transfer.

The value of the method used was investigated by simulating the behavior of the one-level store and learning program on the Mercury computer at Manchester University. This was done for several problems using varying amounts of store in excess of the core store available. One of these was the problem of forming the product A of two 80th order matrices B and C. The three matrices were stored row by row, each one extending over 14 blocks; only 14 pages of core store were assumed to be available. The method of multiplication was

- $B(1,1) \times 1^{st}$ row of C = partial answer to 1st row of A,
- $B(1,2) \times 2^{nd}$ row of C + partial answer = second partial answer, etc.

Thus matrix B was scanned once, matrix C eighty times and each row of matrix A eighty times.

Several machine users were asked to spend a short time writing a program to organize the transfers for a general matrix multiplication problem. When the method was applied to the above problem, in no case were fewer than 357 transfers required. A program written specifically for this problem, which paid great attention to the distribution of the rows of the matrices relative to block divisions, required 234 transfers. The learning program required 274 transfers, the gain over the human programmer was chiefly due to the fact that the learning program could take full advantage of the occasions when the rows of A existed entirely within one block.

Many other problems involving cyclic running of single or multiple sets of data were simulated, and in no case did the learning program require more transfers than an experienced human programmer.

7. Conclusions

The Atlas project was a success in many ways. It was the most powerful computer in the world for about two years in the early sixties. Whilst only three systems were sold it established the reputation of Ferranti, later ICL, as a provider of high performance computer systems. The three installations at Manchester, London and the Rutherford Laboratory provided the UK high performance computer users with world class facilities for several years. Most important of all, it was the first computer to present a multilevel store as a unified address space to the programmer, a feature of computer design which is now the norm from PCs to super-computers.

References

[1] T. Kilburn and R.L. Grimsdale: "A Digital Computer Store with a Very Short Read Time," *Proceedings of the IEE*, 107B (1960), 567.

[2] T. Kilburn, D.B.G. Edwards and D. Aspinall, "A Parallel Arithmetic Unit Using a Saturated Transistor Fast Carry Circuit," *Proceedings of the IEE*, 107B (1960), 573.

[3] T. Kilburn, D.B.G. Edwards, M.J. Lanigan and F.H. Sumner, "One Level Storage," *IRE Transactions,* EC-11 (1962), 223.

[4] F.H. Sumner, G. Haley and T.C.Y. Chen, "The Central Control Unit of the Atlas Computer," *Proceedings of the IFIP Congress,* 1962, 657.

[5] T. Kilburn, D.J. Howarth, R.B. Payne and F.H. Sumner, "The Manchester University Atlas Operating System," *The Computer Journal,* The British Computer Society, October 1961.

FRANK H. SUMNER has just retired from the Department of Computer Science at the University of Manchester. He wrote his first program for the Mark 1 computer early in 1952 after meeting Alan Turing who gave him a copy of his programming manual. After completing his Ph.D. in 1954 he joined the staff of the Computer Laboratory and worked on the Mark 1 and the Ferranti Mercury on a wide range of topics including Pattern Recognition and Artificial Intelligence. He then joined the Atlas computer team where he worked with Tom Kilburn on the design and implementation of the "one level store" and on the design of the central processor. In the 1960s, again with Tom Kilburn, he initiated and designed the MU5 computer which lead to the ICL 2900 series. After the MU5 project he was more involved in establishing and running the undergraduate school in Computer Science. He subsequently became the Director of the National Computer Center at Manchester University. He is a member of the British Computer Society and was its President in 1978.

Past into Present: The EDSAC Simulator

Martin Campbell-Kelly

Abstract. The EDSAC was the world's first stored-program computer to operate a regular computing service. Designed and built at Cambridge University, the EDSAC performed its first fully automatic calculation on May 6, 1949. The simulator written at Warwick University is a faithful emulation of the EDSAC designed to run on a personal computer. The user interface has all the controls and displays of the original machine, and the system includes a library of original programs, subroutines, debugging software, and program documentation. This paper describes the reasons for creating the simulator and the steps taken to achieve a historically authentic emulation.

1. Introduction: an EDSAC player

In his book *The Past is a Foreign Country*, the architectural historian David Lowenthal distinguishes five styles of recreating historical artifacts and experiences, which he terms *duplicates*, *re-enactments*, *copies*, *emulations*, and *commemorations*.[1] Lowenthal's taxonomy is complex and his examples are largely drawn form the world of buildings and architecture. Lowenthal's classification, in the context of the recent computer "rebuild" projects, has been ably discussed by Jon Agar.[2] For the purpose of this paper, I want to distinguish between the two terms of Lowenthal that most closely relate to the present conference: *duplicates* and *emulations*.

By *duplicates* Lowenthal means reproductions that "aim simply to duplicate admired relics."[3] Clearly the recent British projects to rebuild Babbage's Difference Engine, the Colossus, and the Manchester "Baby" fall into this category. In all three cases the project leaders have gone to extraordinary and admirable lengths to recreate not only the architectural and logical details of the artifact, but also its physical characteristics – by seeking out authentic components and materials, for example. Above all, the re-creations have had

---

[1] Lowenthal, D. *The Past is a Foreign Country*, Cambridge University Press, (1985).

[2] Agar, J.. "Digital Patina: Texts, Spirit and the First Computer," *History and Technology*, forthcoming.

[3] Lowenthal, Ibid, p. 290.

Figure 1: A photograph of the EDSAC taken shortly after its completion in May 1949. The left three-quarters of the picture shows the main racks of the arithmetic unit, control and memory. The input-output equipment (a paper-tape reader and teleprinter) can be seen on the table on the right. Three of the monitor tubes can be seen at the rear and right of the picture. The EDSAC operated at a speed of approximately 600 operations per second.

the goal of producing a working facsimile – in the sense of executing an original program or routine. It is notable that all three projects are museum-based, and when completed they will become permanent static exhibits, with just occasional use by a trained docent. Thus the projects fall comfortably into the museum tradition of functional replicas such as Stephenson's *Rocket*.

The EDSAC simulator[4] (Figs. 1 and 2) is a very different kind of re-creation that fits much more closely with what Lowenthal terms *emulations*. These he describes as "self-conscious period revivals" and "respectful yet creative reworkings of earlier forms and styles [that] transcend mere copy-ing" (p. 301). Lowenthal views an emulation as an on-going and evolving activity that is always of its time – thus a Tudor revival of the 1990s would be quite different to one of the 1930s, and that, in turn, would be quite differ-ent to an original Tudor building of the 16th century. Yet an essential Tudor-ness persists in all the variations and derivatives.

The EDSAC simulator is likewise intended to be fluid and of its time, while always capturing the essential EDSAC. Since its original creation in 1978, the simulator has evolved through three distinct reinterpretations and

---

[4] The EDSAC simulator can be downloaded from the Internet address at the University of Warwick: http://www.dcs.warwick.ac.uk/~edsac.

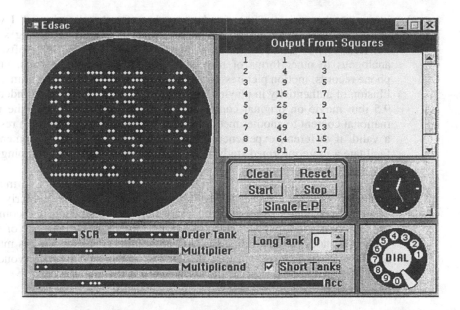

Figure 2: Screen shot of the EDSAC simulator, shown executing the "Squares" program, first run in May 1949. The circular display (top left) shows the main memory monitor tube. The register panel (bottom left) displays the accumulator and other registers. The clock (middle right) gives the elapsed EDSAC time. The output panel (top right) shows the teleprinter print-out. The central panel contains the five user-accessible controls: Clear, Start, Reset, Stop and Single E.P. The dial (bottom right), added in 1951, enabled a single decimal digit to be entered into the machine.

five versions, and I expect this periodic reincarnation to continue into the future. The reason for the fluidity of the EDSAC simulator is that its purpose is to enable present-day computer users to understand what it was "like" to write programs for the EDSAC. The simulator has consciously sought to bridge the ever changing gulf between past and present, by enabling the user to bring the leverage of his or her current programming knowledge and experience to bear on the task of understanding the EDSAC.

Two metaphors have constantly informed the design of the simulator. The first is the analogy between a computer program and a musical score – once described as "frozen music" needing only an orchestra to melt it. Consequently, the EDSAC simulator is textual rather than artifactual in spirit. This has several important implications. For example, the attention that other projects have given to physical authenticity has been directed at obtaining authentic program texts. However, as with musical scholarship, this textual approach permits the informed and explicit filling in of lost textual fragments where this will produce a richer experience for the user. The second metaphor informing the design of the simulator is that of a "media player." I would argue that the informational content of the EDSAC programming system transcends its original physical environment. In this way, EDSAC software is analogous to other forms of recorded information – pianola rolls, gramophone records, motion pictures, etc. While there is an enormous charm and an illusion of authenticity in playing a pianola roll, an Edison wax cylinder, or a 9.5 mm movie on original contemporary equipment, transferring the informational content to another media for use on modern equipment can result in a valid, if different, experience. Indeed, the use of modern equipment can facilitate a more informed study – for example by starting and stopping mid-session, replaying significant passages for closer study, and so on.

Thus the primary aim of the EDSAC simulator has been not so much to create an "EDSAC experience" as to afford a vehicle for the scholarly study of an early programming system by today's university students and computer professionals. As it happens, many users have reported a real sense of walking in the shoes of the original EDSAC programmers. I suppose this must be an illusion – and since I was not an EDSAC user myself I cannot vouch for the authenticity of the experience. I will return to this point later.

2. The EDSAC texts

The EDSAC (Electronic Delay Storage Automatic Calculator) contained some 3000 vacuum tubes and consumed about 12 kW of power. It was designed with a delay-line memory of 32 mercury-filled "tanks," each of which held 32 18-bit words, giving a total capacity of 1024 words with a cycle time of approximately 1 ms. The machine executed about 600 operations per second. The Appendix describes the EDSAC instruction set.[5]

The key text for the EDSAC simulator project is the first edition of the classic textbook on programming, *The Preparation of Programs for an Electronic Digital Computer* by M. V. Wilkes, D. J. Wheeler, and S. Gill (1951) – usually known as *Wilkes, Wheeler and Gill*, or simply *WWG*.[6] This book contains the detailed coding of a number of sample programs, and an appen-

---

[5] A more complete description can be found in: Wilkes M. V. and W. Renwick. "The E.D.S.A.C.," *Report of a Conference on High-Speed Automatic Calculating Machines,* (University of Cambridge, 1949), pp. 9–12. Reprinted in B. Randell, *Origins of Digital Computers,* Springer-Verlag, (Berlin, 1982), 417–21. Also in: Campbell-Kelly, M. "Programming the EDSAC: Early Programming Activity at the University of Cambridge," *Annals of the History of Computing* 2 (1980), 7–36. Reprinted in: *Annals of the History of Computing* 20 (1998), 46–67.

[6] Wilkes, M. V., D. J. Wheeler and S. Gill. *The Preparation of Programs for an Electronic Digital Computer,* Addison-Wesley, (1951). Reprinted as Vol. 1 of the *Charles Babbage Institute Reprint Series for the History of Computing,* Tomash Publishers, Los Angeles, and MIT Press, (Cambridge Ma, 1982).

dix contains specifications and coding for much of the subroutine library. *Wilkes, Wheeler and Gill* is widely available in academic libraries (three thousand copies were originally printed in 1951), and a reprint edition was published as the first volume of the *Charles Babbage Institute Reprint Series for the History of Computing* in 1982. This textbook is uniquely important in making EDSAC programming tangible – it adds a credibility that transcends the informally published programming manuals of other historic computers. Incidentally, the fact that *Wilkes, Wheeler and Gill* contains large fragments of code means that the frozen-music metaphor fairly leaps off the page. About half of the code supplied with the simulator comes directly from *Wilkes, Wheeler and Gill*, the remainder coming from the Cambridge University Archives with a small amount recreated (see Tables 1, 2, and 3).

Table 1: Routines for Initial Orders 1[7]

Routine	Status	Sources
Initial Orders 1	Extant	Wheeler 1950, Renwick & Worsley 1949
Programs		
Squares	Lost	
Print Squares	Extant	Renwick & Worsley 1949
Print Primes	Extant	Renwick & Worsley 1949
The Airy Tape	Partially reconstructed	Personal communication

9	A 46 S	Put 046S in S(65)
50	T 65 S	
1	T 129 S	Clear S(129)
2	A 35 S	Put power of 10
3	T 34 S	in S(34)
4	E 61 S	Jump to 61
5	T 48 S	
6	A 47 S	

Figure 3: Code fragment from Print Squares

---

[7] Wheeler, D. J. 1950. "Programme Organisation and Initial Orders for the EDSAC," *Proc. Roy. Soc.* (A) 202 (1950), 573–589. Renwick, W. and B. H. Worsley. "The E.D.S.A.C Demonstration," *Report of a Conference on High-Speed Automatic Calculating Machines,* (University of Cambridge, 1949), 12–16. Reprinted in Randell 1982, n. 3, 423–9.

CAMBRIDGE
EDSAC.
FIRST ACHIEVEMENT
MAY 7th 1949.

```
0000  0001  0004  0009  0016  0025  0036  0049  0064  0081
0100  0121  0144  0169  0196  0225  0256  0289  0324  0361
0400  0441  0484  0529  0576  0625  0676  0729  0784  0841
0900  0961  1024  1089  1156  1225  1296  1369  1444  1521
1600  1681  1764  1849  1936  2025  2116  2209  2304  2401
2500  2601  2704  2809  2916  3025  3136  3249  3364  3481
3600  3721  3844  3969  4096  4225  4356  4489  4624  4761
4900  5041  5184  5329  5476  5625  5776  5929  6084  6241
6400  6561  6724  6889  7056  7225  7396  7569  7744  7921
8100  8281  8464  8649  8836  9025  9216  9409  9604  9801
```

Figure 4: Output from the "first" program

To understand these tables, it is necessary to know something about the evolution of the EDSAC programming system. This system was based on a set of "initial orders" and a subroutine library. The initial orders combined in a rudimentary fashion the functions performed by a bootstrap loader and an assembler in later computer systems.

The initial orders existed in three versions. The first version, *Initial Orders 1*, was devised by David Wheeler, then a research student, in 1949. The initial orders resided in locations 0 to 30, and loaded a program tape into locations 31 upwards. The program was punched directly onto tape in a symbolic form using mnemonic operation codes and decimal addresses, foreshadowing in a remarkable way much later assembly systems (Fig. 3). Just three programs survive from this period. The first two are demonstration programs, to print primes and squares-and-differences respectively, which were published in the *Report of a Conference on High-Speed Automatic Calculating Machines* held at Cambridge in June 1949 to celebrate the completion of the EDSAC. The third program is a draft version of a program, written by Wilkes in the summer of 1949, to compute the Airy integral. Most tantalizingly, the historic *first* program, which was run on May 6, 1949 and printed a table of squares, has been lost, although there are several extant copies of the output (Fig. 4). Re-creating this program has been a popular exercise for present-day EDSAC programmers, and one that requires insight and empathy to achieve in a way that is historically authentic.

Table 2: Routines for Initial Orders 2

Routine	Status	Sources
Initial Orders 2	Extant	Wheeler 1950; WWG 1951
Subroutine Library		
Approx. 90 subroutines	Extant	Cambridge University Archives WWG 1951, ca. 30 subroutines
Postmortem Routines		
P0	Extant	Cambridge University Archives
P1-P5	Reconstructed	
Programs		
Chapman's Integral	Extant	WWG 1951
Noughts & Crosses	Partial reconstruction	Douglas 1954
Glennie Program	Extant	Dodd and Glennie 1951

In September 1949, the first form of the initial orders was replaced by a new version. Again written by Wheeler, *Initial Orders 2* was a *tour de force* of programming that combined a surprisingly sophisticated assembler and relocating loader in just 41 instructions. The initial orders read in a master routine (main program) in symbolic form, converted it to binary and placed it in the main memory; this could be followed by any number of subroutines, which would be relocated and packed end-to-end so that there were none of the memory allocation problems associated with less sophisticated early attempts to organize a subroutine library. During 1949-1951, some 90 library subroutines were developed. In a typical program, two-thirds of the code would come from the subroutine library, and only one-third would be directly written by the user. Library subroutines existed for input and output in many different formats, for the common mathematical functions (square root, sine, etc.), and for more complex numerical processes such as interpolation and numerical integration.

There are few extant programs from this period, and only one that can be genuinely described as compelling – this is Sandy Douglas's program to play Noughts and Crosses (tic-tac-toe in the U.S.) The program was written while Douglas was a PhD student and was reproduced in his PhD thesis.[8] Other extant programs include a demonstration program for the numerical integration of a differential equation written by A. E. Glennie, and several programming examples in *Wilkes, Wheeler and Gill*, all of a mathematical nature. The mathematical orientation of EDSAC applications points up one of

---

[8] Douglas, A.S. 1954. *Some Computations in Theoretical Physics*, PhD Dissertation 2478, (Cambridge University, 1954).

Table 3: Routines for Initial Orders 3

Routine	Status	Sources
Initial Orders 3	Extant	Cambridge University Archives
Test programs	All lost	

the problems of making the programming system accessible to present-day programmers who lack the sophisticated mathematical training of the pioneers – almost all of whom were mathematicians or engineers with a numerical aptitude.

A third set of initial orders, written by Stanley Gill, was introduced in late 1951. These initial orders loaded programs punched in a form very close to binary and were used solely for running hardware test programs. *Initial Orders 3* used only the most basic, single-length instructions (such as load and store, add and subtract, and the basic branch orders) so that they would usually work even if the multiplier, for example, was being temperamental. Unfortunately, none of the original test programs has survived so these initial orders are essentially a historical curiosity. *Initial Orders 3* are not currently supplied with the simulator as a conscious editorial decision to avoid confusing the learning experience with items of marginal significance.

3. Interpreting the EDSAC

There is no known complete architectural description of the EDSAC at the gate level, still less at the level of electronic components. There are, however, a number of descriptions of the EDSAC at the systems level. For example, an article "The E.D.S.A.C." in the proceedings of the *Report of a Conference on High-Speed Automatic Calculating Machines*[9] gives both a narrative and diagrammatic account of the instruction fetch-execute cycle. Wilkes's 1956 textbook *Automatic Digital Computers* gives a much fuller description, including some parts, such as the serial adder and multiplier, which are described at the gate level.[10] The most complete description is an unpublished report produced by a visiting Australian academic.[11] Impressively detailed as this report is, it is a description of work in progress that predates completion of the machine and is therefore incorrect in many details.

---

[9] Wilkes and Renwick, see n. 4 above.

[10] Wilkes, M. V. *Automatic Digital Computers*, Methuen (London, 1956).

[11] Anonymous, *The EDSAC*, unpublished report, Cambridge University Mathematical Laboratory, May 1948.

In fact, although this literature has occasionally been used to clarify my own understanding of the EDSAC architecture, it has been little used in any tangible way in developing the simulator. The EDSAC "player" is defined by the software, not the hardware. For example, although in principle the simulator could emulate the original machine at the gate level, there would be very little to be gained by doing so and a great deal to be lost: the EDSAC was a serial machine, and performing a bit-by-bit simulation on a contemporary byte-organized machine would produce such a massive computational overhead that the simulator would be reduced to a crawl. Instead, advantage has been taken of the fact that the EDSAC was a fairly conventional word-based two's complement machine. Thus, for example, adding a pair of numbers using the built-in hardware circuits of a present-day machine gives exactly the right result with no time penalty and no effort on the part of the author. This approach has been found to be effective for interpreting the various arithmetical and logical instructions, most of which are identical to those in present-day machines. This remains true whether the implementation language is assembler, C, Pascal or Java – all of which have been used for various incarnations of the simulator.

As noted earlier, the EDSAC simulator has existed in three distinct forms and five deployed versions. The first form of the simulator, 1978–82, was a batch oriented program, reflecting the available technology of the day. EDSAC programs were prepared on punched cards, delivered to the computer center, and the printed results collected later. The system was also used on a Unix time-sharing system, which enabled the use of files rather than punched cards, and achieved a faster turnaround, but was nevertheless batch-oriented in principle. As a result users of the simulator were not able to develop very exciting programs in the few hours they were able to use the system. A typical user program would be to sum the first hundred integers and print the total.

The second version of the simulator, 1982–87, was produced by a research student and it provided a dynamic display of the EDSAC store and registers similar to that on the original machine. This provided a much more evocative experience and was enthusiastically used by students. However, the program operated in a time-shared Unix environment and was so computationally intensive that it was possible to get a real-time simulation only at the dead of night when the system was unencumbered with competing users. This problem was alleviated by the arrival of inexpensive workstations and personal computers in the late 1980s.

The present simulator is based on one designed for the Macintosh computer in 1988. The system provides an integrated editor for program texts, dynamically updated displays, and press-buttons for the EDSAC controls. The development was quite labor intensive, involving about six months of student effort. The simulator was consciously modeled on the Interactive Development Environments (IDEs) that were then becoming the norm for programmers, so that they could easily transfer their existing programming expertise to the task of understanding the EDSAC. The system was quite

Figure 5: EDSAC "console"

widely distributed, being made available as freeware in Macintosh-related magazines via CD-ROMs or in one case by mail order. It was not until 1996, with the arrival of the Windows 95 operating system, that the IBM-compatible computer finally caught up with the Macintosh by providing a fast, programmer-accessible graphical user interface. At the same time, the development of Rapid Application Development (RAD) systems had reduced the task of developing the simulator to about two weeks. The PC-based simulator was released in 1996. The documentation for the system has now been converted to Adobe Acrobat PDF format for platform independent web-based delivery. A Java-based version is currently under development, which should allow universal platform-independent deployment. One outcome of the Macintosh and PC versions of the simulator is that users have been able to study the EDSAC programming system far more deeply than was ever possible with a batch or time-shared system. As a result, there have been some very interesting programs created that could never have been envisioned by the original EDSAC programmers – of which more later.

The notion of an EDSAC "player" has freed the design of the simulator from the obligation to produce an authentic console for the user. Such a console is a distinctive feature of some other simulators of historic computers, such as Chris Burton's Pegasus emulator or Keith Reed-Green's IBM 650 simulator. In the case of the EDSAC the decision to forgo a naturalistic console was not least due to the fact that there *was* no console. The closest that EDSAC came to a console was collecting together all the displays and controls onto a wooden table (Fig. 5). The face presented to the user of the EDSAC simulator is a heavily sanitized and stylized version of this equipment.

The most critical part of the EDSAC display was to represent the monitor tubes of the original machine. There was a total of six monitor tubes on the EDSAC. The most important of these displayed the dynamically changing contents of one of the 32-word main memory delay lines. Which of the 32 "long tanks" was displayed was determined by a rotary switch. This main memory tube is the most distinctive visual feature of the simulator (Fig. 2). The remaining monitor tubes displayed the processor registers or "short tanks" – the accumulator, multiplier registers, instruction decode register, and sequence control register. The short-tank displays can be turned off to enable the simulator to run more quickly – with all the displays turned on the simulator will run somewhat slower than the original EDSAC on an average PC. A clock, which shows elapsed EDSAC time, provides user feedback on the speed of the calculation. The EDSAC was controlled by five push buttons: Start, Stop, Clear, Reset and Single E. P. (single shot). In addition, a telephone dial was provided to input a single decimal digit into the machine. These are all represented by conventional GUI buttons.

To an extent the simulator can be regarded as an ephemeral creation, although efforts have gone into keeping a consistent visual appearance from version to version. Much more persistent are the program texts themselves. At least as much effort has gone into preparing the supporting programs and documentation as in developing the simulator itself. The most important user documentation is the *Tutorial Guide*. This is essentially a present-day programmer's primer for the EDSAC. One could argue that a more authentic experience would be gained by just supplying a copy of *Wilkes, Wheeler and Gill*. However, the *Tutorial Guide* has an overtly pedagogical aim to explain and interpret the nuances of EDSAC programming for the current generation of programmers. The *Tutorial Guide* includes a set of staged programming examples to take the user though the programming process, and concludes with a set of programming problems for the user to undertake. Several of the programming problems were first used at the summer school in programming, which was held annually at Cambridge, for several years from 1950.

In the interests of pedagogy, the program documentation and program code supplied with the simulator is not an uncritical reproduction of the contents of the EDSAC library. For example, a selection of just 18 of the 90 extant library subroutines has been included. These routines are sufficient to attempt all of the sample problems given in the *Tutorial Guide*. The subroutines that have been omitted are primarily mathematical – such as variants of the basic mathematical functions, and advanced numerical procedures, such as interpolation and numerical integration. This conscious editorial excision was made purely to make the simulator more accessible and less overwhelming. It should be noted that the summer schools, designed to introduce mathematical and engineering professionals to computing, were of two full weeks duration. The aim of the simulator is to enable the students to develop a significant appreciation of the EDSAC in a half-day, although there is enough material for at least a week's intensive study. For the truly dedicated student (and there are some) access to *Wilkes, Wheeler and Gill* will provide more depth again.

4. Authentication and validation

Because the EDSAC simulator is relatively abstract, not being defined by an underlying hardware implementation, there was a need for positive proof that the representation of EDSAC was correct – both in terms of the few hardware features emulated as well as all the software features.

In terms of the physical hardware environment, the most important source has been *The EDSAC Film*. This movie was made for the Joint Computer Conference in Philadelphia in December 1951. Although of only 5 minutes duration, the film gives a very detailed account of the creation and running of a complete EDSAC program. The movie includes a sequence showing the main memory monitor tube, on which the simulator display is very closely based. Incidentally, *The EDSAC Film* also demonstrates that it is possible to be misled by one's sources. According to the authority of all the textual sources, the monitor tube displayed words so that binary numbers appeared with their most significant bit on the left and least significant on the right. This made it much easier for humans, accustomed to writing decimal numbers left-to-right, to read data from the face of the tube. This innovation was non-obvious: in a serial machine in which the least significant bit of a word is generated first, numbers would appear with the least significant digit on the left when using the conventional left-to-right sweep of a CRT time-base. This was the case in almost all prototype computers. Wilkes and his co-workers were rather proud of their simple innovation, which just involved inter-changing the connections to the X-plates of the CRT monitor tube, and they were somewhat scathing of Manchester University, which forced program-mers into using "base-32 backwards" numbers. However, *The EDSAC Film* contradicted the textual sources, because the monitor tube displayed numbers with their least significant digit on the *left*, as on other machines. The expla-nation turned out to be that the original film – shot on 16 mm film stock, which has a left-right symmetry – had been reversed in the editing process. (Additional photographic evidence, noted below, subsequently confirmed this hypothesis.)

It is not possible to verify all the behaviors of the various controls pro-vided in the EDSAC simulator, and there may well be some minor flaws. For example, when the original EDSAC obeyed a *Stop* instruction, the machine halted and sounded a bell; however, if an attempt was made to execute an undefined operation code, the machine halted *without* sounding the bell. This is a subtle distinction which appears in only one place in all the EDSAC literature. Undoubtedly, the simulator will have failed to observe some other minor subtleties of this type. However, it has been used by a number of EDSAC pioneers, including Wilkes and Wheeler, and they vouch that it captures the spirit of EDSAC; but they could not be expected to pick up subtle flaws at this distance in time.

The most important aid for authenticating the software behaviors of the simulator are five original programs, for which both the original program and a physical copy or a photograph of the output exists. These programs are:

- Print Squares (Wilkes, June 1949)
- Print Primes (Wheeler, June 1949)
- The Glennie Program (Glennie, July 1951)
- Noughts & Crosses (Douglas, c.1952)
- The Airy Tape (Wilkes, July 1949)

The Print Squares and Print Primes programs are the two programs written for the EDSAC demonstration at the *Conference on High-Speed Automatic Calculating Machines* in June 1949. The published proceedings have bound into them a spirit-duplicated copy of the output of these two programs and it has been possible to verify the correctness of the simulator results by a character by character comparison. However, these programs only performed single length arithmetic and so are a less than exacting test.

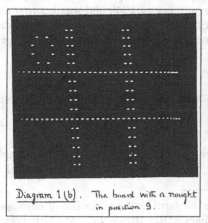
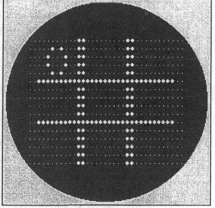

Figure 6: Verification of the EDSAC Simulator. *Noughts & Crosses* program display: original *left*; simulated *right*

The Glennie Program is much more demanding. This is a demonstration program written by Alick Glennie, who was a visitor at Cambridge from the Armament Research Establishment during 1949. The program appears in the report that Glennie and his colleague K. N. Dodd wrote for their employers,[12] and it includes both a program listing and a photograph of the printed output. The program solves the recurrence relation:

$$\Omega_n = v \sum_{m=0}^{n=1} \frac{1}{\sqrt{n-m}} (1-\Omega_m)$$

This is a complex program of over 400 instructions that exercises all the double-length arithmetic instructions, and would be sensitive to any faults in the numerical processes. The report states that the program took 7 minutes to execute. The EDSAC simulator produces visually identical output to that in the report, and shows an elapsed time of 7 minutes 25 seconds. Incidentally, the Glennie program is not included with the demonstration programs supplied with the simulator as it is visually unexciting and is too mathematically arcane for the intended users of the simulator.

The fourth program used to test the simulator is Sandy Douglas's *Noughts & Crosses* program. The significance of this program is that the output is produced on the face of the CRT monitor tube rather than the teleprinter. It is the only extant original program to produce its output in this way. There are Polaroid photographs of the CRT monitor in Douglas's PhD dissertation, which match exactly the simulator display (Fig. 6). (Incidentally, the program also confirms the right-left sweep of the CRT time-base.)

The *Airy Tape* was a more problematic program.[13] This program was written by M. V. Wilkes in the summer of 1949 and was his first attempt to write a "real" program for the EDSAC. The program is another mathematical one, the integration of the Airy Integral, which is the solution of the second-order differential equation:

$$y'' = xy$$

The program used a step-by-step integration process with the central difference formula:

$$\delta^2 y = (\delta x)^2 (y'' + \frac{1}{12}\delta^2 y'')$$

The printed results of the program were reproduced in photographic form in an article that Wilkes wrote for *Nature* in 1949.[14] The original program

---

[12] Dodd, K. N. and A. E. Glennie. *An Introduction to the Use of High-Speed Automatic Digital Computing Machines*, ARE Memo No. 7/51, ARE Fort Halstead, Sevenoaks, (Kent, 1951).

[13] Campbell-Kelly, M. "The Airy Tape: An Early Chapter on the History of Debugging," *Annals of the History of Computing* 14 (1992), 18–28.

[14] Wilkes, M. V. "Electronic Calculating Machine Development in Cambridge," *Nature*, 164 (1949), 557–558

was thought lost, but in 1979, shortly before his retirement from Cambridge University, Wilkes found a paper tape "in a rather fragile condition" with the single word "Airy" written on it in pencil. He sent me a copy of the tape to see what I could make of it.

The tape turned out to be not the final version of the program but an early version with a number of errors. I duly "corrected" the tape to the best of my ability but was unable to reproduce exactly the results that appeared in *Nature*. The program was transcribed into Fortran, which *did* produce the correct results. This suggested that either my EDSAC code had a very subtle error, or there was an error in the multiplication or rounding orders. As a result, although it seemed a long shot, I rewrote the multiplication routine simulating exactly the shift-and-add algorithm given by Wilkes in his textbook *Automatic Digital Computers* (1956, p. 55). The results were unchanged. There the matter lay, unresolved for several years. It was only when the Macintosh version of the simulator became available that I was able to reexamine the program, now with the advantage of being able to "peep" into the store and observe the numerical processes at very close hand over an extended period of time. The error was eventually tracked down to the nonobvious need for a double-length arithmetic constant in place of a single-length one. The simulator then produced results that exactly matched those in the *Nature* paper. Personal programming shortcomings apart, I think this experience shows how very difficult it is to completely step into the shoes of the EDSAC programmers – my inability to discover this programming error for over a decade was fundamentally due to the fact that I have never had any personal experience of hand-machine computation and therefore do not have the "feel" for numbers that almost very EDSAC programmer had. One notes that Wilkes managed to fix the program in a couple of weeks.

5. Conclusions: the "So What?" question

Historians often ask of a piece of scholarship: So what? It's always disarming, but usually the question deserves to be asked. What is the simulator for? Is it an "exhibit," a celebration, or a piece of recreational software? Or is there some deeper, academic purpose?

The EDSAC simulator is certainly an exhibit in a small way, but plainly not in the same league as the physical rebuild projects. It is also recreational software of a sort, and can be viewed as a kind of video game for computer programmers and geeks. However, I believe that even casual users take away some insight from using the simulator. The following appeared in a semipopular article by Brian Hayes in *The Sciences* in 1993:

> On May 6, 1949, a length of punched paper tape was threaded into a machine at the University of Cambridge; a few seconds later a nearby teleprinter began tapping out a list of numbers: 1, 4, 9, 16, 25, [...] It

was the first time any full-scale computer in the modern sense of the term had successfully run a program. [...]

The EDSAC is long gone; most of its parts were sold for scrap in 1958. No one will ever build another machine like it. Nevertheless, it is still possible to write programs for the EDSAC, to load those programs into the paper tape reader and then to see (and hear) the results come ticking out of the teleprinter. The time machine that offers this transport of delight is a simulator created by Martin Campbell-Kelly, an historian of computing at the University of Warwick.[15]

The following unsolicited e-mail from a female journalist is another typical, if unusually articulate, reaction:

Being 37 years old, I was brought up to think of computers as being these big mysterious boxes cooled by liquid nitrogen and manned by cool-voiced guys who all looked like Nicholas Negroponte. Big spools of magnetic tape would spin at improbable rates, manila cards and greasy paper tape would display enigmatic designs, and everywhere there would be big banks of twinkling lights. When I was 13, I was taught how to work with a PDP-8, but it didn't feel like a computer. Neither did the ZX-81 I learned BASIC on or finally, this Macintosh. I was born, I feel, somewhat too late. I'm less than a great programmer with BASIC, indifferent to COBOL and FORTRAN, and C++ sounds like a great idea, but hard to put into practice.

FINALLY, I have a computer [the EDSAC simulator] I can UNDERSTAND! No longer is there a problem showing how binary numbers work in calculation; I can SEE them. This may not be a great thing in the overall course of computing, but for my own satisfaction, this is a major breakthrough.

Perhaps reactions such as the above are justification enough for the simulator. However, the simulator does have two serious academic uses: as a research tool, and as a pedagogical device.

The EDSAC simulator grew out of the research that eventually appeared in my paper "Foundations of Computer Programming in Britain."[16] This research aimed to make a comparative study of the development of programming on the first three British computers: the EDSAC at Cambridge, the Mark I at Manchester, and the Pilot ACE at the National Physical Laboratory. Simulators were developed for all three machines and the TPK algorithm was coded for each of them. (The TPK algorithm is a universal bench-mark pro-

---

[15] Hayes, B. "The Discovery of Debugging," *The Sciences*, July/August, 1993, 10–13.
[16] Campbell-Kelly, M. 1982. "Foundations of Computer Programming in Britain 1945–1955," *Annals of the History of Computing* 4 (1982), 121–131.

Table 4: Comparative Performance of EDSAC, Manchester Mark I and Pilot ACE

Time to Run TPK (sec)	EDSAC	Mark I	Pilot ACE
Processor	24	37	5
Input-output	70	44	17
Total	94	82	21

gram devised by Knuth and Trabb Pardo.)[17] The results of this exercise are summarized in Table 4.

Two significant conclusions can be drawn from the table:

1. The Pilot ACE, which was based on Turing's unique ACE design, was overwhelmingly more powerful and cost-effective than its competitors in raw hardware terms. It was 5 to 7 times faster, yet had a tube count of less than a third of either the Cambridge or Manchester machines.[18]

2. Although the Manchester Williams Tube memory was a random access device, which was said to be a great improvement over the cyclic mercury delay lines of the EDSAC, the performances of the two machines were roughly comparable. This was because both were serial machines in which the speed of arithmetic instructions was determined by the basic pulse rate rather than the memory cycle time.

There was some controversy over both of these issues in the early 1950s. The simulators added a degree of quantitative analysis to the rhetoric of the contemporary debate. Beside these major findings, the act of running original software on the simulators provided many small insights, and afforded a degree of scholarly intensity that could never have been achieved with the texts alone.

A second piece of research[19] concerned Wilkes's *Airy Tape*. The process of turning Wilkes original tape into a working program using the EDSAC simulator again produced many insights that could not have been gained by textual study alone. In particular, the exercise gave a real appreciation of the unanticipated nature of the debugging problem, and provided dramatic contemporary evidence to support Wilkes's famous assertion in his *Memoirs*:

---

[17] Knuth, D. E. and L. Trabb Pardo. "The Early Development of Programming Languages," in N. Metropolis et al (ed.), *A History of Computing in the Twentieth Century*, Academic Press, (New York, 1980), 197–213.

[18] See in this volume: Harry D. Huskey, "Hardware Components and Computer Design."

[19] Campbell-Kelly, see n. 13 above.

By June 1949 (...) I was trying to get working my first non-trivial program, which was for the numerical integration of Airy's differential equation. It was on one on my journeys between the EDSAC room and the punching equipment that "hesitating at the angles of the stairs" the realization came over me that a good part of the remainder of my life was going to be spent in finding the errors in my own programs.[20]

The second purpose of the simulator is pedagogical. A number of universities have used the simulator as a visual aid in introductory computer systems and software engineering classes. However, the main use of the simulator is in undergraduate courses on the history of computing. Many science students prefer writing a program to writing an essay. The simulator offers such students an opportunity to demonstrate their historical understanding in a way that compensates for shortcomings in essay writing. A surprising result we have noticed at Warwick University is that the quality of undergraduate submissions has as great, or even greater, variance than traditional essays. The tariff of grades is roughly as follows:

A: An excellent solution to a challenging or original problem. Program methodology and documentation fully empathetic to the Cambridge style.
B: A good program that makes full use of the EDSAC programming system, including subroutines and pseudo-operations. Good or acceptable documentation.
C: A good solution to a simple problem – such as the easier summer school examples. Acceptable documentation.
D: A "hack" that works. Documentation not in keeping with the EDSAC style.
E: A "hack" that does not work. Flawed documentation.

The scale effectively runs from a program that is almost indistinguishable from a contemporary routine in the EDSAC library, down to a machine code "hack" that shows little or no historical empathy.

One surprising development during the last two years we have been using the simulator at Warwick University, is that of students devising programs which could not have been written in EDSAC's time because the theory postdates machines of the EDSAC's vintage. One example of such a program was the computation of the *Mandelbrot Set* and another was a simulation of Conway's *Game of Life*. Both programs used the CRT monitor to visualize their results. The phenomenon is similar to hearing a Beatle's melody played on a pianola: a captivating sense of temporal dislocation. I'm not sure of the historical significance, but it gives pause for thought.

---

[20] Wilkes, M. V. *Memoirs of a Computer Pioneer*, MIT Press, (Cambridge, Ma., 1985), 145.

Appendix: The EDSAC Instruction Set

The EDSAC used a single-address instruction format, as shown in Fig. A1. Although the EDSAC was based on an 18-bit word, only 17 bits were used, the leading bit being unusable for reasons connected with circuit set up time. The opcode (or "function") was specified in 5 bits and the address in 10 bits. A further bit specified the operand length: most instructions could operate on either a 17-bit short word or a 35-bit double-length word; the length indicator specified which.

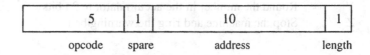

Figure A1: Instruction format

Table A1 shows the EDSAC instruction set as it existed in 1949. Operations were represented by letters of the alphabet, some of which suggested the function they denoted (for example, A for Add, S for subtract, and so on). Average instruction times were 1.5 ms, although multiplication was longer, taking 6 ms; input/output times were determined by the basic speeds of the peripheral equipment – a 50 character per second tape reader and a 6 $\frac{2}{3}$ character per second teleprinter.

Table A1: The EDSAC Instruction Set (1949)

A   n   Add the number in storage location n into the accumulator

S   n   Subtract the number in storage location n from the accumulator

H   n   Copy the number in storage location n into the multiplier register

V   n   Multiply the number in storage location n by the number in the
            multiplier register and add the product into the accumulator

N   n   Multiply the number in storage location n by the number in the
            multiplier register and subtract the product from the accumulator

T   n   Transfer the contents of the accumulator to storage location n and
            clear the accumulator

U   n   Transfer the contents of the accumulator to storage location n and
            do not clear the accumulator

C   n   Collate [logical *and*] the number in storage location n with the
            number in the multiplier register and add the result into the accu-
            mulator

R 2^{n-2} Shift the number in the accumulator n places to the right

L 2^{n-2} Shift the number in the accumulator n places to the left

E	n	If the sign of the accumulator is positive, jump to location n; otherwise proceed serially
G	n	If the sign of the accumulator is negative, jump to location n; otherwise proceed serially
I	n	Read the next character from paper tape, and store it as the least significant 5 bits of location n
O	n	Print the character represented by the most significant 5 bits of storage location n
F	n	Read the last character output for verification
X		No operation
Y		Round the number in the accumulator to 34 bits
Z		Stop the machine and ring the warning bell

MARTIN CAMPBELL-KELLY is Reader in Computer Science at the University of Warwick where he specializes in the history of computing. He has a BSc in computer science from the University of Manchester and a Ph.D. in the history of science. His Ph.D. thesis explored the development of the programming systems for the EDSAC and other early British computers. His books include: *Computer: A History of the Information Machine*, co-authored with William Aspray (1996); *ICL: A Business and Technical History* (1989), which won the Wadsworth Prize for British Business History; and the eleven volume *Works of Babbage* (1989). He is currently working on a business history of the international software industry.

Part V: Early Japanese Computers

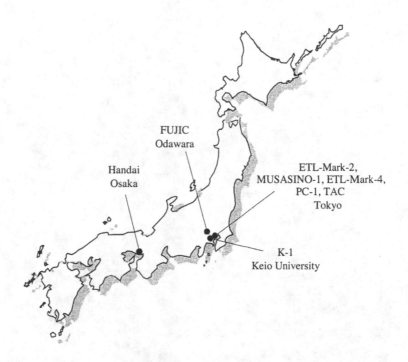

FUJIC
Odawara

Handai
Osaka

ETL-Mark-2,
MUSASINO-1, ETL-Mark-4,
PC-1, TAC
Tokyo

K-1
Keio University

Part VI. Early Japanese Computers

The First Japanese Computers and Their Software Simulators

Seiichi Okoma

Abstract. In the second half of the 1950s, many experimental computers were designed and produced by Japanese national laboratories, universities and private companies. The researchers had only a limited amount of literature and poor materials at their disposal. In those days, many experiments were carried out using various electronic and mechanical techniques and materials such as relays, vacuum tubes, parametrons, transistors, mercury delay lines, cathode ray tubes, magnetic cores and magnetic drums. These endeavors provided a solid basis for the development of electronics in Japan. The paper presents nine early Japanese computers, six of which have been emulated in software in order to test and validate their instruction sets.

1. Introduction

In the second half of the 1950s, many experimental computers were designed and produced by national laboratories, universities and private companies in Japan. Some of them became models for mass-produced computers, but mostly they were once-off constructions. In those days, many experiments were conducted using a range of electronic and mechanical techniques and materials, such as relays, vacuum tubes, parametrons, transistors, mercury delay lines, cathode ray tubes, magnetic cores and magnetic drums. Furthermore, single-address, two-address, three-address instructions, the binary or decimal number system, instruction traps and interrupts, and other new ideas, were being tested and adopted. These endeavors provided a solid basis for the development of electronics in Japan.

Table 1 shows a list of the first Japanese computers, not including commercially developed machines. At the time, there were other experimental machines which could be called "computers," but they were never put into use nor had they been fully completed.

I have prepared software simulators for six early Japanese computers (marked with an asterisk in Table 1) in order to test and validate their instruction sets. We will discuss the nine computers listed in Table 1. Only the

Table 1: The First Japanese Computers

	Computer	Completion	Produced by	Comments
1*	ETL-Mark-2	Autumn 1955	Electrotechnical Laboratory (ETL)	Not a stored-program computer; made of relays
2*	FUJIC	March 1956	Fuji Photo Film Company	First stored-program computer
3	MUSASINO-1	March 1957	Nippon Telegram and Telephone Public Corporation	Second electronic computer, compatible with ILLIAC-1
4*	ETL-Mark-4	November 1957	ETL	First transistor computer·
5*	PC-1	March 1958	Tokyo University	Parametron computer, its initial order was very famous
6*	ETL-Mark-4a	August 1958	ETL	Revised machine from ETL-Mark-4
7	TAC	March 1959	Tokyo University	Last vacuum tube computer, compatible with EDSAC
8	Handai-Computer	Incomplete	Osaka University	Visionary but incomplete computer
9*	K-1	June 1959	Keio University	1 address and 1+1 address

FUJIC and the ETL-Mark-4 will be presented in detail, including their instruction sets.

2. The ETL-Mark-2

The ETL-Mark-2 was built in 1955, using a huge number of relays. It was not a stored-program computer, but it was the first programmable computer in Japan. Programs for the ETL-Mark 2 were punched on paper tape, which was read and executed instruction by instruction. To obtain a program loop, both ends of the tape were pasted together. The ETL-Mark-2 was basically a floating-point machine, but additions and subtractions could be done using fixed-point arithmetic. In those days, vacuum tubes were not very reliable, and transistors had not been yet put to any practical use. Therefore, the use of relays was an obvious choice. After completion, the ETL-Mark-2 was used for a period of 10 years as a stable, programmable computer.

The ETL-Mark-1 was produced in 1952, as a prototype of a relay computer, and provided data and know-how for the construction of the next machine, the ETL-Mark-2. The ETL-Mark-1 was only used for tests and experiments.

Table 2: The ETL-Mark-2

Completion date	1955
Number of relays	22,253
Memory	200 words (relays)
	250 words (relay ROM, constants such as e or π)
Word length	42 bits
Input	60-unit paper tape reader × 6
Output	60-unit paper tape punch × 6
Addition/subtraction time	0.11 s (fixed-point), 0.2 s (floating-point)
Multiplication/division time	0.14-1.39 s (floating-point)
Decimal to binary conversion time	0.1-0.8 s
Binary to decimal conversion time	1.7 s

3. The FUJIC

The FUJIC, a monumental computer, was completed in March 1956. It was the first stored-program computer made in Japan. The FUJIC was a binary, fixed-point, three-address machine with vacuum tubes as the basic elements, and a mercury delay line for the memory. It had been built almost entirely by one person, Dr. Bunji Okazaki, with some outside help for the wiring and soldering. He even built the card reader with his own hands. It was originally designed to perform calculations for optical lenses for the Fuji Photo Film Company. Later, it was used for other purposes by other companies and universities. The FUJIC is part of the collection at Japan's National Science Museum.

Table 3: The FUJIC

Completion date	March 1956
Number of vacuum tubes	Approx. 1,700
Memory	Mercury delay line, 255 words, access 0.5 ms (average)
Word length	33 bits
Input	Hand-made card reader (2 cards/sec, 12 instructions or data/card)
Output	Remodeled electric typewriter (10 characters/sec)
Addition/subtraction time	0.1 ms
Multiplication time	1.6 ms (average)
Division time	2.1 ms
Clock frequency	30 kHz (arithmetic unit)
	1,080 kHz (memory)

Internal structure

The FUJIC processor contains a single accumulator. The machine uses the following special hexadecimal notation (for the digits 9 to 15):

Decimal	10	11	12	13	14	15
FUJIC hexadecimal	J	K	L	M	N	P
Modern hexadecimal	a	b	c	d	e	f

The FUJIC computes using fixed-point arithmetic. A data word consists of 1 bit for the sign, 4 bits for the integer part and 28 bits for the fractional part. If during arithmetic computations, the absolute value becomes greater than 16, an overflow occurs. This reportedly provided enough accuracy for the computation of optical light tracking.

Sign 1 bit	Integer part 4 bits	Fractional part 28 bits

Figure 1: Data word

Alphabetic and special characters are not used at all, only numerical values are treated as data.

The instruction set

The FUJIC has 17 simple instructions for addition, subtraction, multiplication, division, move, unconditional/conditional jump, input/output and stop. It has neither shift nor zero test instructions. For each instruction, an 8-bit quadruple is used. The left 8 bits encode the type of instruction, the other three 8-bit fields, called Xx, Yy, and Zz, denote three addresses in the memory.

8 bit	8 bit	8 bit	8 bit
Instruction type	Xx	Yy	Zz

Figure 2: Instruction Format

Machine language instructions are always written in the FUJIC hexadecimal notation, using absolute addresses only.

The following conventions are used in Table 4:

- Xx, Yy, Zz are two-digit hexadecimal numbers.
- [Xx] denotes the content of address Xx.

Table 4: FUJIC Instruction Set

Type	Instruction code	Meaning
	70 Xx Yy Zz	([Acc])+[Xx]+[Yy] → Zz
Add and Subtract	60 Xx Yy Zz	([Acc])-[Xx]+[Yy] → Zz
	50 Xx Yy Zz	([Acc])+[Xx]-[Yy] → Zz
	40 Xx Yy Zz	([Acc])-[Xx]-[Yy]→ Zz
	20 Xx Yy Zz	([Acc])+[Xx] x [Yy]→ Zz
Multiply	15 Xx Yy Zz	([Acc])-[Xx] x [Yy] → Zz
	10 Xx Yy Zz	[Acc] x [Xx] x [Yy] → Zz
	16 Xx Yy	[Acc] x [Xx] → Yy
Divide	30 Xx Yy Zz	(([Acc])+[Xx])/[Yy]→ Zz
Move	19 Xx Yy	[Xx] → Yy
	1L Xx	Unconditional jump to Xx
Jump	1K Xx Yy Zz	Conditional jump: if [Xx] is positive, jump to Yy, otherwise to Zz
	1J Xx Yy	Overflow detection: if the absolute value of [Acc] is less than 16, jump to Yy, otherwise to Xx
Input	1P	Read card[s]
Output	1M Xx Yy Zz	[Xx] print after decimal conversion
	1N Xx Yy Zz	[Xx] print direct with hexadecimal image
Stop	00	Stop

- [Acc] denotes the content of the accumulator.
- 00 for Xx, Yy, Zz specifies the accumulator, its value can be used by the next instruction.
- Yy, Zz for output instruction specify the layout and one character in front of the data.

An example of a FUJIC instruction is:

70 10 2P 1K ; (Acc) + (address 10) + (address 2P) → address 1K

Address 00 denotes accumulator, therefore,

1K 00 20 1J ; jump to address 20 if the content of accumulator is positive, otherwise jump to address 1J.

Input instruction and program loading

The input instruction (1P) of the FUJIC is both unique and powerful, since it continues reading a data triplet from cards until the end of data is detected. Input data in a card must have the following format: (a) address, (b) type and

(c) word. The type is 1, 2 or 3, which corresponds to hexadecimal, decimal, or end of data respectively. For example:

```
(a)  (b)      (c)
30   1  16 1M 50 00 ;    161M5000 is stored in address 30, since type is 1 (hexadecimal).
31   2   05 00 00 00 ;   08000000 is stored in address 31, because the type is 2(decimal).
20   3   00 00 00 00 ;   end of data, i.e., end of 1P instruction execution, because the type
                         is 3 (end of input), then the program control goes to address 20.
```

The end of the input data shows which address is to be executed next. Therefore, the next instruction of an input statement (1P) is written not in the program, but in its input data. A user's source program is loaded using this input instruction, so that a special loading program, such as an initial order, is not required.

4. MUSASINO-1

The MUSASINO-1 was the second electronic computer and the first parametron computer built in Japan; it was completed in March 1957. "Musasino" is the name of the area where the machine was produced by the Nippon Telephone and Telegram Public Corporation. The MUSASINO-1 was a binary, single-address, fixed-point computer, compatible with the ILLIAC-1 of Illinois University. Parametrons were used as its basic elements. In order to use the extensive library of programs for the ILLIAC-1, the MUSASINO-1 implemented a superset of its instructions, but it was not totally compatible with the ILLIAC-1. Because many programs of the ILLIAC-1 used redundant bits of the instruction, this meant that they were

Table 5: The MUSASINO-1

Start	1952
Completion date	March 1957
Retirement date	July 1962
Number of parametrons	5,400
Number of vacuum tubes	519
Memory	Magnetic core, 256 words
Word length	40 bits, 2 instructions/word
Input	Photo electric paper tape reader
Output	Paper tape punch
Addition/subtraction time	1.35 ms
Multiplication time	6.8 ms
Division time	26.1 ms
Clock frequency	6–25 kHz

Table 6: The ETL-Mark-4

Design start	October 1956
Completion date	November 1957
Retirement date	August 1959 (remodeled to ETL-Mark-4a)
Number of transistors	470
Number of diodes	4,600
Memory	Magnetic drum (18,000 rpm), 1000 words Average access time 1.65 ms
Word length	5 decimal digits+sign
Input	Mechanical paper tape reader (10 characters/sec) Photo electric paper tape reader (200 characters/sec)
Output	Electric typewriter (8 characters/sec) Paper tape punch (30 characters/sec)
Addition/subtraction time	3.4 ms
Multiplication time	4.8 ms
Division time	6.4 ms
Comparison time	1.8 ms
Clock frequency	180 kHz

not redundant for MUSASINO-1. In the first version, the memory consisted of only 32 words. A year later, it was extended to 256 words.

5. The ETL-Mark-4

The ETL-Mark-4 was developed by Dr. Sigeru Takahasi and his group at the Electrotechnical Laboratory in November 1957.

This was Japan's first transistorized computer, and the first decimal computer with a magnetic drum memory. Although its word length was only 5 decimal digits, the 1000-word memory was the largest at that time.

Internal structure

An arithmetic word consists of 5 decimal digits and a sign. There is an implicit decimal point at the leftmost position of a word, a convention respected by the arithmetical operations.

Sign	5 decimal digits

Figure 3: Arithmetic Word

Operation part two decimal digits	Address part three decimal digits

Figure 4: Instruction Word

An instruction word consists of 2 decimal digits to encode the operation, and 3 decimal digits for the address. The sign is not used.

One digit consists of 4 bits. The ETL-Mark-4 could not handle the full alphabet – only a few letters, such as D, Q and Y, could be recognized. The processor contains three registers: the accumulator, MDR (keeps a multiplier for multiplication) and MQR (keeps the quotient after division). Their sizes were 10, 5 and 5 decimal digits, plus a bit for the sign.

Source programs for the ETL-Mark-4 were written in machine code and were loaded according to the initial order described below. Neither mnemonic operation codes nor symbolic addresses could be used in a program.

Instruction set

The operation part must be always written using 2 decimal digits, but the address may consist of significant digits only. Absolute addressing or relative addressing may be used, indicated by the letter Y or Q respectively. For instance, "0223Y" denotes the absolute address 02023. On the other hand, "0223Q" denotes a relative address. If this program group is loaded starting at address 500, this becomes the address 02523.

A constant is written in the format of an absolute address instruction. If the value of a constant is 10, "0010Y" or "00010Y" should be used. Writing "10Y" is not allowed. A negative constant cannot be written.

The ETL-Mark-4 provides no address modification facilities such as index registers. In Table 7 the following conventions were used:

- n: address part of an instruction.
- [x]: content of x, x is address n, Acc, MDR or MQR.

Although the ETL-Mark-4 cannot handle alphabets, the "type out" instruction listed above (code 36 n) prints the 10 special characters listed in Table 8 on a typewriter.

Table 7: ETL-Mark-4 Instruction Set

Code	Name	Meaning
02 n	Add	$[n]+[Acc] \rightarrow Acc$
03 n	Clear add	$[n] \rightarrow Acc$
04 n	Subtract	$-[n]+[Acc] \rightarrow Acc$
05 n	Clear subtract	$-[n] \rightarrow Acc$
06 n	Multiply and add	$[n] \times [MDR]+[Acc] \rightarrow Acc$
07 n	Clear multiply and add	$[n] \times [MDR] \rightarrow Acc$
08 n	Divide	$[Acc] / [n] \rightarrow MQR$
10 n	Multiply and subtract	$-[n] \times [MDR]+[Acc] \rightarrow Acc$
11 n	Clear multiply and subtract	$-[n] \times [MDR] \rightarrow Acc$
12 n	Load MD	$[n] \rightarrow MDR$
14 n	Store	$[Acc] \rightarrow n$
15 n	Clear Store	$0 \rightarrow n$
16 n	Store MQ	$[MQR] \rightarrow n$
20 n	Plus jump	Jump if $[Acc] >= 0$
22 n	Stop and jump	Stop, jump when restart
24 n	Minus jump	Jump if $[Acc] < 0$
26	Raise	$[Acc]+10^{-5} \rightarrow Acc$
27	Clear raise	$10^{-5} \rightarrow Acc$
28	Round off	$Sign([Acc]) \times (abs([Acc])+5 \times 10^{-6}) \rightarrow Acc$
30 n	Left shift	$[Acc] \times 10^{n} \rightarrow Acc$
32 n	Right shift	$[Acc] \times 10^{-n} \rightarrow Acc$
34 n	Type out	Type left most n digit from Acc
36 n	Type special character	Type a specified character, see below
40 n	Read in	Read n digits into Acc
41 n	Clear read in	Clear read n digits into Acc

Table 8: Special character output

n	Character
0	space
1	carriage return
2	line feed
3	upper case
4	lower case
5	+
6	-
7	.
8	/
9	=

Load and start address

There are two parameters for the initial order, one is the first address of the program loading or the origin of the relative address, the other is the start address of execution.

- "14nD": *n* indicates the first address for program loading. For example, 14100D means that subsequent instructions will be stored starting at address 100, and the value 100 will be added to relative addresses.
- "20nD00Y": *n* indicates the start address of program execution. The number *n* must always be an absolute address. For example: "20500D00Y" stops program loading and begins program execution from address 500.

Comments could not be added to a source program because the ETL-Mark-4 could not handle letters. However, in my simulator source programs can be annotated. All characters occurring between two semicolons, or between semicolon and a carriage return, are treated as comments.

An initial order is used for ETL-Mark-4 source program loading. The paper tape punched source program is loaded by the initial order which was stored previously from addresses 0 to 31 of the memory. After pressing a start button on the MTL-Mark-4 console, the loader started, the source program was loaded onto the memory and started running. When the stored initial order was damaged in memory, which often occurred, the following three instructions were manually inserted from address 0 to 2. Starting the machine at address 0, after loading the initial order tape on the paper tape reader, the initial order itself was loaded to memory again.

```
address   instruction

   0       40005         ; read 5 digits into Acc
   1       14003         ; store Acc into address 3
   2       40005         ; read 5 digits into Acc
```

Further developments

The ETL-Mark-4 was remodeled into the ETL-Mark-4a (see Section 7) in August 1959. In July 1956, the ETL-Mark-3 was developed as a pilot model of a transistorized computer (memory: optical glass delay line). This was, in fact, the first Japanese transistorized computer, but it was not put to any practical use due to the unreliability of the pin-point contact transistors. Since that time the pin-point contact type transistors have never again been used.

The ETL-Mark-4 and the ETL-Mark-4a were succeeded by the Yamato (Feb. 1959), the ETL-Mark-5 (May 1959), the K-1 (June 1960) and the ETL-Mark-6 (March 1966). Some of the earliest experiments in English to Japanese machine translation were carried out using the Yamato ("Yamato" is an ancient name for Japan). It had a memory capacity of 82 Kbits – huge for computers at that time.

6. The PC-1

The PC-1 (Parametron Computer 1) was developed by Professor Hidetosi Takahasi's Laboratory at the Department of Physics, University of Tokyo, in March 1958. The parametron was invented by Dr. Eiichi Goto when he was a post-graduate student in 1954. Since it was cheap and reliable, it was widely used in Japan, but it was replaced by transistors, because of the difference in speed. The PC-1 was a binary, single-address computer using parametron and magnetic core memory. Its architecture was similar to that of the EDSAC computer built at Cambridge University, but it was not compatible with the EDSAC. Its instruction set and instruction format were very carefully refined.

The PC-1 was very famous for its initial order, which installed a paper tape punched- source program written in symbols into the memory. It took charge of decimal to binary conversion, code conversion, the process of relative and absolute addresses, and so on, in only 68 words. The PC-1 initial order is undoubtedly one of the masterpieces of computer programming in the world. It is worth decoding the initial order.[1]

Table 9: The PC-1

Start	September 1957
Completion date	March 1958
Retirement date	May 1964
Number of parametrons	4,200
Memory	Magnetic core 512 words (short), 256 words (long)
Addition/subtraction time	0.4 ms
Multiplication time	2.6 ms (short), 4.4 ms (long)
Division time	16.1 ms (short/long)
Clock frequency	10 kHz
Input	Photoelectric paper tape reader
Output	Tele-typewriter

---

[1] See in this volume; E. Wada, "The Parametron Computer PC-1 and its Initial Input Routine."

Table 10: The ETL-Mark-4a

Completion date	August 1959
Memory	Magnetic drum, 1,000 words (access time 2 ms) Magnetic core, 1,000 words (access time 0.01 ms)
Word length	7 decimal digits + sign
Input	Photoelectric paper tape reader
Output	Electric typewriter and paper tape punch
Addition/subtraction time	4.2 ms (drum) 0.24 ms (core)
Multiplication/division time	7.2 ms (drum) 3.40 ms (core)
Comparison time	2.2 ms (drum) 0.25 ms (core)

7. The ETL-Mark-4a

The ETL-Mark-4a was remodeled from ETL-Mark-4 in August 1959. The word length was extended from 5 decimal digits to 7 decimal digits. A 1000-word core memory, 2 index registers and many new instructions were added. Its basic architecture, including decimal coding, single-address instruction format, and input/output units were left unchanged.

8. The TAC

The development of the TAC (Todai Automatic Computer) began in 1952 and was completed in 1959, using a huge number of vacuum tubes as basic elements and cathode ray tubes as memories. The TAC was a binary, single-address, fixed-point arithmetic computer, basically compatible with the EDSAC. The following extra functions were incorporated in the TAC:

- division instruction,
- floating-point arithmetic, and
- B-register (index register).

The reason it took almost eight years to develop the TAC, from the start of its design to completion, was not the challenge of regulating the huge number of vacuum tubes, but the tremendous problems involved in regulating the cathode ray tube memories. Subsequently, cathode ray tubes were never again used as memories in Japan.

Table 11: The TAC

Start	1952
Completion date	January 1959
Retirement date	July 1962
Number of vacuum tubes	7,000
Number of diodes	3,000
Memory	16 cathode ray tubes, 512 words
Word length	17 bits (short), 35 bits (long), 70 bits (floating-point)
Input	Paper tape reader
Output	Paper tape punch Electric typewriter
Addition/subtraction time	0.48 ms
Multiplication time	5.04 ms
Division time	9.84 ms
Clock frequency	330 kHz

9. The Handai-Computer

The development of the Handai-computer began in 1954 at Osaka University. "Handai" is an abbreviation of Osaka University in Japanese. The Handai-computer was a binary, single-address, fixed-point arithmetic computer, and vacuum tubes were used as basic elements with crystal delay units as memories.

Although the computer functioned partially, it did not operate as a whole system, and unfortunately, its development was abandoned in 1959. Thus, the Handai-computer was never officially inaugurated, and it is now kept at Osaka University.

Table 12: The Handai Computer

Start date	1953
Abandon date	1959
Number of vacuum tubes	1,500
Number of diodes	4,000
Memory	Crystal delay unit, 512 words
Word length	40 bits (2 instructions/word)
Input	Paper tape reader
Output	Paper tape punch (6 characters/sec)
Addition/subtraction time	0.04 ms
Multiplication time	1.6 ms
Division time	not implemented
Clock frequency	1 MHz

10. The K-1

The K-1 was produced to commemorate the 100th anniversary of the foundation of Keio University in April 1960.

The basic architecture of the K1 (i.e., the use of the decimal system, transistors and magnetic drum memories) was the same as that of the ETL-Mark-4 and the ETL-Mark-4a. Its word length was extended to 11 decimal digits and floating-point arithmetic operations were added. Instructions with both one and 1+1 addresses, could be coded. However, since there was no assembler comparable to IBM 650 SOAP2, its 1+1 address facility could not be used effectively. The K-1 is still kept at the Faculty of Science and Technology, Keio University.

11. Software simulators

I have programmed six software simulators for the first Japanese computers (ETL-Mark-2, FUJIC, ETL-Mark-4, PC-1, ETL-Mark-4a and K-1) which emulate the instructions of these early machines. The source programs of these computers can now be tested and executed. Anyone who is interested in the first Japanese computers can test and use the simulators using either UNIX or MS-DOS.

However, the functions of the simulators are not exactly the same as the original machines. For instance, they use the standard input of the language C instead of a paper tape or card, and the standard output instead of an output typewriter or printer. Moreover, they do not simulate the timing of instructions. Also, a busy jump instruction for I-O-units will never jump, because this information is absorbed by the operating system.

Table 13: The K-1

Completion date	June 1959
Number of transistors	900
Number of diodes	11,500
Memory	Magnetic drum (10,000 rpm), 1,200 words (average access time 3 ms). Magnetic core 1,000 words (access time 0.01 ms)
Word length	11 decimal digits + sign
Input	Photoelectric paper tape reader
Output	Electric typewriter and paper tape punch
Addition/subtraction time	0.36 ms
Multiplication/division time	5.50 ms
Clock frequency	200 kHz

More information, sample programs and simulator programs written in C are available at http//www.comp.ae.keio.ac.jp/pub/simulator. This web site is updated whenever new information or materials are obtained.

12. Conclusions

The pioneers of computing in Japan produced their own machines soon after the end of World War II, under difficult circumstances and with small budgets. We, as their successors, need to preserve their achievements as much as possible. However, due to the non-availability of old vacuum tubes and paper tape punchers, it is nearly impossible to operate or to reconstruct the machines in their original form. However, using software simulators of the early machines, we can easily test and operate the old programs, while keeping maintenance costs low. Moreover, they can be very easily transferred to any other computer.

Several papers and documents have been written about the hardware, but, unfortunately very few documents exist that deal with the program libraries or software. For instance, "1000-digit calculation of e or π which required only a few seconds" or "a partial differential equation which was solved numerically" were reported, but the programs themselves were not kept. One exception is the program libraries of PC-1, which were collected systematically but are only available in private edition.

References written in English are rare and difficult to obtain. References written in Japanese are available on my web site.

References

[1] M. Goto, Y. Komamiya, R. Suekane, M. Takagi and S. Kuwabara, "Theory and Structure of the Automatic Relay Computer ETL-Mark-2," *Research of the Electrotechnical Laboratory*, No. 556, Sep. 1956.

[2] H. Takahasi, "Parametron Computers," Iwanami-shoten, 1968 (in Japanese).

[3] H. Takahasi, "The Birth of Electronic Computers," Chuuou-kouronsha, 1972 (in Japanese).

[4] E. Wada, "Initial Order R0 of PC-1," bit, kyouritsu-shuppan, Vol. 4, No.12, pp.1149-1162, 1972 (in Japanese).

[5] H. Takahasi, "Some Important Computers of Japanese Design," *Annals of the History of Computing*, 2:4 (1980), p. 330.

[6] E. Wada, "PC-1 Program Library," private edition, 1983.

[7] S. Takahashi, "Early Transistor Computers in Japan," *Annals of the History of Computing*, 8:2 (1986), 144–154.

[8]   IPSJ Special History Committee (ed.) "History of Japanese Computers,"
       OHM-sha, 1985 (in Japanese).
[9]   S. Takahashi, "Computer Chronicle," OHM-sha, 1996 (in Japanese).
[10] S. Okoma, "The Software Simulators for Japanese Computers in the
       Cradle," *Proceedings of the 37th Programming Symposium*, Informa-
       tion Processing Society of Japan(IPSJ), Jan. 1996 (in Japanese).
[11] S. Okoma, "Second report of the Software Simulators for Japanese
       Computers in the Cradle," *Proceedings for History of Computing,
       Summer Programming Symposium*, Information Processing Society of
       Japan(IPSJ), July 1996 (in Japanese).

SEIICHI OKOMA received a B.E. degree in 1959 from the Faculty of Engineering, and a Ph.D. in Engineering in 1977 from the Faculty of Science and
Technology, both at Keio University. From 1959 to 1964 he worked as a
programmer at the Onoda Cement Co. Ltd. In 1964 he became an assistant at
the Faculty of Engineering, Keio University. He is now a professor of the
Faculty of Science and Technology of this university. His earlier subjects of
interest were programming languages, compiling techniques and algorithms.
He took part in the standardization of COBOL as a member of the ISO
COBOL standardization committee of the Information Processing Society of
Japan (IPSJ). His current research interests are natural language processing
and the history of computing. He has written 16 books on such topics as
COBOL, FORTRAN and C. He is a member of the IPSJ, the Japanese Society for Software Science and Technology, the Mathematical Linguistic Soci-
ety of Japan, and the Association for Computing Machinery.

The Parametron Computer PC-1 and Its Initial Input Routine

Eiiti Wada

Abstract. The Parametron Computer 1 (PC-1) was born at Professor Hidetosi Takahasi's Laboratory in 1958. The logical components of the PC-1 were parametrons, elements which could compute majority logic. The memory system operated using a dual frequency read/write scheme. The address decoding mechanism applied an error correcting scheme in order to decrease the number of components necessary. Most of the hardware technology was conceived by Eiichi Goto.

We studied the EDSAC computer carefully at the time, but we developed the architecture and programming system based upon our own philosophy. The instruction set was chosen to ease programming. Conventional teletypes were employed, leaving the burden of code conversion to software, which seemed to us to have almost infinite abilities.

However, the memory capacity was small and we had to invent clever ways of implementing some operations. In this contribution, I describe the initial input routine and then: (a) the code conversion table contained in the program body, and (b) the magic number method to control the number of multiplications during decimal-binary conversion.

The PC-1 was one of the first computers which implemented interrupts. The peripheral devices interrupted the running program by saving the address of the next instruction to be executed, and then jumping to a fixed location in memory. As a simple experiment in multiprogramming, we implemented concurrently the binary to decimal conversion program and a printer control routine which used a circular buffer.

1. Introduction

The Parametron Computer 1 (PC-1) was a binary, single-address computer developed at Professor Hidetosi Takahasi's Laboratory at the Department of Physics, University of Tokyo, and was one of the first general purpose computers that used parametron components and dual frequency magnetic core memory.[1] Construction started in September 1957 and was completed on March 26, 1958. The PC-1 was used at Takahasi's Laboratory for research related both to hardware and software and the researchers in the Faculty of Science also used it for scientific computing. The PC-1 was retired in May 1964.

The British EDSAC computer had a great influence on the design and implementation of all aspects of the PC-1 because it was the only computer described in a textbook at the time.[2]

The arithmetic and control circuits of the PC-1 consisted of 4200 parametrons. Binary numbers were coded using the two's complement representation; a short number was coded using 18 bits and a long one using 36. The single-address instructions were 18 bits long and there were about 20 of them. The memory consisted of 512 short words. The clock frequency was 15 KHz. One addition or subtraction required 4 clock cycles; one multiplication 26 cycles for a short multiplier, or 44 cycles for a long multiplier. Division consumed 161 cycles and a store operation 8. The power consumption was 3 Kw and the floor area required was 8 square meters. The input was done using a photoelectric paper tape reader; the output was provided by a teletype.

The PC-1 seems to be the first computer which implemented interrupts, which means that we were experimenting with multiprogramming as early as 1959. The research on modular computations mentioned by Knuth was conducted on this machine.[3]

This contribution reviews the parametron and memory circuits in sections 2 and 3, then the structure of memory and registers in section 4. The teletype code used by the PC-1 is given in section 5, so that the readers can understand the details of the input/output routines. Section 6 gives an overview of the instruction set. A few other instructions implemented later for experimental use are not included here. The operation of interrupts is described briefly in section 7. Section 8 sketches the initial input routine R0.[4] The program listing can be found in Appendix A.

---

[1] H. Takahasi (ed.), *Parametron Computers*, *Iwanami Shoten*, 1968 (in Japanese).

[2] M. V. Wilkes, D. J. Wheeler, and S. Gill: *The Preparation of Programs for an Electronic Digital Computer*, Addison-Wesley Press, Inc. 1951.

[3] D. Knuth, *The Art of Computer Programming*, Vol. 2, *Seminumerical Algorithms*, 3rd ed. p. 291.

[4] The editors of this volume suggested using the word "assembler" instead of " initial input routine". However, from the viewpoint of systems programming, they are quite different things. The initial input routine is different from the assembler since the initial input routine is: 1) very primitive; it normally coexists in the storage with the user's program; 2) the parameters must be explicitly present; 3) parameters are not symbolic addresses; they cannot refer forward; 4) the programmer may use parts of the initial input routine if he fully understands it; 5) special functions may be achieved by combining the directives; 6) interleaving can be used to save storage space. The assembly program is on the contrary: 1) very large; assembly and pro-

2. Parametrons

The parametron was invented by Eiichi Goto in 1954 when he was a graduate student. A parametron is a resonant circuit of frequency f energized by parametric excitation. When the circuit parameter is excited repeatedly with twice the resonance frequency, the circuit is energized. Fig. 1 shows a parametron circuit. Here, the reactance is modulated by a driver line. With respect to the excitation frequency $2f$ (gray line), the circuit might be energized in one of

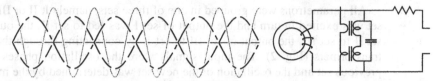

Figure 1: Parametron circuit

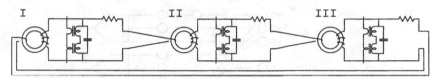

Figure 2: Parametron connection

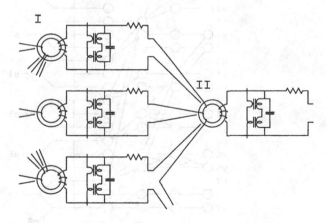

Figure 3: Parametron connection for majority logic

duction run are distinct phases; 2) symbolic addresses may be used; forward and backward references are automatically resolved; 3) special functions are designed as fundamental from the beginning; they are expressed with names and parameters;

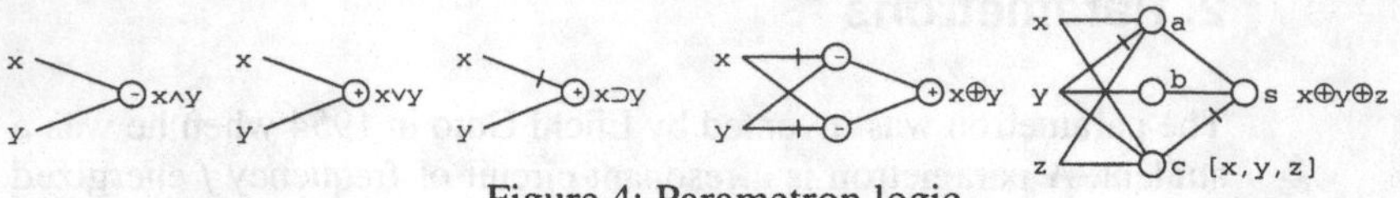

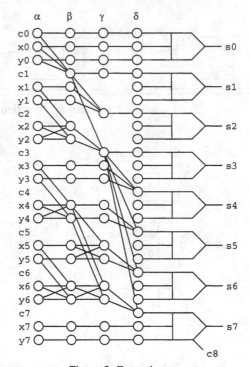

Figure 4: Parametron logic

the two possible phases, 0 and π (broken lines). One of the two phases is considered to represent 0 and the other 1, thus it is possible to represent one bit of information.

All parametrons were grouped in one of three sets, namely I, II or III. Each set was excited in turn and the output of set I was fed to set II, the output of set II to set III, the output of set III to set I, by connecting through the input transformers (Fig. 2). The input signals were the oscillation phases of the previous set and the oscillation of the new set was determined by the majority

Figure 5: Carry detector

---

4) inside of the assembly program is the black box for the programmer; program segment cannot be used by the user.

of input signals (phases). Therefore parametrons implemented majority logic (Fig. 3). Negation was achieved by reverse coupling.

Fig. 4 shows the typical operations with the possible constant input 0 or 1. (0) is represented as −, 1 as + in the circle. (The small cross on the input line indicates negation.) In Fig. 4, the rightmost circuit is a full adder ([x,y,z] means majority of x, y, z). One of the interesting circuits was a carry detector which detects n digit carries before performing additions in $\log_2 n$ logical steps.

Fig. 5 shows an array of the full adders with the carry detector. The carry detector works as follows: the parametrons in the figure are subscripted in each column α, β, etc., from the left. The input carry to the highest full adder, $c_{7\delta}$, is 0 if both $x_{6\alpha}$ and $y_{6\alpha}$ are 0 and 1 if both $x_{6\alpha}$ and $y_{6\alpha}$ are 1. If $x_{6\alpha}$ and $y_{6\alpha}$ are 1 then $x_{6\beta}$ and $y_{6\beta}$ become 1 and $x_{6\gamma}$ and $y_{6\gamma}$ also become 1, thus $c_{7\delta}$ is 1. In case both $x_{6\alpha}$ and $y_{6\alpha}$ are 0, $c_{7\delta}$ is 0 in the same way. However, if $x_{6\alpha}$ and $y_{6\alpha}$ are 0 and 1, then two inputs to $x_{6\beta}$ and $y_{6\beta}$ cancel out and $x_{5\alpha}$ and $y_{5\alpha}$ will have the casting vote. If $x_{5\alpha}$ and $y_{5\alpha}$ are again 0 and 1, then $c_{7\delta}$ is determined by $x_{4\alpha}$ and $y_{4\alpha}$ at $x_{6\gamma}$ and $y_{6\gamma}$. In this fashion, carry inputs $c_{4\delta}$ to $c_{7\delta}$ are determined at the third parametron column (δ). The carry inputs to $c_{2\delta}$ and $c_{3\delta}$ are determined at the second parametron column (γ). Thus, in general, the steps needed to detect carries are $\log_2 n$. Although not shown in Fig. 5, the usual carry paths from one full adder to the next exist. They are used for repeated additions during multiplication. At the final stage of multiplication, carries are detected using this circuit. The addends together with the detected carries are fed to the full adders to obtain the sums.

3. Memory

Goto also invented most of the technology for the memory of the PC-1. The magnetic core memory of the PC-1 used sinusoidal waves rather than pulses for a write/read operation. The core matrix consisted of a 36×256 rectangular array. In each writing operation, a sinusoidal wave of frequency $\frac{1}{2} f$ went through the wire selected from one of the 256 rows, and the 36 information bits were applied to the 36 column wires in the form of a sinusoidal wave of frequency f, where the phase of the latter wave represented each information bit. The cores on the cross points of both wires were subjected to a magnetizing force of the form

$$I_0 \cos \pi f t \pm I_1 \cos 2\pi f t,$$

and the asymmetry of this wave caused magnetization of the core in one or the other direction (Fig. 6).

The read out signals were obtained from the second harmonic waves of the column wires (Fig. 7).

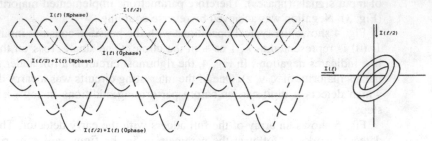

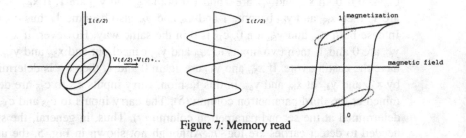

Figure 6: Memory write

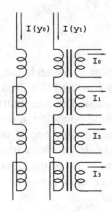

Figure 7: Memory read

The address selection mechanism was based on an error-correcting code developed by Takahasi and Goto.[5]

Figure 8: Word selector

---

[5] H. Takahasi and E. Goto: "Application of Error Correcting Codes to Multiway Switching", *UNESCO International Conference on Information Processing*, Paris 1959, G 2.9.

Table 1: Output currents

input		output			
$I(y_0)$	$I(y_1)$	I_0	I_1	I_2	I_3
$-I$	$-I$	$-2I$	0	0	$+2I$
$+I$	$-I$	0	$-2I$	$+2I$	0
$-I$	$+I$	0	$+2I$	$-2I$	0
$+I$	$+I$	$+2I$	0	0	$-2I$

Selection logic

The transformers shown in Fig. 8 will produce the output current shown in Table 1, when driven by input currents $I(y_0)$ and $I(y_1)$ of the same frequency and the same amplitude but in either phase, 0 or π (indicated by $+I$ and $-I$). The parametrons will oscillate when driven with full amplitude, but will not oscillate under a weaker current. So, in this case, of the word selection parametrons driven by the output current, only those that are fed with $\pm 2I$ oscillate, the others remain inactive. However, the difference between the maximal amplitude and the next one is only $2I$ and the discrimination power is not enough. However, by employing, for example, the 7 bit Hamming error-correcting code for 4 input lines, the discrimination power is $7I$ vs. $\pm I$, which is enough for the purpose of selection. The PC-1 used 18-bit input lines and excited only one address selection parametron out of 256.

4. Structure of memory and registers

The structure of the PC-1 memory is shown on the left of Fig. 9. It consists of 512 short words. Like the EDSAC, two short words at addresses $2n$ and $2n+1$ can be combined into one long word. Instructions are coded using short words, but numeric data may be 18 or 36-bit long. Instructions which refer to long word operands must use an even numbered address and have a 1 in the l/s bit of the instruction. Some instructions, however, used the l/s bit for other purposes. Of three arithmetic registers contained in the arithmetic unit, the accumulator and the R register were used for programming. The memory register was used to hold the multiplier and divisor. The contents of the arithmetic register were assumed to be fractional, i.e. numbers are represented in the range of $-1 \leq n < 1$.

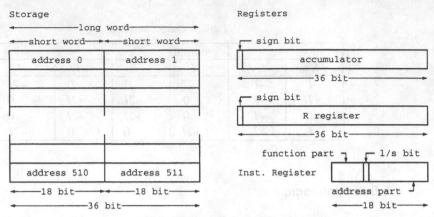

Figure 9: Storage and registers

5. Teletype code

The PC-1 used a conventional teletype for input/output without modifying the code. In those days, the teletype code used in Japan consisted of 6 bits. The lower 5 bits are similar to the 5 bits of the international teletype code. Digits and some special characters have a 1 in the most significant bit. However, since the codes of the digits are based on the character codes of the third row of the teletype, the code patterns and the numerical values of the digits are quite independent. Therefore the input and output routines need a code conversion table.

Teletype code table

000000	Blank	010000	e	100000	Signal	110000	2
000001	t	010001	z	100001	4	110001	
000010	CR	010010	d	100010	-	110010	
000011	o	010011	b	100011	8	110011	
000100	SP	010100	s	100100		110100	
000101	h	010101	y	100101		110101	5
000110	n	010110	f	100110	,	110110	
000111	m	010111	x	100111	.	110111	=
001000	LF	011000	a	101000	+	111000	0
001001	l	011001	w	101001		111001	1
001010	r	011010	j	101010	3	111010	
001011	g	011011	FGRS	101011		111011	
001100	i	011100	u	101100	7	111100	6
001101	p	011101	q	101101	9	111101	
001110	c	011110	k	101110		111110	
001111	v	011111	LTRS	101111	:	111111	Erase

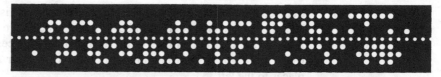

abcdefghijklmnopqrstuvwxyz 0123456789 +-=,.:

Figure 10: Tape image

6. The PC-1 Instruction Set

The PC-1 implemented the following list of instructions. The suffix "l" makes the instruction refer to a long word or indicates variations.

a n, al n	Add the number in storage location n (nL) to the accumulator.
b n, bl n	Replace the accumulator with the bitwise XOR of the numbers in storage location n (nL) and the accumulator.
c n, cl n	Replace the accumulator with the bitwise logical AND of the numbers in storage location n (nL) and the accumulator.
d n, dl n	Divide the number in the accumulator and R register by the number in storage location n (nL) and place the quotient in the accumulator, the remainder in the R register. The remainder is always positive.
i n	If the tape reader is ready, read a character and place it in the bits 0–5 of the accumulator and clear the bits 6–35; if not ready, jump to n.
jl n	Jump to n.
k n	If the number in the accumulator is smaller than 0, jump to n.
kl n	If the number in the accumulator is positive, jump to n.
l n	If n < 1024, shift the accumulator n places to the left; if n ≥ 1024, shift the accumulator logically 2048 − n places to the right.
ll n	Same as l n except that it shifts the accumulator and the R register.
n n, nl n	Clear the accumulator and subtract the number in storage location n (nL) from the accumulator.
o n	If the teletype is ready, place bits 0–5 of the accumulator into the teletype; if not ready, jump to n.
p n, pl n	Clear the accumulator and add the number in storage location n (nL) to the accumulator.

q n, ql n	Replace the accumulator with the number in the R register and load the R register with the number in storage location n (nL).
r n	If n < 1024, shift the number in the accumulator n places to the right; if n ≥ 1024, shift the number in the accumulator 2048–n places to the left.
rl n	Same as r n except that it shifts the accumulator and the R register.
s n, sl n	Subtract the number in storage location n (nL) from the accumulator.
t n, tl n	Store the contents of the accumulator to storage location n (nL).
v n, vl n	Multiply the number of the accumulator and the number in storage location n (nL) and place the product in the accumulator and the R register.
w n, wl n	Do the same as vn (v ln) and add the original contents of the R register multiplied by 2^{-17} (for w n), or by 2^{-35} (for wl n), to the product.
X n	Store bits 7–17 of the accumulator to the address part of storage location n.
z n	Jump to n if bits 7–17 of the accumulator contain 0.
zl n	Jump to n if the content of the accumulator is 0.

7. Interrupts

From the beginning, the input/output instructions were designed to have busy jump facilities, that is, when a device was not ready, instead of waiting for the completion of the operation, the programmer could choose another path by jumping to the alternative context. However, after one year's experience we concluded that the busy jump facilities were hard to use effectively.

Therefore, in the summer of 1959, another approach was adopted: interrupts. The devices were designed to interrupt the program whenever they were ready for use. The interrupt worked as follows: •

1. When the current peripheral operation was completed, the program counter holding the location of the next instruction was stored in the address part of location 510 and control was sent to 511 (last storage location).
2. At the same time, further interrupts were disabled by setting a flip-flop, because otherwise the return address could be overwritten by a subsequent interrupt.
3. The jump instruction in 511 led the control to the interrupt handling routine.

4. At the end of interrupt handling and after resetting the interrupt disable flip-flop, the program returned to the former routine with the information in 510.

In the print routine digits were printed after doing the binary-decimal conversion and code conversion to the teletype codes. However, a digit could be converted faster than it could be printed with the teletype. Accordingly, the output conversion program placed the teletype code in the circular buffer without waiting for completion of the previous printing. Later, when the teletype finished printing one digit, it interrupted a running program, and the interrupt handling, taking control, started the teletype again with the next code taken from the circular buffer.

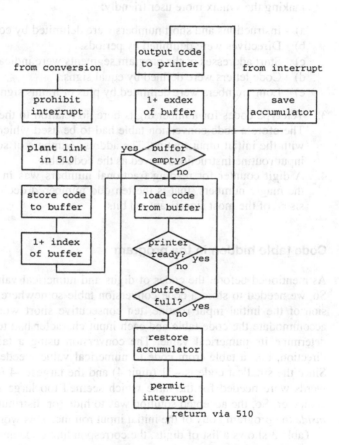

Figure 11: Interrupt handling routine

8. The initial input routine R0

The initial input routine R0 of the PC-1 used an approach similar to the initial orders of the EDSAC or the DOI (decimal order input) of the ILLIAC I, insofar as instructions are denoted by a mnemonic code and addresses by decimal numbers etc. In detail, there are several differences.

1.   Because of the use of 6 bit teletype code, there were more characters available for programming. Without shifting to upper case, we could use all the small alphabetic characters, decimal digits, and 6 special characters (comma, period, colon, equal, plus and minus).
2.   Instructions and directives were terminated with special characters, making the syntax more user friendly:

     a)   Instructions and short numbers were delimited by commas.
     b)   Directives were delimited by periods.
     c)   Start addresses of the program segments were indicated by colons.
     d)   Code letters were defined by equal signs.
     e)   Long numbers were delimited by plus or minus signs.

3.   Teletype codes for number digits bore no relation to the numeric values. Therefore a code conversion table had to be used which was intermixed with the initial input routine. The address numbers of some of the initial input routine instructions served as the code table.
4.   A digit counter for reading fractional numbers was implemented using the magic numbers, that is the ten odd numbered decimal fraction consisted of the most significant ten bits.

Code table hidden in the program

As mentioned before, the codes of digits and numerical values are different. So, we needed to store a code conversion table somewhere. In the first version of the initial input routine, ten consecutive short words were used to accommodate the code table and each input character had to be compared to determine its numerical value. The conversion using a table for the other direction, i.e., a table from code to numerical value, needed a larger space. Since the smallest code is –31 (digit 4) and the largest –4 (digit 6), 28 short words were needed for the table which seemed too large for the 512-word computer. So, the accepted solution was to hide (or distribute) the code table *inside* the program body of the initial input routine. This worked very well.

Table 3 shows a list of digits, the corresponding code and the value of the codes. Each of the values, increased by 56, was the address at which an instruction with the number in its address part had to be stored. The instructions of these locations are shown in the last column of the table.

Table 3: Conversion table for digits

digit	code	value	address	content
0	111000	–8	48	r 0
1	111001	–7	49	l 1
2	110000	–16	40	p 2
3	101010	–22	34	x 3
4	100001	–31	25	jl 4
5	110101	–11	45	kl 5
6	111100	–4	52	jl 6
7	101100	–20	36	rl 7
8	100011	–29	27	Space 8
9	101101	–19	37	ll 9

Table 4: Base orders for special symbols

symbol	code	value	address	content	directive
,	100110	–26	30	*s 45	jl 57
,	100111	–25	31	*s 3	jl 15
:	101111	–17	39	*el 54	x 66
=	110111	–9	47	*nl 12	p 24
+	101000	–24	32	*s 6	jl 18
-	101000	–30	26	*s 6	jl 18

Similarly, other terminating characters were treated by assigning the pseudo instructions (in the program of R0, this is referred to as the base orders) in the appropriate locations. The results are shown in Table 4. The stars in the content column indicate 1 in the sign bit. To obtain the instructions (most of them are jump instructions), the content was added to *nl 12 (instruction in 14 of the initial input routine).

This was possible because the initial input routine was stored in a fixed place and the addresses were unchanged (compare Table 3 with R0, Appendix A).

The "magic number" for binary conversion

In the days of the PC-1, memory was the most crucial resource. One of the techniques used in the input routine was to count the number of characters by means of the strobe, the magic numbers, in the course of reading a fractional number up to 10 decimal digits.

When reading a fractional number, for instance 0.25, the fractional part was first read as an integer. Thus 0.25 appeared in the accumulator simply as 25. The number of characters already read was 2. Accordingly, the content of the accumulator was multiplied by 10^8 and then divided by 10^{10}.

When the decimal point was read, a "magic number" was loaded into the accumulator where the decimal to binary conversion was performed (the period was identified as the decimal point if the working area was cleared; otherwise, it served as the directive delimiter).

The magic number was 1.193359375 in decimal, and 1.001100011 in binary representation. Since all digits are odd, in the course of multiplication by 10 during the decimal to binary conversion, the sign bit always remained as 1, i.e., the number seemed negative. And the least significant bit of the magic number was being shifted to the left 1 bit each time. At the same time, the result of decimal to binary conversion crept up from the right.

The input of a long fractional number is terminated by + or −. When the input was terminated, the dummy multiplication by 10 was repeated until the accumulator became positive, which meant the multiplication by 10 was executed exactly 10 times.

The magic number for 6 digits is shown as an example in the diagram below. The 6 digits magic number is calculated like this. Write down the numbers 2^0, 2^{-1}, 2^{-2}, ..., 2^{-5}. Summing up proceeds from the bottom checking if the result of addition produces an odd digit in the last position of that number. If an odd digit is produced, add that number, otherwise skip the addition of the number by crossing it out. This brings up the number 1.59375 as shown below. The magic number for 6 digits is:

```
1.0
0.5
0.25
0.125
0.0625
0.03125  +
1.59375  (decimal)
1.10011  (binary)
```

9. Conclusion

The PC-1 played a remarkable role in our group's research of computer architecture and program libraries and influenced the work in computational physics and chemistry in the neighboring laboratories. The preparation of the system program was essential for all activities within and outside of the group. In this paper, some typical techniques used in our system programs were explained.

Judging the research activity of that period retrospectively, the members of the laboratory benefited immensely from having their own computer. In a relatively few years, we could learn the whole life cycle of computer development from the hardware elements to the program library. Our homemade computer enabled us to make all the experiments in hardware or software that came to mind. The scarcity of computers around us led to original research in this field.

After the fundamental system programs were prepared, more advanced programming systems had been developed, though I have to say, the PC-1 was too small for such ambitious projects. For instance, we tried to introduce symbolic addresses in the initial input routine. This idea was implemented by colleagues in the chemistry department. They used lists to keep the unsolved symbols until symbol/location correspondence was settled. The modular arithmetic system was an idea of Takahasi. In this implementation, he used the *wl* *n* operation to obtain the remainder of division by a large prime number.

One day, a flip-flop of the PC-1 was connected to a loudspeaker and a program made the loudspeaker oscillate, thus generating a sound. The pitch was controlled by adjusting the shift number of the shift instruction and the sound duration was controlled by a busy jump in the output instruction.

A few years later, work was started on the design of the next parametron computer. Construction of this new machine, the PC-2, was in the hands of Fujitsu Ltd., so it didn't have the same impact on us as the PC-1. The PC-2 was installed in the computer room of the Faculty of Science, University of Tokyo and was used by the community.

Every five years, the members of the group who built the PC-1 and maintained the library meet on March 26 to celebrate the birthday of our beloved machine.

Appendix A: the initial input routine R0

The PC-1 is a stored program computer and uses the binary system within the machine for the representation of numbers and addresses. In using the PC-1, the program and the numerical data should first be stored in the machine's memory, before computation starts. This may be done by preparing a tape in which all the instructions and numbers are represented in binary form. The tape is read by pressing down the "initial load switch". However, writing down the instructions and numbers in binary notation is by no means simple.

The basic input routine "R0" enables the instructions and numbers punched on tape (in decimal and alphanumeric notation) to be read and placed in the PC-1 memory. R0 decodes the teleprinter code of the PC-1 perforator, converts the decimal numbers into binary form, adds the operation codes and places the assembled words in specified locations in the memory. When all the instructions and numbers have been stored, R0 causes the machine to start the program by transferring control to a specified word in the memory. R0 provides further facilities for turning the relative addresses on tape into absolute addresses, and adding one or more parameters to the words before they are stored in the memory. The input routine R0 itself occupies the locations 0–67 of the PC-1 memory and is stored there by placing the binary tape of R0 in the tape reader and pressing down the "initial load switch".

Each PC-1 instruction is punched in exactly the same form as it is written in the text, that is, the operation code consisting of a letter, followed or not by a letter „l", followed by a decimal integer denoting the address, and terminated by a comma. Non significant zeros at the beginning of the address may be omitted. A sequence of instructions is normally placed in consecutive memory locations. The location of the first instruction in a sequence must be specified by a "directive" in front of the sequence.

Example: The tape 100: pl150, vl152, sl154, tl156, jl130, causes the memory locations 100–104 to be loaded with the following object code:

location	instruction	contents of the memory
100	pl 150	0011011000100 10110
101	vl 152	0011111000100 11000
102	sl 154	0101001000100 11010
103	tl 156	0000011000100 11100
104	jl 130	0110101000100 00010

A program tape will consist of one or more sections of such sequences, or subprograms. A blank section of some ten centimeters should be left at the beginning of the program tape, and the program should begin with a "carriage return and line feed", which clears the working positions of R0 prior to reading essential information.

The complete program should end with the control code "jl M" (terminated by a period), which stops the operation of R0 and starts the program by causing the control to be transferred to the location M.

Listing of the initial input routine, R0

Location	Order		Notes
38 → 0	al	28	Decimal to binary conversion
1	jl	49	
2		0	Constant
3		(0)	Temporary storage for binary number
25 → 4	a	64	Add function
45 → 5	t	64	Store function and parameter
52,7 → 6	i	6	Read code, number or symbol
7	zl	6	Jump if blank
46 → 8	rl	12	Shift teletype code to address part
9	kl	20	Jump if code letter
10	a	33	Add address base
11	x	12	Assemble load order
12	p	(0)	Load number or base order
13	kl	34	Jump if number
14	a	47	Modify base order into switch order
18 → 15	t	18	Set switch order
16	p	29	Load address
17	a	64	Add function and parameter
18		0	switch order (jl 18, jl 57, jl 15, x 66, p 24)
19	jl	55	Jump to "set transfer order"
9 → 20	a	33	Add address base
21	x	24	Assemble load order
22	s	4	Examine whether code letter is LF
23	z	40	Jump to clear order in case of LF
24	p	(0)	Load parameter
25	jl	4	(4) Jump to "add function" order
26	*s	6	(-) Base order for "jl 18" (* means 1 in sign bit)
27	SP	8	(8) Constant, function part is SP
28		(0)	Working space for
29		(0)	decimal to binary conversion
30	*s	45	(,) Base order for "jl 57"
31	*s	3	(.) Base order for "jl 15"
32	*s	6	(+) Base order for "jl 18"
33	CR	56	Address base, function part is CR
13 → 34	x	3	(3) Store binary number
35	pl	28	(7) Decimal to binary conversion
36	rl	7	
37	ll	9	(9)

38	jl 0	
39	*el 54	(:) Base order for "x 66"
56,23 → 40	p 2	(2) Load 0
41	tl 28	Clear working space for conversion
42	t 64	
44 → 43	i 43	Read function, number or symbol
44	zl 43	Jump if blank
45	kl 5	(5) Jump if function letter
46	jl 8	
47	*nl 12	(=) Base order for "p 24"
48	r 0	(0)
1 → 49	l 1	(1)
50	al 2	Decimal to binary conversion
51	tl 28	
52	jl 6	(6)
58 → 53	p 57	
54	a 49	Increase "transfer order"
19 → 55	x 57	Set "transfer order"
56	jl 40	(Blank) Jump to clear order
18 → 57	t (67)	(t) Transfer order
58	jl 53	(CR)
59	28	(o) o-parameter
60	0	(SP)
61	0	(h) h-parameter
62	parity	(n) n-parameter
63	digit	(m) m-parameter
64	(0)	(LF) Working space for function and parameters
65	2048	(l) l-parameter for long word order
66	(0)	(r) r-parameter
67	68	(g) g-parameter, End of tape

May 1, 1958
E. Wada

EIITI WADA received his B.S. degree from the University of Tokyo. He joined Professor Hidetosi Takahasi's Laboratory at the Physics Department, University of Tokyo, as a graduate student. He worked there with Dr. Eiichi Goto to develop parametron logic and parametron computers. Dr. Wada constructed a copying and calculating machine, which performs computations while translating the arithmetic expressions contained in the paper tape. In 1958, when the parametron computer PC-1 was completed, Dr. Wada developed the software library. In 1964, he moved to the Faculty of Engineering at the University of Tokyo. In 1973 he spent one year at MIT, Project MAC, as a visiting professor. Dr. Wada is the Japanese member of the IFIP WG2.1. Dr. Wada is a Professor Emeritus of the University of Tokyo and Executive Advisor of Fujitsu Laboratories.

Index

Printed in the United States
by Baker & Taylor Publisher Services

Printed in the United States
by Baker & Taylor Publisher Services